Degenerate and other problems

Pitman Monographs and Surveys in Pure and Applied Mathematics 61

Degenerate and other problems

A Dzhuraev
Tajik Academy of Sciences

CRC Press
Taylor & Francis Group
Boca Raton London New York

CRC Press is an imprint of the
Taylor & Francis Group, an **informa** business
A CHAPMAN & HALL BOOK

First published 1992 by Longman Group

Published 2019 by CRC Press
Taylor & Francis Group
6000 Broken Sound Parkway NW, Suite 300
Boca Raton, FL 33487-2742

First issued in paperback 2019

ISBN-13: 978-0-367-45016-8

Visit the Taylor & Francis Web site at
http://www.taylorandfrancis.com

and the CRC Press Web site at
http://www.crcpress.com

ISSN 0269-3666

British Library Cataloguing in Publication Data

A catalogue record for this book is
available from the British Library

Library of Congress Cataloging-in-Publication Data

Dzhuraev, A. (Abdukhamid), 1932–
Degenerate and other problems / A. Dzhuraev.
p. cm. -- (Pitman monographs and surveys in pure and applied mathematics, ISSN 0269-3666 ;)
Includes bibliographical references (p.).
1. Differential equations, Elliptic. 2. Boundary value problems.
3. Integral equations. I. Title. II. Series.
QA377.D95 1992
515'.353--dc20 92-4762
CIP

Contents

Preface

In this book we have collected various problems of partial differential equations which possess degeneration properties, so they do not fit into the framework to which general methods are applicable. These problems are elliptic systems degenerate on the boundary or on part of it, overdetermined boundary value problems for elliptic systems and initial-boundary value problems for nonstrictly hyperbolic systems. Elliptic systems that are degenerate on the boundary possess different properties: some of them are like pure elliptic systems, and the determinacy of their solution depends on the boundary conditions, but some of them are such that for the solution to be determinate it is even necessary to cancel the boundary conditions. The Cauchy problem for elliptic systems are degenerate problems too, because they are 'unsolvable' problems, i.e. the necessary and sufficient solvability conditions on the right-hand side are not finite, and so the problem arises of describing those sets of right-hand sides for which the Cauchy problem is solvable.

It turns out that this set is described by means of the Bergman kernel function of the domain and its generalizations, introduced here for consideration.

Chapters 1 and 2 of this book are dedicated to the analysis of boundary-value problems for elliptic systems on bounded multiply connected plane domains on the basis of the method of singular integral equations over two-dimensional multiply connected domains, elaborated by the author, and of boundary-value problems for some first-order elliptic systems such as Moisil-Theodorescu and also some second-order elliptic systems, not necessarily strongly elliptic, in the multidimensional case. Chapter 3 consists of degenerate problems. In Chapter 4 the initial-boundary-value problems for hyperbolic (including nonstrictly hyperbolic) systems are stated and investigated as well as the problems for some mixed type systems. In Chapter 5 the theory of first-order elliptic and non elliptic overdetermined systems are discussed. Finally, in Chapter 6 the well-posed problems for the systems with degenerate principal symbols are studied.

These analytical tools have greatly improved the fluency of the analysis presented in this book on the above-mentioned problems. I make no claims for completeness; my intention is to sketch some of those properties arising in this subject which seem to me interesting and to deserve analysis.

I wish to thank Professor A. Jeffrey of the Department of Engineering Mathematics, University of Newcastle upon Tyne for his interest and for ensuring the prompt receipt of my manuscript by the publisher. I would also like to acknowledge Longman Group UK Ltd for the rapid publication of this book and Professor H. Begehr of the Free University of Berlin for his interest in the topic presented here and for support in the publication of this book.

A. Dzhuraev

Dushanbe/Berlin
July 1992

1 Elliptic systems in the plane

1.1 First-order elliptic systems in the plane

1.1.1. Introduction

Any linear first-order system of two real equations in the plane

$$a_{11}(x,y)\frac{\partial u_1}{\partial x}+a_{12}\frac{\partial u_2}{\partial x}+b_{11}(x,y)\frac{\partial u_1}{\partial y}+b_{12}(x,y)\frac{\partial u_2}{\partial y}$$

$$+\,c_{11}(x,y)u_1+c_{12}(x,y)u_2=f_1(x,y),$$

$$a_{21}(x,y)\frac{\partial u_1}{\partial x}+a_{22}\frac{\partial u_2}{\partial x}+b_{21}(x,y)\frac{\partial u_1}{\partial y}+b_{22}(x,y)\frac{\partial u_2}{\partial y}$$

$$+\,c_{21}(x,y)u_1+c_{22}(x,y)u_2=f_2(x,y)$$

may be put in the complex form

$$a(z)w_{\bar z}+b(z)\bar w_z+c(z)w_z\ d(z)\bar w_{\bar z}+a_0(z)w+b_0(z)\bar w=f(z), \tag{1.1}$$

where all quantities are now complex valued.

The characteristic form of (1.1) is

$$\chi(z,\zeta)\equiv|\,a(z)\zeta+c(z)\bar\zeta\,|^2-|\,d(z)\zeta+b(z)\bar\zeta\,|^2$$

$$=(|\,a(z)|^2+|\,c(z)|^2-|\,b(z)|^2-|\,d(z)|^2)\,|\,\zeta\,|^2+2\,\mathrm{Re}\,(\alpha(z)\zeta^2), \tag{1.2}$$

where $\zeta=\xi_1+i\xi_2$,

$$\alpha(z)=a(z)\overline{c(z)}-\overline{b(z)}\,d(z). \tag{1.3}$$

The ellipticity of (1.1) at the point z means the definiteness of the form $\chi(z,\zeta)$ at

z, which leads to the inequality

$$\sigma_0(z) = (|a(z)|^2 + |c(z)|^2 - |b(z)|^2 - |d(z)|^2)^2 - 4|\alpha(z)|^2 > 0. \tag{1.4}$$

Since

$$|\beta(z)|^2 - |\alpha(z)|^2 = A(z) \cdot B(z),$$

where

$$A(z) = |a(z)|^2 - |b(z)|^2,\ B(z) = |d(z)|^2 - |c(z)|^2,\ \beta(z) = a(z)\overline{d(z)} - \overline{b(z)}c(z), \tag{1.5}$$

then

$$\sigma_0 = |A|^2 - 2(|\alpha|^2 + |\beta|^2) + |B|^2 = |A|^2 - (|\alpha| + |\beta|)^2 + |B|^2 - (|\alpha| - |\beta|)^2$$

$$= |A|^2 - (|\alpha| + |\beta|)^2 + |B|^2 - \frac{|A|^2 |B|^2}{(|\alpha| + |\beta|)^2}$$

$$= |A|^2 - (|\alpha| + |\beta|)^2 + |B|^2 \left(1 - \frac{|A|^2}{(|\alpha| + |\beta|)^2}\right)$$

$$= (|A|^2 - (|\alpha| + |\beta|)^2) \left(1 - \frac{|B|^2}{(|\alpha| + |\beta|)^2}\right)$$

$$= \frac{1}{(|\alpha| + |\beta|)^2} (|A|^2 - (|\alpha| + |\beta|)^2)((|\alpha| + |\beta|)^2 - |B|^2)$$

$$= \frac{1}{(|\alpha| + |\beta|)^2} (|A| + |\alpha| + |\beta|)(|\alpha| + |\beta| + |B|))(A - |\alpha| - |\beta|)(|\alpha| + |\beta| - |B|)$$

$$= k(|A| - |\alpha| - |\beta|(|\alpha| + |\beta| - |B|),\quad k > 0.$$

Hence, (1.1) is elliptic at the point z if and only if at this point

(a) $\qquad |A(z)| > |\alpha(z)| + |\beta(z)|,\quad |B(z)| < |\alpha(z)| + |\beta(z)|,$

or

(b) $$|B(z)|>|\alpha(z)|+|\beta(z)|,\quad |A(z)|<|\alpha(z)|+|\beta(z)|.$$

In case (a) eliminating $\bar{w}_z$ (1.1) may be put in the form

$$w_{\bar{z}}-q_1(z)w_z-q_2(z)\overline{w_z}+a(z)w+b(z)\bar{w}=g_1(z) \tag{1.6}$$

with condition

$$|q_1(z)|+|q_2(z)|<1 \tag{1.7}$$

and in case (b) eliminating $\bar{w}_{\bar{z}}$ it may be put in the form

$$w_z-p_1(z)w_{\bar{z}}-p_2(z)\overline{w_{\bar{z}}}+a(z)w+b(z)\bar{w}=g_2(z) \tag{1.8}$$

with condition

$$|p_1(z)|+|p_2(z)|<1. \tag{1.9}$$

Note that by conjugation and transformation $\bar{w}\to w$ (1.8) reduces to (1.6).

Further simplification of the (1.6) with $C^{1,\alpha}$ coefficients connected with the linear transformation of the unknown function $w(z)$. To do this we rewrite (1.6) as

$$w_x+\frac{i(1-|q_1|^2+|q_2|^2+2i\,\mathrm{Im}\,q_1)}{\delta}w_y-\frac{2i\,q_2}{\delta}\bar{w}_y+2a^1w+b^1\bar{w}=2g,$$

where

$$\delta(z)=|1-q_1(z)|^2-|q_2(z)|^2, \tag{1.10}$$

$$a^1=\frac{(1-\overline{q_1})a_1+q_2\overline{b_1}}{\delta},\quad b^1=\frac{(1-\overline{q_1})b_1+q_2\overline{a_1}}{\delta}, \tag{1.11}$$

$$g=\frac{(1-\bar{q}_1)g_1+q_2\bar{g}_1}{\delta}.$$

The linear transformation

$$\varphi=\frac{1-|q_1|^2+|q_2|^2+\sqrt{\sigma}}{2(1-|q_1|^2)}w-\frac{q_2}{1-|q_1|^2}\bar{w}, \tag{1.12}$$

where

$$\sigma = (1 - |q_1|^2 + |q_2|^2)^2 - 4|q_2|^2 \tag{1.13}$$

leads to

$$\varphi_x = -\frac{1}{(1-|q_1|^2)\delta}\left[\frac{i(1-|q_1|^2+|q_2|^2+2\operatorname{Im} q_1)(1-|q_1|^2+|q_2|^2+\sqrt{\sigma})}{2} - 2i|q_2|^2\right]w_y$$

$$+\frac{q_2(i\sqrt{\sigma}-2\operatorname{Im} q_1)}{(1-|q_1|^2)\delta}\bar{w}_y - \left[\frac{(1-|q_1|^2+|q_2|^2+\sqrt{\sigma})a^1 - 2q_2\bar{b}^1}{1-|q_1|^2} - \left(\frac{1}{2}\frac{1-|q_1|^2+|q_2|^2+\sqrt{\sigma}}{1-|q_1|^2}\right)_x\right]w$$

$$-\left[\frac{(1-|q_1|^2+|q_2|^2+\sqrt{\sigma})b^1 - 2q_2\bar{a}'}{1-|q_1|^2} + \left(\frac{q_2}{1-|q_1|^2}\right)_x\right]\bar{w} + \frac{(1-|q_1|^2+|q_2|^2+\sqrt{\sigma})g - 2q_2\bar{g}}{1-|q_1|^2}, \tag{1.14}$$

$$\varphi_y = \frac{1-|q_1|^2+|q_2|^2+\sqrt{\sigma}}{2(1-|q_1|^2)}w_y - \frac{q_2}{1-|q_1|^2}\bar{w}_y + \frac{1}{2}\left(\frac{1-|q_1|^2+|q_2|^2+\sqrt{\sigma}}{1-|q_1|^2}\right)_y w - \left(\frac{q_2}{1-|q_1|^2}\right)_y \bar{w}. \tag{1.15}$$

But since

$$i(1-|q_1|^2+|q_2|^2+\sqrt{\sigma}) = \frac{i(1-|q_1|^2+|q_2|^2-2i\operatorname{Im} q_1)}{\delta} + \frac{i\sqrt{\sigma}}{\delta} - \frac{2\operatorname{Im} q_1}{\delta},$$

$$\frac{i(1-|q_1|^2+|q_2|^2+2i\operatorname{Im} q_1)(1-|q_1|^2+|q_2|^2+\sqrt{\sigma})}{2\delta(1-|q_1|^2)} - \frac{2i|q_2|^2}{1-|q_1|^2}$$

$$= -\left(\frac{i\sqrt{\sigma}}{\delta} - \frac{2\operatorname{Im} q_1}{\delta}\right)\frac{(1-|q_1|^2+|q_2|^2+\sqrt{\sigma})}{2(1-|q_1|^2)},$$

then (1.14) takes the form

$$\varphi_x = \frac{2\,\mathrm{Im}\,q_1 - \mathrm{i}\sqrt{\sigma}}{\delta}\,\frac{1-|q_1|^2+|q_2|^2+\sqrt{\sigma}}{2(1-|q_1|^2)}w_y - \frac{q_2(2\,\mathrm{Im}\,q_1-\mathrm{i}\sqrt{\sigma})}{\delta}\overline{w}_y$$

$$+ a^0 w + b^0\overline{w} + g^0. \tag{1.14'}$$

Now from (1.12), (1.15) and (1.14′) it follows that

$$\varphi_x - \frac{2\,\mathrm{Im}\,q_1 - \mathrm{i}\sqrt{\sigma}}{\delta}\varphi_y = A^0\varphi + B^0 + g^0, \tag{1.16}$$

where

$$A^0 = \frac{1}{(1-|q_1|^2)\Delta}\Big(\frac{\tilde{A}^0(1-|q_1|^2+|q_2|^2+\sqrt{\sigma}}{2} + \tilde{B}^0\overline{q}_2\Big),$$

$$B^0 = \frac{1}{(1-|q_1|^2)\Delta}\Big(\tilde{A}^0 q_2 + \tilde{B}^0\frac{(1-|q_1|^2+|q_2|^2+\sqrt{\sigma})}{2}\Big), \tag{1.17}$$

$$\Delta = \frac{\sqrt{\sigma}(\sqrt{\sigma}+1-|q_1|^2+|q_2|^2)}{2(1-|q_1|^2)},$$

$$\tilde{A}^0 = a^0 - \frac{2\,\mathrm{Im}\,q_1 - \mathrm{i}\sqrt{\sigma}}{4\,\delta}\Big(\frac{1-|q_1|^2+|q_2|^2+\sqrt{\sigma}}{1-|q_1|^2}\Big)_y,$$

$$\tilde{B}^0 = b^0 - \frac{2\,\mathrm{Im}\,q_1 - \mathrm{i}\sqrt{\sigma}}{2\,\delta}\Big(\frac{q_2}{1-|q_1|^2}\Big)_y.$$

We can write (1.16) also as

$$\varphi_{\overline{z}} - q(z)\varphi_z + A_0(z)\varphi + B_0(z)\overline{\varphi} = f_0(z), \tag{1.18}$$

with

$$q(z) = \frac{2\,\mathrm{Im}\,q_1(z) - \mathrm{i}(\sqrt{\sigma} - \delta)}{2\,\mathrm{Im}\,q_1(z) - \mathrm{i}(\sqrt{\sigma} + \delta)}, \tag{1.19}$$

$$A_0 = \frac{A^0 \cdot \delta}{\delta + \sqrt{\sigma} + 2\mathrm{i}\,\mathrm{Im}\,q_1}, \quad B_0 = \frac{B^0 \cdot \delta}{\delta + \sqrt{\sigma} + 2\mathrm{i}\,\mathrm{Im}\,q_1}, \quad f_0 = \frac{g^0 \cdot \delta}{\delta + \sqrt{\sigma} + 2\mathrm{i}\,\mathrm{Im}\,q_1}.$$

By (1.7) it follows that $\delta = |1 - q_1|^2 - |q_2|^2 > 0$ and by (1.19)

$$|q(z)| < 1. \tag{1.20}$$

The equation

$$B\varphi \equiv \varphi_{\bar{z}} - q(z)\varphi_z = 0 \tag{1.21}$$

is called the Beltrami equation. By means of a homeomorphism of the Beltrami equation the equation (1.28) can be simplified to the equation (see [58]):

$$\varphi_{\bar{z}} + A(z)\varphi + B(z)\bar{\varphi} = f_0(z). \tag{1.22}$$

When $f_0(z) \equiv 0$ solutions of (1.22) called generalized analytic or pseudoanalytic functions, must have isolated zeros and singularities and it is meaningful to consider the order of a zero. Also entire generalized analytic functions satisfy Liouville's theorem. The above results can also be extended to the more general equation (1.6) with $g_1 \equiv 0$. In particular a Liouville theorem for the equation

$$w_{\bar{z}} - q_1(z)w_z - q_2(z)\overline{w_z} + a_1(z)w + b_1(z)\bar{w} = 0 \tag{1.6^0}$$

with bounded measurable $q_k(z)$ satisfying (1.7) and with $a_1(z)$, $b_1(z)$ belonging to the space $L_{p,2}(\mathbb{C})$ of all complex valued functions $f(z)$ given on whole complex plane $\mathbb{C}$ with finite norm (see [51])

$$\|f\|_{L_{p,2}} = \Big(\iint_{|z|<1} |f(z)|^p\,dx\,dy\Big)^{1/p} + \Big(\iint_{|z|<1} \Big|z^{-2} f\Big(\frac{1}{z}\Big)\Big|^p\,dx\,dy\Big)^{1/p}$$

asserts: solution (1.6^0), which is continuous on the whole plane $\mathbb{C}$ and vanishes at the infinity, is identically zero. This follows from a similarity principle for solutions (1.6^0):

$$w(z) = \Phi[\,W(z)\,]\,e^{s(z)} \tag{1.23}$$

where Φ is an analytic function, $W(z)$ is some homeomorphism of a z-plane to a W-plane and $s(z)$ is bounded and continuous. All this and many other results can be found in [58].

1.1.2. The Riemann problem for the general first-order elliptic equation

In this section we prove a theorem on the Riemann problem for the general equation (1.6) as sharp as in the case of the classical Riemann problem for analytical functions.

Let G^+ be a multiply connected domain in $\mathbb{C}$ that is bounded by a finite number of closed, noninteresecting $C^{1,\alpha}$ curves Γ_k, $k = 0,1,..,m$, with Γ_0 containing the remainder in its interior. We denote the domains composing the complement of G^+ by G_k^-, $k = 0,...,m$ with G_0^- being unbounded. Let G^- denote $\bigcup_{k=0}^{m} G_k^-$ (see Fig. 1.1):

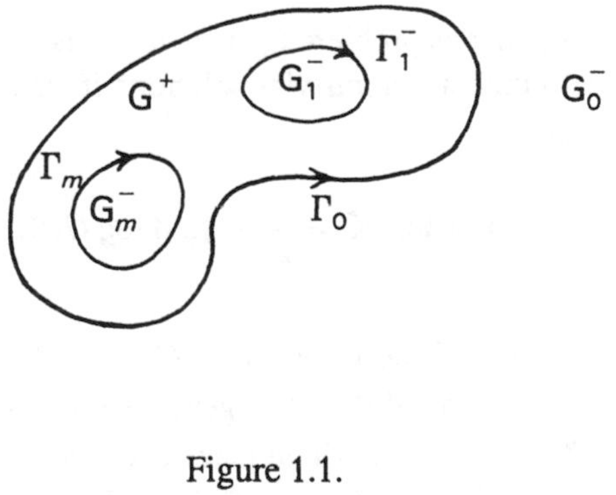

Figure 1.1.

As usual, we orient Γ_0 so that counterclockwise is positive, and thus clockwise is positive for the other Γ_k. If $w(z)$ is a function defined on $\mathbb{C} \setminus \Gamma$, then for $t \in \Gamma$ we denote by $w^+(t)(w^-(t))$ the limit of $w(z)$ (if it exists) as $z \to t$ from within $G^+(G^-)$. Let us now state the Riemann problem that will be considered.

Riemann Problem (R). Let there be given Hölder continuous functions $G(t) \neq 0$, $g(t)$ on Γ. Find a solution to (1.6) in $G^+ \cup G^-$ that is Hölder continuous in $\bar{G}^+ = G^+ + \Gamma$ and $\bar{G}^- = G^- + \Gamma$ with $\bar{z}$-derivative in $L_{p,2}(\mathbb{C})$, $p > 2$, vanishing at infinity and satisfying the jump condition

$$w^+ - G w^- = g \tag{1.24}$$

on Γ.

We recall a classical result, known in the analytical case, i.e. when $q_1 = q_2 = a_1 = b_1 = g_1 \equiv 0$ (see [52]): if $\kappa \geq 0$ then the problem (R) is solvable unconditionally and the general solution is represented by the formula

$$w(z) = \frac{X(z)}{2\pi i}\int_\Gamma \frac{g(t)\,dt}{X^+(t)(t-z)} + X(z)P_{\kappa-1}(z), \tag{1.25}$$

but if $\kappa < 0$ then the problem (R) is solvable if and only if its right-hand side $g(t)$ satisfies the conditions

$$\int_\Gamma g(t)\frac{t^j\,dt}{X^+(t)} = 0, \quad j = 0,\ldots, -\kappa - 1 \tag{1.26}$$

and the solution is represented by the formula

$$w(z) = \frac{X(z)}{2\pi i}\int_\Gamma \frac{g(t)\,dt}{X^+(t)(t-z)}. \tag{1.27}$$

In particular, the homogeneous problem ($g \equiv 0$) has exactly 2κ linearly independent (over the field of real numbers) nontrivial solutions if $\kappa > 0$ and has no nontrivial solutions if $\kappa \leq 0$, where

$$\kappa = \mathrm{Ind}_\Gamma G = \frac{1}{2\pi i}\int_\Gamma d\log G(t),$$

$P_{\kappa-1}(z)$ is the polynomial of the order $\kappa-1$ ($P_{\kappa-1} \equiv 0$ for $\kappa = 0$), $X(z)$ is a piecewise holomorphic solution of the homogeneous problem (R), which is called the canonical solution. $X(z) \neq 0$ on the finite part of $\mathbb{C}$, has a zero of order κ if $\kappa > 0$ and a pole of order $-\kappa$ if $\kappa < 0$ at infinity. First we derive the formulae for solutions of the problem (R) with $q_1 = q_2 = a = b = g \equiv 0$:

$$w_{\bar{z}} = f(z) \tag{1.28a}$$

in $\mathbb{C} \setminus \Gamma$ and

$$w^+ - Gw^- = 0 \tag{1.28b}$$

on Γ.

Lemma 1.1 *If* $\kappa \geq 0$ *then the problem (1.28a,b) is solvable for any* $f(z) \in L_{p,2}(\mathbb{C})$, $p > 2$, *solutions represents as*

$$w(z) = Tf + X(z)\,(T\theta f - \theta(z)Tf) + X(z)P_{k-1}(z)$$
$$\equiv T_0 f + \sum_{j=0}^{\kappa-1} a_j\, z^j X(z), \tag{1.29}$$

where

$$Tf = -\frac{1}{\pi}\iint_{\mathbb{C}} \frac{f(\zeta)\,d\xi\, d\eta}{\zeta - z},\quad \theta(z) = \frac{1}{2\pi i}\int_\Gamma \frac{(1-G(t))\,dt}{X^+(t)\,(t-z)}.$$

If $\kappa < 0$ then the problem (1.28a,b) is solvable if and only if

$$\iint_{\mathbb{C}} f(z)\frac{z^j}{X(z)}\,dx\,dy = 0,\quad j = 0,\dots,-\kappa-1. \tag{1.30}$$

If this condition is satisfied then the solution represents as

$$w(z) = X(z)\,T\frac{f}{X}. \tag{1.31}$$

Proof Integrating (1.27) we have

$$w(z) = \varphi(z) + Tf \tag{1.32}$$

with piecewise holomorphic φ satisfying (according to (1.28)) the boundary condition

$$\varphi^+ - G\varphi = (G-1)\,Tf \tag{1.33}$$

on Γ.

Formula (1.29) follows from (1.25), (1.32). If $\kappa > 0$ then conditions (1.26) are

$$\iint_{\mathbb{C}}\left(\frac{1}{2\pi i}\int_\Gamma \frac{(1-G(t))\,t^j dt}{X^+(t)\,(t-\zeta)}\right) f(\zeta)\,d\xi\,d\eta = 0,\ j = 0,\dots,-\kappa-1. \tag{1.34}$$

But

$$\frac{1}{2\pi i}\int_\Gamma \frac{t^j\,dt}{X^+(t)\,(t-\zeta)} = \begin{cases} \dfrac{\zeta^j}{X(\zeta)}, & \zeta\in G^+ \\ 0, & \zeta \in G^-, \end{cases}$$

$$\frac{1}{2\pi i}\int_\Gamma \frac{G(t)\,t^j\,dt}{X^+(t)(t-\zeta)} = \frac{1}{2\pi i}\int_\Gamma \frac{t^j\,dt}{X^-(t)(t-\zeta)} = \begin{cases} 0 \ , & \zeta \in G^+, \\ -\dfrac{\zeta^j}{X(\zeta)}, & \xi \in G^-, \end{cases}$$

and thus (1.34) reduces to (1.30). Moreover, since

$$\frac{1}{2\pi i}\int_\Gamma \frac{(G(t)-1)Tf}{t-z}\,dt = -\frac{1}{\pi}\iint_{\mathbb{C}}\Big(\frac{1}{2\pi i}\int_\Gamma \frac{(1-G(t))\,dt}{(t-\zeta)(t-z)}\Big) f(\zeta)\,d\xi\,d\eta$$

$$= T\theta f - \theta(z)Tf,$$

and in case $\kappa > 0$:

$$\theta(z) = \frac{1}{2\pi i}\int_\Gamma \Big(\frac{1}{X^+(t)} - \frac{1}{X^-(t)}\Big)\frac{dt}{t-z} = \frac{1}{X(z)},$$

then (1.31) follows from (1.27), (1.32).

Without loss of generality we assume that $g \equiv 0$.

Theorem 1.1 *Let* $q_1(z)$, $q_2(z)$ *be bounded measurable functions, vanishing* at *infinity of the order* $\varepsilon > 0$ *satisfying the condition*

$$\text{l.u.b}_{\mathbb{C}}\,(|q_1(z)| + |q_2(z)|) < 1, \tag{1.35}$$

$a_1(z)$, $b_1(z)$, $g_1(z)$ *belongs to the space* $L_{p,2}(\mathbb{C})$, $p > 2$. *Then in case* $\kappa \geq 0$ *the Riemann problem* (R) *is solvable unconditionally. In case* $\kappa < 0$ *the problem* (R) *has a solution if and only if*

$$\iint_{\mathbb{C}} g_1(z) v_j(z)\,dx\,dy = 0, \quad j = 1,\dots,2\kappa, \tag{1.36}$$

where $\{v_j(z)\}$ *forms a complete system of linearly independent solutions of the homogeneous adjoint problem*:

$$v_{\bar z} - (q_1(z)v)_z - \overline{(q_2(z)v_{\bar z}} - a_1(z)v - \overline{b_1(z)v} = 0$$

in $\mathbb{C}\setminus\Gamma$ *and*

$$Gv^+ - v^- = 0$$

on Γ. *The homogeneous Riemann problem* $(g_1 \equiv 0)$ *has exactly* 2κ *linearly independent nontrivial solutions if* $\kappa > 0$, *and has no nontrivial solution if* $\kappa \le 0$.

Proof The solution $w(z)$ is represented in the form (1.29) if $\kappa \ge 0$ with

$$f = w_{\bar z} \in L_{p,2}(\mathbb{C}), \quad p > 2 \tag{1.37}$$

and in the form (1.31) if $\kappa < 0$, provided the function f satisfies condition (1.30).

Let us assume $\kappa \ge 0$. Then differentiating (1.29), we obtain

$$w_{\bar z} = f(z), \; w_z = Sf + X(z)(S\theta f - \theta(z) Sf) + X'(z)(T\theta f - \theta(z) Tf)$$
$$- X(z)\theta'(z)\, Tf + \sum_{j=0}^{\kappa-1} a_j \varphi_j'(z) \equiv Sf + T_1 f + \sum_{j=0}^{\kappa-1} a_j \varphi_j', \tag{1.38}$$

where

$$Sf = -\frac{1}{\pi}\iint_{\mathbb{C}} \frac{f(\zeta)\, d\xi\, d\eta}{(\zeta - z)^2}, \quad \varphi_j(z) = z^j X(z). \tag{1.39}$$

Substituting (1.29), (1.38) into (1.6) we obtain the integral equation for f:

$$f - K_+ f = g(z) + \sum_{j=0}^{\kappa-1} a_j g_j^{(1)} + \bar a_j g_j^{(2)}, \tag{1.40}$$

where K_+ is linear continuous operator in $L_{p,2}(\mathbb{C})$, $p > 2$:

$$K_+ f \equiv q_1 Sf + q_2 \overline{Sf} - (a_1 T_0 f - q_1 T_1 f) - (b_1 \overline{T_0 f} - q_2 \overline{T_1 f}), \tag{1.41}$$

$$g_j^{(1)} = -a_1 \varphi_j + q_1 \varphi_j', \quad g_j^{(2)} = -b\overline{\varphi_j} + q_2 \overline{\varphi_j'}.$$

Since the L_2-norm of the singular integral operator S is equal to one, then by means of (1.35) we see that the operator

$$K_+^0 f \equiv q_1 Sf + q_2 \overline{Sf}$$

will be contractive in $L_{p,2}(\mathbb{C})$ with some $2 < p < 2 + \varepsilon'$ and so $K - K_+^0$ will be invertible in this space.

Multiplying both sides (1.40) to $(I - K_+^0)^{-1}$ we reduce it to Fredholm's equation, which is equivalent to (1.40). Let $f^0(z) \in L_{p,2}(\mathbb{C})$ be solution of the corresponding homogeneous equation

$$f^0 - K_+ f^0 = 0. \tag{1.42}$$

Then the function

$$\overset{\circ}{w}(z) = Tf^0 + X(z)\,(T\theta f^0 - \theta(z)\,Tf^0) \tag{1.43}$$

will be a solution of the homogeneous problem (R).

Since $G = -X^+/X^-$, then the function $\overset{\circ}{v}(z) = \overset{\circ}{w}(z)/X(z)$ will be continuous in finite part of the plane $\mathbb{C}$ solution of the homogeneous equation

$$\overset{\circ}{v}_{\bar z} - q_1\,\overset{\circ}{v}_z - q_1^0\,\overline{\overset{\circ}{v}_z} + A^0(z)\overset{\circ}{v} + B^0(z)\,\overline{\overset{\circ}{v}} = 0, \tag{1.44}$$

growing at infinity of the order $\kappa - 1$, where

$$q_2^0(z) = q_2(z)X(z)/\overline{X(z)}, \; A^0 = a_1 - q_1 X'/X, \; B^0 = b_1\bar{X}/X - q_2\overline{X'}/X.$$

Since

$$\text{l.u.}_{\mathbb{C}}\text{.b}\,(\,|\,q_1(z))\,|+|\,q_2^0(z)\,|\,) = \text{l.u.}_{\mathbb{C}}\text{.b}\,(\,|\,q_1(z)\,|+|\,q_2(z)\,|\,) < 1$$

and evidently $A^0(z), B^0(z) \in L_{p,2,\,p>2}$ because $X/(z)/X(z)$ vanishes at infinity and so A^0, B^0 vanishes at infinity at least of the order $1+\varepsilon, \varepsilon > 0$, then it follows by the

Liouville thoerem that $\overset{\circ}{v}(z) \equiv 0$ for $\kappa = 0$ and by (1.23)

$$\overset{\circ}{v}(z) = \sum_{k=0}^{\kappa-1} C_k W^k e^s$$

for $\kappa > 0$, where $W = W(z)$ is a homeomorphism of a z-plane to the W-plane. Consequently $\overset{\circ}{w}(z) \equiv 0$ for $\kappa = 0$ and

$$\overset{\circ}{w}(z) = \sum_{k=0}^{\kappa-1} C_k W^k(z)\, e^{s(z)} \tag{1.45}$$

for $\kappa > 0$. Since Tf^0 is continuous on the whole plane for $f^0(z) \in L_{p,2}(\mathbb{C})$, $p > 2$, then it follows from (1.43) that

$$\overset{\circ}{w}^+ - \overset{\circ}{w} = X^+(T\theta f^0 - \theta^+ Tf^0) - X^-(T\theta f^0 - \theta^- Tf^0)$$

and by means of Cauchy's theorem and formula we obtain for $z \in G^-$:

$$\frac{1}{2\pi i}\int_\Gamma \frac{\overset{\circ}{w}{}^+(t) - \overset{\circ}{w}{}^-(t)}{t-z}\,dt = X(z)\,(T\theta f^0 - \theta(z)\,Tf^0). \tag{1.46}$$

The right-hand side of this equality vanishes at infinity of the order $\kappa + 1$, so the same must be so for the left-hand side, i.e.

$$\int_\Gamma (\overset{\circ}{w}{}^+(t) - \overset{\circ}{w}{}^-(t))\, t^j\, dt = 0, \quad j = 0,\dots,\kappa-1. \tag{1.47}$$

Substituting (1.45) into (1.47) we obtain

$$\sum_{k=0}^{\kappa-1}\left(\int_\Gamma (G(t) - 1)\, t^j\, W^k(t) X^-(t)\, e^{s(t)}\, dt\right) C_k = 0$$

and since $\kappa > 0$ it follows then that $C_k = 0$ and so $\overset{\circ}{w}(z) \equiv 0$, because

$$\det\int_\Gamma (G(t) - 1)\, t^j\, W^k(t) X^-(t)\, e^{s(t)}\, dt \neq 0,$$

i.e. the right-hand side of (1.43) is identically zero, from which we obtain $f^0(z) \equiv 0$ by differentiation with respect to $\bar{z}$. This means that the equation (1.40) is uniquely solvable. Since the homogeneous Riemann problem (R) is equivalent to the nonhomogeneous integral equation (1.40) with $g_1 \equiv 0$ the right-hand side of which is the linear combination with 2κ real coefficients, then it has exactly 2κ nontrivial linearly independent solutions. If $\kappa < 0$, then differentiating (1.31) we obtain

$$w_{\bar{z}} = f, \quad w_z = X(z) S\frac{f}{X} + X'(z)\, T\frac{f}{X}.$$

Substituting this expression and (1.31) into (1.6) we obtain the integral equation

$$f - K_- f = g_1, \tag{1.48}$$

where K_- is linear continuous operator in $L_{p,2}(\mathbb{C})$, $p > 2$:

$$K_- f \equiv q_1 X s\frac{f}{X} + q_2 \bar{X}\, \overline{S\frac{f}{X}} - (a_1 X - q_1 X')T\frac{f}{X} - (b_1 \bar{X}')\overline{T\frac{f}{X}}.$$

In terms of function $f_1 = f/X$ the equation (1.48) has the form

$$f_1 - \hat{K}_- f_1 = g_1/X, \tag{1.48'}$$

where

$$\hat{K}_- = \hat{K}_-^0 + \hat{K}_-^1,$$

$$\hat{K}_-^0 \equiv q_1 Sf_1 + q_2^0 \overline{Sf_1}\ , \ \hat{K}_-^1 f_1 \equiv (q_1 \frac{X'}{X} - a_1) Tf_1 + (q_2 \frac{\overline{X'}}{X} - b_1 \frac{\overline{X}}{X}) \overline{Tf_1}\ .$$

As above it follows that $I - \hat{K}_-^0$ is invertible in $L_{p,2}(\mathbb{C}), p > 2$, and so we may apply Fredholm's theorem to the euqation (1.48′). Let us denote by $f_1^0(z) \in L_{p,2}(\mathbb{C})$, $p > 2$ a solution of the homogeneous equation

$$f_1^0 - \hat{K}_- f_1^0 = 0.$$

Then the function $v^1(z) = Tf_1^0$ will be a solution of the homogeneous equation

$$v_{\bar{z}}^1 - q_1 v_z^1 - q_2^0 \overline{v_z^1} + (a_1 - q_1 \frac{X'}{X}) v^1 + (b_1 \frac{\overline{X}}{X} - q_2 \frac{\overline{X'}}{X}) \bar{v}^1 = 0$$

continuous on the whole plane $\mathbb{C}$ and vanishing at infinity. Therefore by Liouville's theorem,

$$Tf_1^0 = v^1(z) \equiv 0,$$

and so $f_1^0 \equiv 0$ by differentiation with respect to $\bar{z}$. Consequently (1.48) is uniquely solvable and its solution may be written as

$$f = g_1 + R_- g_1, \tag{1.49}$$

where R_- is some continuous linear integral operator. Since the equation (1.48) is equivalent to the Riemann problem provided the complementary conditions (1.30) are satisfied, then of (1.49) implies (1.36)

Let us instead (1.24) consider now the general jump condition

$$w^+ - Gw^- - H\overline{w^-} = 0 \tag{1.24′}$$

with $G(t), H(t)$ Hölder continuous on Γ.

If $G(t) \neq 0$ on Γ, then the problem (1.6), (1.24′) can be investigated by means of the theory of singular integral equations on Γ, which allows us to prove the Noetherian property of this problem and to calculate its index equal to $\mathrm{Ind}_\Gamma G(t)$, but does not allow us to get sharp results as in the case of the Riemann problem.

Nevertheless for the case $|G|>|H|$ on Γ we have

Corollary 1.1. *If* $|G(t)|>|H(t)|$ on Γ, *then the results of Theorem 2.1 hold also for the general problem (1.6), (1.24').*

Proof. Indeed in this case we can write (1.24') as

$$w^+ - G(w^- - H/G\cdot \overline{w^-}) = 0. \tag{1.50}$$

Now we extend the function $p = H/G$ into G^- such that $p(z) \in C^1(G^-)$ and $|p|<1$ in G^-. Then we introduce ψ by

$$\psi(z) = \begin{cases} w(z), & z\in G^+, \\ w(z)-p(z)\overline{w(z)}, & z\in G^- \end{cases}$$

and obtain for ψ the Riemann problem

$$\psi_{\bar z} - q_1^*(z)\psi_z - q_2^*(z)\overline{\psi_z} + a^*(z)\psi + b^*(z)\bar\psi = f^*(z) \text{ in } \mathbb{C}\setminus\Gamma$$

with

$$|q_1^*(z)|+|q_2^*(z)|<1,\ a^*, b^*, f^* \in L_{p,2,},\ p>2$$

and

$$\psi^+ - G\psi^- = 0 \text{ on } \Gamma$$

and then we apply Theorem 2.1.

1.1.3. The Schwartz problem in a multiply connected domain

In this section we investigate the Schwartz problem for the general first-order elliptic equation (1.6) by means of a singular integral equation on a multiply connected domain G^+. The Schwartz problem is to find a solution $w(z)$ to (1.6) in G^+ that is Hölder continuous in $\overline{G^+}$ with $\bar z$-derivative in $L_p(\overline{G^+})$, $p>2$, and satisfying the condition[1]

$$\operatorname{Re} w = 0 \tag{1.51}$$

on Γ. We assume that $q_k(z)$ are bounded measurable and satisfy (1.7) in $\overline{G^+}$, and

1 For simplicity we take a homogeneous boundary condition without loss of generality.

$a_1, b_1, g_1 \in L_p(\overline{G^+})$ with $p>2$.

To solve this problem, let us denote by ρ the $\bar{z}$-derivative

$$w_{\bar{z}} = \rho(z). \tag{1.52}$$

The problem (1.51), (1.52) is solvable if and only if (see [25]).

$$\operatorname{Re}\iint_{G^+} \rho(z)\,\Phi_j(z)\,dx\,dy = 0,\ j = 1,\dots,m, \tag{s}$$

where $\Phi_j(z) = 2\partial u_j/\partial z$, u_j is the harmonic measure of the curve Γ_j with respect to the domain G^+. In this case all solutions of the problem (1.51), (1.52) are given by (see [25])

$$w(z) = T^0_{G^+}\rho + iC, \tag{1.53}$$

where C is an arbitrary real constant and

$$T^0_{G^+}\rho \equiv \iint_{G^+} \left[\left(\tilde{\ell}(z,\zeta) - \frac{1}{\pi(\zeta - z)}\right)\rho(\zeta) - \tilde{K}(z,\bar{\zeta})\overline{\rho(\zeta)}\right] dG^+_\zeta, \tag{1.54}$$

$$\tilde{\ell}(z,\zeta) = \int_0^z \ell(z,\zeta)\,dz,\quad \tilde{K}(z,\bar{\zeta}) = \int_0^z K(z,\zeta)\,dz,$$

$K(z,\bar{\zeta})$ is Bergman's kernel function of the domain G^+, the function $\ell(z,\zeta)$ is holomorphic in z, ζ and continuous in the closure $\overline{G^+}$. Differentiating (1.53) with respect to $\bar{z}$ and z we obtain

$$w_{\bar{z}} = \rho,\ w_z = S^0_{G^+}\rho \tag{1.55}$$

where $S^0_{G^+}$ is the singular integral operator

$$S^0_{G^+}\rho \equiv \iint_{G^+} \left[\left(\tilde{\ell}(z,\zeta) - \frac{1}{\pi(\zeta - z)^2}\right)\rho(\zeta) - K(z,\bar{\zeta})\,\overline{\rho(\zeta)}\right] dG^+_\zeta. \tag{1.56}$$

Substituting (1.53), (1.55) into (1.6) we obtain the following singular integral equation for $\rho \in L_p(\overline{G^+})$, $p>2$:

$$\rho - q_1 S^0_{G^+}\rho - q_2 \overline{S^0_{G^+}\rho} + a_1 T^0_{G^+}\rho + b_1 \overline{T^0_{G^+}\rho}$$

$$= g_1 - \mathrm{i}C(a_1 - b_1). \tag{1.57}$$

According to the Calderon-Zygmund theorem the operator $S^0_{G^+}$ is linear and continuous in $L_p(\bar{G}^+)$, $p > 1$, with Λ_p norm continuous in p.

Lemma 1.2. *The operator* $S^0_{G^+}$ *has an* L_2*-norm equal to one.*

Proof. Let $\rho(z)$ be a C^∞ function with compact support in G^+. Since

$$S^0_{G^+}\rho = \frac{\partial T^1_{G^+}\rho}{\partial z} - K\bar{\rho},$$

where

$$T^1_{G^+}\rho \equiv \iint_{G^+} \left[\tilde{\ell}(z,\zeta) - \frac{1}{\pi(\zeta - z)}\right]\rho(\zeta)\, \mathrm{d}G^+_\zeta,$$

$$K\rho \equiv \iint_{G^+} K(z,\bar{\zeta})\,\rho(\zeta)\,\mathrm{d}G^+_\zeta,$$

then by means of the Green formula

$$\| S^0_{G^+}\rho \|^2_{L_2(\bar{G}^+)} = \iint_{G^+} \frac{\partial T^1_{G^+}\rho}{\partial z}\,\frac{\partial \overline{T^1_{G^+}\rho}}{\partial \bar{z}}\,\mathrm{d}G_z - \iint_{G^+} K\bar{\rho}\,\frac{\partial \overline{T^1_{G^+}\rho}}{\partial \bar{z}}\,\mathrm{d}G^+_z$$

$$- \iint_{G^+} \frac{\partial T^1_{G^+}\rho}{\partial z}\,\overline{K\bar{\rho}}\,\mathrm{d}G^+_z + \| K\bar{\rho} \|^2_{L_2(\bar{G}^+)}$$

$$= \frac{1}{2\mathrm{i}} \int_\Gamma \overline{T^1_{G^+}\rho}\,\frac{\partial T^1_{G^+}\rho}{\partial t}\,\mathrm{d}t - \iint_{G^+} \frac{\partial \rho}{\partial z}\,\overline{T^1_{G^+}\rho}\,\mathrm{d}G^+_z$$

$$- \mathrm{Im} \int_\Gamma K\bar{\rho}\,\overline{T^1_{G^+}\rho}\,\mathrm{d}t + \| K\bar{\rho} \|^2_{L_2(\bar{G}^+)}$$

$$= \|\rho\|^2_{L_2(\bar{G}^+)} + \|K\bar{\rho}\|^2_{L_2(\bar{G}^+)} + \iint_{G^+} \rho(z)\,dG_z^+ \iint_{G^+} \overline{\rho(\zeta)}\,dG_\zeta^+$$

$$\times \Big\{ \frac{-1}{2\pi^2 i} \int_\Gamma \frac{dt}{(\bar{t}-\bar{\zeta})(t-z)^2} + \frac{1}{2i} \int_\Gamma \overline{\tilde{\ell}(t,\zeta)}\,\ell(t,\zeta)\,dt$$

$$+ \frac{1}{2\pi i} \int_\Gamma \frac{\ell(t,z)}{\bar{t}-\bar{\zeta}} - \frac{1}{2\pi i} \int_\Gamma \frac{\overline{\tilde{\ell}(t,\zeta)}\,dt}{(t-z)^2} \Big\}$$

$$+ \operatorname{Im} \iint_{G^+} \overline{\rho(\zeta)}\,dG_\zeta^+ \iint_{G^+} \rho(z)\,dG_z \Big[-\frac{1}{\pi} \int_\Gamma \frac{K(t,\bar{\zeta})\,dt}{\bar{t}-\bar{z}} - \int_\Gamma \overline{\tilde{\ell}(t,z)}\, K(t,\bar{\zeta})\,dt \Big].$$

Let $G_\varepsilon^+ = G^+ - (|\tau - z| < \varepsilon)$. Then by Green's formula

$$-\frac{1}{\pi} \iint_{G_\varepsilon^+} \frac{\ell(\tau,z)\,dG_\tau^+}{(\bar{\tau}-\bar{\zeta})^2} = \frac{1}{2\pi i} \int_\Gamma \frac{\ell(t,z)\,dt}{\bar{t}-\bar{\zeta}} + O(\varepsilon),$$

$$\frac{1}{\pi} \iint_{G_\varepsilon^+} \frac{K(\tau,\bar{\zeta})\,dG_\tau^+}{(\bar{\tau}-\bar{z})^2} - \iint_{G_\varepsilon^+} \overline{\ell(\tau,z)}\,K(\tau,\bar{\zeta})\,dG_\tau^+$$

$$= \frac{1}{2\pi i} \int_\Gamma \Big(\frac{1}{\bar{t}-\bar{z}} + \pi i\, \overline{\tilde{\ell}(t,z)} \Big) K(t,\bar{\zeta})\,dt + O(\varepsilon)$$

and so as $\varepsilon \to 0$:

$$-\frac{1}{\pi} \iint_{G^+} \frac{\ell(\tau,z)\,dG_\tau^+}{(\bar{\tau}-\bar{\zeta})^2} = \frac{1}{2\pi i} \int_\Gamma \frac{\ell(t,z)\,dt}{\bar{t}-\bar{\zeta}},$$

$$\frac{1}{\pi} \iint_{G^+} \frac{K(\tau,\bar{\zeta})\,dG_\tau^+}{(\bar{\tau}-\bar{z})^2} - \iint_{G_\varepsilon^+} \overline{\ell(\tau,z)}\,K(\tau,\bar{\zeta})\,dG_\tau^+$$

$$= \frac{1}{2\pi i} \int_\Gamma \Big(\frac{1}{\bar{t}-\bar{z}} + \pi i\, \overline{\tilde{\ell}(t,z)} \Big) K(t,\bar{\zeta})\,dt.$$

Taking into account an integral representation of Bergman's kernel (see [6]):

$$K(z,\bar\zeta) = \frac{1}{2\pi^2 i}\int_\Gamma \frac{dt}{(\bar t - \bar\zeta)(t-z)^2} + \iint_{G^+} \overline{\ell(\tau,\zeta)}\,\ell(\tau, z)\, dG_\tau^+$$

and that

$$\frac{1}{\pi}\iint_{G^+} \frac{K(\tau,\bar\zeta)\, dG_\tau^+}{(\bar t - \bar\zeta)(t-z)^2} = \iint_{G^+} \overline{\ell(\tau,z)}\, K(\tau,\bar\zeta)\, dG_\tau^+$$

we obtain

$$\| S^0_{G^+}\rho \|^2_{L_2(\overline{G^+})} = \| \rho \|^2_{L_2(\overline{G^+})} + \| K\bar\rho \|^2_{L_2(\overline{G^+})} - \iint_{G^+}\iint_{G^+} K(z,\bar\zeta)\,\rho(z)\,\overline{\rho(\zeta)}\, dG_z^+\, dG_\zeta^+$$

$$= \| \rho \|^2_{L_2(\overline{G^+})} + \| K\bar\rho \|^2_{L_2(\overline{G^+})} - \sum_{k=1}^{\infty} |c_k|^2,$$

where $c_k = (\bar f, \omega_k)$ are Fourier coefficients of the function $\overline{f(z)}$, because of

$$K(z,\bar\zeta) = \sum_{k=1}^{\infty} \omega_k(z)\,\overline{\omega_k(\zeta)},$$

$\{\omega_k\}$ is a complete orthonormal system of analytic functions in G^+.

Finally since

$$\| K\bar\rho \|^2_{L_2(\overline{G^+})} = \iint_{G^+} \sum_{k=1}^{\infty} c_k\omega_k(z)\cdot \sum_{k=1}^{\infty} \bar c_k \omega_k(z)\, dG_z^+ = \sum_{k=1}^{\infty} |c_k|^2,$$

then it follows that $\| S^0_{G^+}\rho \|^2_{L_2(\overline{G^+})} = \| \rho \|^2_{L_2(\overline{G^+})}$, i.e. $\| S^0_{G^+} \|_{L_2(\overline{G^+})} = 1$.

According Lemma 1.2 an L_p-norm of the operator

$$\hat S_{G^+}\rho \equiv q_1 S^0_{G^+}\rho + q_2 \overline{S^0_{G^+}\rho}$$

satisfy

$$\| \hat S_{G^+} \|_{L_p(G^+)} \le q\,\Lambda_p < 1$$

for some $0 < p - 2 < \varepsilon$.

Hence for some fixed $p > 2$ the operator $I - \hat{S}_{G^+}$ is invertible in $L_p(G^+)$, so we may apply Fredholm's theorem to (1.57).

Let $\rho^0 \in L_p(G^+)$, $p > 2$ be a solution of the homogeneous equation, corresponding to (1.57).

Then the function

$$\overset{\circ}{w}(z) = T^0_{G^+} \rho^0 \tag{1.58}$$

will be a solution of the Schwarz's problem (1.51) for the homogeneous equation (1.5′).

Since the function $T_{G^+}\rho$ is holomorphic in G^- and vanishes at infinity, then by Cauchy's theorem we obtain from (1.58),

$$\frac{1}{2\pi i}\int_\Gamma \frac{\overset{\circ}{w}(t)\,dt}{t - z} = \iint_{G^+} \left[\tilde{\ell}(z,\zeta)\,\rho(\zeta) - \tilde{K}(z,\bar{\zeta})\,\overline{\rho(\zeta)}\right] dG^+_\zeta$$

for $z \in G^+$, because the right-hand side of this expression is holomorphic in G^+. Moreover this right-hand side vanishes at the origin $z = 0$, so the left-hand side must also vanish:

$$\int_\Gamma \frac{\overset{\circ}{w}(t)\,dt}{t} = 0. \tag{1.59}$$

From (1.6′) by the similarity principle (see [51]) we write

$$\overset{\circ}{w}(z) = \Phi[\,W(z)\,]\,e^{s(z)}$$

with holomorphic Φ and continuous $s(z)$: $\operatorname{Im} s \mid_T = 0$, where $W(z)$ is some homeomorphism of the z-plane to the W-plane. And then from (1.51) it follows that $\operatorname{Re}\Phi = 0$, so

$$\overset{\circ}{w}(z) = i\,C^0 e^s, \tag{1.60}$$

where C^0 is an arbitrary real constant. Now from (1.60) and (1.59) it follows that $\overset{\circ}{w}(z) \equiv 0$, i.e.

$$T^0_{G^+} \rho^0 \equiv 0.$$

Differentiating this identity with respect to $\bar{z}$ we obtain $\rho^0 \equiv 0$. This means that (1.57) has a unique solution, which may be put in the form

$$\rho = g_1 - \mathrm{i}C(a_1 - b_1) + \tilde{R}g_1 - \mathrm{i}C\tilde{R}(a_1 - b_1), \tag{1.61}$$

where $\tilde{R}$ is a linear continuous (resolvent) operator.

Substituting (1.61) into equations (Φ), we obtain the conditions

$$\mathrm{Re}\iint\limits_{G^+}\left[(a_1 - b_1) + \tilde{R}(a_1 - b_1)\right]\Phi_j(z)\,\mathrm{d}G_z^+ + \mathrm{Re}\iint\limits_{G^+}(g_1 + \tilde{R}g_1)\,\Phi_j\,\mathrm{d}G_z^+ = 0,$$

$$j = 1, \dots, m.$$

If, for all j,

$$\mathrm{Re}\iint\limits_{G^+}\left[(a_1 - b_1) + \tilde{R}(a_1 - b_1)\right]\Phi_j\,\mathrm{d}G_z^+ = 0, \tag{1.62}$$

then we get exactly m solvability conditions for the right-hand side g_1 of the Schwartz problem

$$\mathrm{Re}\iint\limits_{G^+} g_1 \cdot V_j\,\mathrm{d}G_z = 0, \quad j = 1, \dots, m, \tag{1.63}$$

where $v_j = \Phi_j + \tilde{R}^*\Phi_j$, $\tilde{R}^*$ being the operator adjoint to $\tilde{R}$, but if for some j^0 $(0 \le j^0 \le m)$:

$$\mathrm{Re}\iint\limits_{G^+}\left[(a_1 - b_1) + \tilde{R}(a_1 - b_1)\right]\Phi_j\,\mathrm{d}G_z^+ \neq 0,$$

then we get $m - 1$ solvability conditions of the form (1.63), because in this case the real constant C is determined automatically. Thus we have proved the following:

Theorem 1.2. *If the equalities (1.62) hold for all* $j = 1, .., m$ *(which is realized when* $a_1 = b_1 = 0$*) then the Schwartz problem is solvable if and only if its right-hand side satisfies exactly* m *orthogonal conditions, but if for some* j^0 *the equality fails, then Schwartz problem is solvable if and only if* $m - 1$ $(m \ge 1)$ *orthogonal conditions are fulfilled for the right-hand side. In the first case the homogeneous* $(g_1 \equiv 0)$ *Schwartz problem has one nontrivial solution and in the second case it has none.*

1.1.4 The Riemann-Hilbert problem in a multiply connected domain

The boundary condition (1.51) is a particular case of the general condition on Γ

$$\operatorname{Re} \bar{\lambda} w = 0 \tag{1.64}$$

with given Hölder continuous function $\lambda(t) \neq 0$, $t \in \Gamma$. The Riemann-Hilbert (RH) boundary-value problem is to find a solution to (1.6) in G^+ that is Hölder continuous in $\bar{G}^+$ with $\bar{z}$-derivative in $L_p(G^+)$ and satisfies the condition (1.64). We assume that $z = 0$ is the inner point of G^+ and denote $\kappa = \kappa_0 + \kappa_1 + \cdots + \kappa_m$, where

$$\kappa_j = \frac{1}{2\pi} \Delta_{\Gamma_j} \arg \lambda, \quad j = 0, \ldots, m. \tag{1.65}$$

As in the previous section we shall investigate the RH problem by means of a singular integral equation on a multiply connected domain G^+. To do this we need an integral representation formula for functions which satisfy the boundary condition (1.64). To derive this representation we may, without loss of generality, take $t^{-\kappa}$ instead of $\bar{\lambda}$, i.e. write (1.64) as

$$\operatorname{Re} t^{-\kappa} w = 0 \tag{1.66}$$

on Γ (see [58]). Evidently the case $\kappa = 0$ is the Schwartz problem.

First we assume $\kappa > 0$. If we denote the $\bar{z}$-derivative of w by ρ, then we have

$$w = \varphi + T_{G^+} \rho, \tag{1.67}$$

where φ is holomorphic in G^+ and Hölder continuous in $\bar{G}^+$. Substituting (1.67) into (1.66) we obtain the RH problem for a holomorphic function:

$$\operatorname{Re} t^{-\kappa} \varphi = h_0(t) = -\frac{1}{2}\left(t^{-\kappa} T_{G^+} \rho + \bar{t}^{-\kappa} \overline{T_{G^+} \rho}\right), \tag{1.68}$$

which is solvable if and only if

$$\int_\Gamma h_0(t)\, \bar{t}^{-\kappa} \psi_j \, dt = 0, \quad j = 1, \ldots, m - \tau, \tag{1.69}$$

where ψ_j are the holomorphic solutions of the adjoint problem

$$\operatorname{Re} t'(s) t^{\kappa} \psi = 0, \tag{1.70}$$

where $t'(s) = dt/ds$, $t = t(s)$ is parametrical equation of the curve Γ.

If (1.69) is fulfilled, then a solution of the problem is represented by the formula (see [25]):

$$\varphi = \frac{z^{\kappa}}{i}\int_{\Gamma}\tilde{L}(z,t)h_0(t)\,dt + \sum_{j=1}^{2\kappa+1-\tau} c_j\,\Phi^{(j)}, \tag{1.71}$$

where

$$\tilde{L}(z,\zeta) = \int_0^z L(z,\zeta)\,dz,$$

$$L(z,\zeta) = -\frac{2}{\pi}\frac{\partial^2 g(z,\zeta)}{\partial z\,\partial\zeta} = \frac{1}{\pi(\zeta - z)^2} - \ell(z,\zeta),$$

$g(z,\zeta)$ is the Green function of the domain G^+ and c_j are arbitrary real constants.

It is clear that

$$\tilde{L}(z,t) = \frac{1}{\pi(t-z)} - \frac{1}{\pi t} - \tilde{\ell}(z,t).$$

Since

$$t^{-\kappa}T_{G^+}\rho = -\frac{1}{\pi t^{\kappa}}\iint_{G^+}\frac{\rho(\zeta)\,dG^+_{\zeta}}{\zeta - t}$$

is the boundary value of the function holomorphic in G^-, vanishing at infinity, then by Cauchy's theorem for $z \in G^+$,

$$\int_{\Gamma}\frac{T_{G^+}\rho\,dt}{t^{\kappa}(t-z)} = 0.$$

Hence we obtain

$$\frac{1}{i}\int_{\Gamma}\tilde{L}(z,t)h_0(t)\,dt = -\frac{1}{2i}\int_{\Gamma}\tilde{L}(z,t)\frac{\overline{T_{G^+}\rho}}{\bar{t}^{\kappa}}\,dt$$

$$+\frac{1}{2i}\int_{\Gamma}\tilde{\ell}(z,t)\frac{T_{G^+}\rho}{t^{\kappa}}\,dt + \frac{1}{2i}\int_{\Gamma}\tilde{\ell}(z,t)\frac{\overline{T_{G^+}\rho}}{\bar{t}^{\kappa}}\,dt$$

$$= \iint_{G^+} \left(\frac{-1}{2\pi i} \int_\Gamma \frac{\tilde{L}(z,t)\,dt}{\bar{t}^{\kappa}(\bar{t}-\bar{\zeta})} \right) \overline{\rho(\zeta)}\, dG_\zeta^+ + \iint_{G^+} \left(\frac{1}{2\pi i} \int_\Gamma \frac{\tilde{\ell}(z,t)\,dt}{t^{\kappa}(t-\zeta)} \right) \rho(\zeta)\, dG_\zeta^+$$

$$+ \iint_{G^+} \left(\frac{1}{2\pi i} \int_\Gamma \frac{\tilde{\ell}(z,t)\,dt}{\bar{t}^{\kappa}(\bar{t}-\bar{\zeta})} \right) \overline{\rho(\zeta)}\, dG_\zeta^+ .$$

But as it follows from the identity

$$\frac{\partial^2 g(z,t)}{\partial z\, \partial t}\, t'(s) + \frac{\partial^2 g(z,t)}{\partial z\, \partial \bar{t}}\, \overline{t'(s)} = 0,. \quad z \in G^+,\ t \in \Gamma$$

and the representation of the functions $K(z,\bar{\zeta})$, $L(z,\zeta)$ through the Green function

$$K(z,\bar{\zeta}) = -\frac{2}{\pi}\frac{\partial^2 g(z,\zeta)}{\partial z\, \partial \bar{\zeta}}, \quad L(z,\zeta) = -\frac{2}{\pi}\frac{\partial^2 g(z,\zeta)}{\partial z\, \partial \zeta},$$

we have, for $t \in \Gamma$,

$$\tilde{L}(z,t)t'(s) = -\tilde{K}(z,\bar{t})\, \overline{t'(s)} .$$

Taking into account the Hermitian symmetry of the Bergman kernel $\overline{K(\zeta,\bar{z})} = K(z,\bar{\zeta})$, we obtain

$$-\frac{1}{2\pi i} \int_\Gamma \frac{\tilde{L}(z,t)\,dt}{\bar{t}^{\kappa}(\bar{t}-\bar{\zeta})} = \frac{1}{2\pi i} \int_\Gamma \frac{\tilde{K}(z,\bar{t})\,d\bar{t}}{\bar{t}^{\kappa}(\bar{t}-\bar{\zeta})}$$

$$= \frac{1}{2\pi i} \int_\Gamma \frac{\overline{\tilde{K}(t,\bar{z})}\, d\bar{t}}{\bar{t}^{\kappa}(\bar{t}-\bar{\zeta})} = -\overline{\frac{1}{2\pi i} \int_\Gamma \frac{\tilde{K}(t,\bar{z})\,dt}{t^{\kappa}(t-\zeta)}} .$$

Now using the identity

$$\frac{1}{t-\zeta} = \frac{\zeta^{\kappa}}{t^{\kappa}(t-\zeta)} + \sum_{j=0}^{\kappa-1} \frac{\zeta^j}{t^{j+1}}$$

and Cauchy's formula, we obtain

$$\frac{1}{2\pi i}\int_\Gamma \frac{\tilde{K}(t,\bar{z})\,dt}{t^\kappa(t-\zeta)} = \frac{1}{\zeta^\kappa}\Big(\frac{1}{2\pi i}\int_\Gamma \frac{\tilde{K}(t,\bar{z})\,dt}{t-\zeta}$$

$$-\sum_{j=0}^{\kappa-1}\frac{\zeta^j}{2\pi i}\int_\Gamma \frac{\tilde{K}(t,\bar{z})\,dt}{t^{j+1}}\Big) = \frac{1}{\zeta^\kappa}\Big(\tilde{K}(\zeta,\bar{z}) - \sum_{j=0}^{\kappa-1}\frac{d^j\tilde{K}(\zeta,\bar{z})}{d\zeta^j}\Big|_{\zeta=0}\cdot\frac{\zeta^j}{j!}\Big).$$

Hence

$$\frac{1}{2\pi i}\int_\Gamma \frac{\tilde{L}(z,t)\,dt}{\bar{t}^\kappa(\bar{t}-\bar{\zeta})} = \frac{1}{\bar{\zeta}^\kappa}\Big(\tilde{K}(z,\bar{\zeta}) - \sum_{j=0}^{\kappa-1}\frac{d^j\tilde{K}(z,\bar{\zeta})}{d\bar{\zeta}^j}\Big|_{\bar{\zeta}=0}\cdot\frac{\bar{\zeta}^j}{j!}\Big)$$

and consequently

$$\varphi(z) = -z^\kappa\iint_{G^+}\Big(\tilde{K}(z,\bar{\zeta}) - \sum_{j=0}^{\kappa-1}\frac{d^j\tilde{K}(z,\bar{\zeta})}{d\bar{\zeta}^j}\Big|_{\bar{\zeta}=0}\cdot\frac{\bar{\zeta}^j}{j!}\Big)\frac{\overline{\rho(\zeta)}}{\bar{\zeta}^\kappa}\,dG_\zeta^+$$

$$+z^\kappa\iint_{G^+}\Big\{\Big(\frac{1}{2\pi i}\int_\Gamma\frac{\tilde{\ell}(z,t)\,dt}{t^\kappa(t-\zeta)}\Big)\rho(\zeta) + \Big(\frac{1}{2\pi i}\int_\Gamma\frac{\tilde{\ell}(z,t)\,dt}{\bar{t}^\kappa(\bar{t}-\bar{\zeta})}\Big)\overline{\rho(\zeta)}\Big\}\,dG_\zeta^+$$

$$+\sum_{j=0}^{2\kappa+1-r} c_j\varphi_j(z) = -z^\kappa\varphi_0(z) + \varphi_1(z) + \sum_{j=0}^{2\kappa+1-r} c_j\varphi_j(z). \tag{1.71'}$$

Differentiating (1.67) with respect to $\bar{z}$ and z we obtain

$$w_{\bar{z}} = \rho,\quad w_z = S_{G^+}\rho + \varphi'(z), \tag{1.72}$$

where $\varphi'(z)$ is a continuous linear operator over the function $\rho \in L_p(G^+)$ with principal singular part $K_\kappa\rho$:

$$K_\kappa\rho = -z^{2\kappa}\frac{\partial}{\partial z}\iint_{G^+} K_\kappa(z,\bar{\zeta})\,\overline{\rho(\zeta)}\,dG_\zeta^+, \tag{1.73}$$

$$K_\kappa(z,\bar{\zeta}) = \Big(\tilde{K}(z,\bar{\zeta}) - \sum_{j=0}^{\kappa-1}\frac{\partial\tilde{K}(z,\bar{\zeta})}{\partial\bar{\zeta}^j}\Big|_{\bar{\zeta}=0}\cdot\frac{\bar{\zeta}^j}{j!}\Big)\frac{1}{(\bar{\zeta}z)^\kappa}.$$

Since

$$\tilde{K}(z,\bar{\zeta}) = \sum_{k=1}^{\infty} \tilde{\omega}_k(z)\overline{\omega_k(\zeta)}, \quad \tilde{\omega}_k(z) = \int_0^z \omega_k(z)dz,$$

then we have

$$K_\kappa(z,\bar{\zeta}) = \sum_{k=1}^{\infty} \frac{\tilde{\omega}_k(z)}{(\bar{\zeta}z)^\kappa} \sum_{j=\kappa}^{\infty} \frac{\omega_l^{(j)}(0)}{j!} \bar{\zeta}^j. \tag{1.74}$$

Substituting (1.67) and (1.72) into (1.6) we obtain the following singular integral equation in case $\kappa > 0$:

$$\rho - K_+^{(\kappa)}\rho = g_1 + \sum_{j=1}^{2\kappa+1-\tau} c_j(g_j^{(1)} + g_j^{(2)}), \tag{1.75}$$

where $K_+^{(\kappa)}$ is linear continuous operator in $L_p(G^+)$:

$$K_+^{(\kappa)}\rho \equiv q_1(S_{G^+}\rho + K_\kappa\rho) + q_2(\overline{S_{G^+}\rho} + \overline{K_\kappa\rho}) + \tilde{K}_\rho^{(\kappa)},$$

$$\tilde{K}_\rho^{(\kappa)}\rho \equiv -\kappa z^{\kappa-1} q_1 \varphi_0 + q_1\varphi_1 - \kappa \bar{z}^{\kappa-1} q_2 \overline{\varphi_0} + q_2\overline{\varphi_1} + z^\kappa a_1\varphi_0$$
$$+ \bar{z}^\kappa b_1 \overline{\varphi_0} - a_1\varphi_1 - b_1\overline{\varphi_1}.$$

Since

$$S_{G^+}\rho \equiv \frac{\partial T_1\rho}{\partial z} + K_\kappa\rho - \iint_{G^+} \ell(z,\zeta)\rho(\zeta)\,dG_\zeta^+,$$

where

$$T_1\rho \equiv \iint_{G^+} \left(\tilde{\ell}(z,\zeta) - \frac{1}{\pi(\zeta - z)}\right)\rho(\zeta)\,dG_\zeta^+,$$

then as above for $\rho \in C_0^\infty(G^+)$ we have, for the L_2-norm of the operator

$$S_{G^+}^0\rho = \frac{\partial T_1\rho}{\partial z} + K_\kappa\rho,$$

the following equality:

$$\| S^0_{G^+}\rho \|^2_{L_2(\overline{G^+})} = \| \rho \|^2_{L_2(\overline{G^+})} + \| K_\kappa \rho \|^2_{L_2(\overline{G^+})} - \sum_{k=1}^{\infty} | c_k |^2. \tag{1.76}$$

If G^+ is a unit disc, then there exists a unique system

$$\omega_k(z) = \sqrt{\left(\frac{k}{\pi}\right)}\, z^{k-1},\quad \tilde{\omega}_k(z) = \frac{z^k}{\sqrt{(\pi k)}},\quad \omega_k^{(j)}(0) = 0$$

for all k except for $k = j+1$: $\omega^{(j)}_{j+1}(0) = \sqrt{\left(\frac{j+1}{\pi}\right)}\, j!$ Hence in this case

$$K_\kappa(z, \zeta) = \sum_{j=\kappa}^{\infty} \frac{z^{j+1}\bar{\zeta}^j}{\pi(\bar{\zeta} z)^\kappa} = \frac{z}{1-\bar{\zeta} z}$$

and

$$K_\kappa \rho \equiv -\frac{z^{2\kappa}}{\pi} \iint_{|\zeta|<1} \frac{\overline{\rho(\zeta)}\, d\xi\, d\eta}{(1-\bar{\zeta} z)^2}.$$

Therefore

$$\| K_\kappa \rho \|^2_{L_2} = \iint_{|z|<1} | z |^{4\kappa} \left| \sum_{k=1}^{\infty} \frac{k z^{k-1}}{\pi} \iint_{|\zeta|<1} \zeta^{k-1}\, \overline{\rho(\zeta)}\, d\xi\, d\eta \right|^2 dx\, dy$$

$$= \iint_{|z|<1} | z |^{4\kappa} \left| \sum_{k=1}^{\infty} \sqrt{\frac{k}{\pi}}\, \bar{c}_k z^{k-1} \right|^2 dx\, dy = \sum_{k=1}^{\infty} \frac{|c_k|^2}{2\kappa + k}.$$

Hence in case G^+ is the unit disc $| z | < 1$,

$$\| S^0_{G^+}\rho \|^2_{L_2} = \| \rho \|^2_{L_2} - 2\kappa \sum_{k=1}^{\infty} \frac{|c_k|^2}{2\kappa + k}$$

and

$$\| S^0_{G^+}\rho \|^2_{L_2} = \| \rho \|^2_{L_2}$$

for such $\rho \in C_0^\infty$ which satisfy the conditions

$$\iint\limits_{|z|<1} \rho(z)\, z^k \,dx\,dy = 0, \quad k = 1,2,\dots .$$

It means that $\| S^0_{G^+} \|_{L_2(\overline{G^+})} = 1$.

By the same reason as above we apply Fredholm's theorem to (1.75).

Let $\rho^0 \in L_p(G^+)$, $p > 2$ be a solution of the corresponding homogeneous equation

$$\rho^0 - K_+^{(\kappa)} \rho^0 = 0. \tag{1.77}$$

Then the function

$$\overset{\circ}{w}(z) = T_{G^+}\rho^0 - z^\kappa \iint\limits_{G^+} K_\kappa(z,\zeta)\, \overline{\rho^0(\zeta)}\, dG^+_\zeta$$

$$+ z^\kappa \iint\limits_{G^+} \Big\{ \Big(\frac{1}{2\pi i} \int_\Gamma \frac{\tilde{\ell}(z,t)\,dt}{t^\kappa (t-\zeta)} \Big) \rho(\zeta) + \Big(\frac{1}{2\pi i} \int_\Gamma \frac{\tilde{\ell}(z,t)\,dt}{\bar{t}^\kappa (\bar{t}-\bar{\zeta})} \Big) \overline{\rho(\zeta)} \Big\} dG^+_\zeta \tag{1.78}$$

will be a solution of the homogeneous equation (1.6′), satisfying the boundary condition (1.66).

Moreover, since the first term on the right-hand side is holomorphic in G^- and vanishes at infinity, and the remaining term of the right-hand side is holomorphic in G^+ by Cauchy's theorem, we obtain, for $z \in G^+$,

$$\frac{1}{2\pi i} \int_\Gamma \frac{\overset{\circ}{w}(t)\,dt}{t - z} = z^\kappa \iint\limits_{G^+} K_\kappa(z,\zeta)\, \overline{\rho^0(\zeta)}\, dG^+_\zeta$$

$$-z^\kappa \iint\limits_{G^+} \Big\{ \Big(\frac{1}{2\pi i} \int_\Gamma \frac{\tilde{\ell}(z,t)\,dt}{t^\kappa (t-z)} \Big) \rho(\zeta) + \Big(\frac{1}{2\pi i} \int_\Gamma \frac{\tilde{\ell}(z,t)\,dt}{\bar{t}^\kappa (\bar{t}-\bar{\zeta})} \Big) \overline{\rho(\zeta)} \Big\} dG^+_\zeta.$$

But since the right-hand side of this expression has a zero of the order $2\kappa + 1$ at point $z = 0$, then the left-hand side must have a zero of this order, i.e.

$$\int_\Gamma \frac{\overset{\circ}{w}(t)\,dt}{t^{j+1}} = 0, \quad j = 0,\dots, 2\kappa. \tag{1.79}$$

By the similarity principle we may represent $\overset{\circ}{w}(z)$ as

$$\overset{\circ}{w}(z) = \Phi[W(z)]\, e^{s(z)} \tag{1.80}$$

with holomorphic Φ and continuous s: $\operatorname{Im} s \mid_\Gamma = 0$, where $W(z)$ is some homeomorphism of the z-plane to the W-plane.Now the boundary condition (1.66) reduces to the condition for a holomorphic function: $\operatorname{Re} W^{-\kappa}\Phi(W) = 0$. Hence by (1.79) we obtain $\Phi \equiv 0$, i.e. $\overset{\circ}{W}(z) \equiv 0$.

From this identity by differentiation with respect to $\bar{z}$ we obtain $\rho^0 \equiv 0$. This means that the equation (1.75) is solvable uniquely and its solutions can be represented by

$$\rho = g_1 + R_{\kappa =} g_1 + \sum_{j=1}^{2\kappa+1-r} C_j\, g^{(j)}, \tag{1.81}$$

where R_κ is some linear continuous operator in $L_p(G^+)$, $p > 2$. Substituting for $h_0(t)$ its expression through ρ, by means of the Green formula we obtain

$$\operatorname{Re} \iint_{G^+} \rho(z)\, \psi_j(z)\, dx\, dy = 0, \quad j = 1,\dots, m - r$$

or by (1.81) we obtain the inhomogeneous algebraic equation for c_k:

$$\sum_{k=1}^{2\kappa+1-r} \left(\operatorname{Re} \iint_{G^+} g^{(k)} \psi_j\, dG^+\right) c_k = \operatorname{Re} \iint_{G^+} g_1\, \tilde{\psi}_j^*\, dG^+, \; j = 1,\dots, m - r. \tag{1.82}$$

If

$$\operatorname{rank}(a_{kj}) = r',$$

$$a_{kj} = \operatorname{Re} \iint_{G^+} g^{(k)} \psi_j\, dG^+,$$

then (1.82) becomes

$$\operatorname{Re} \iint_{G^+} g_1 \psi_j^* \, dG^+ = 0, \quad j = 1, \ldots, m - r - r' \tag{1.83}$$

and in the right-hand side (1.81) $2\kappa + 1 - r - r'$ coefficients remain arbitrary, i.e. the homogeneous (RH) problem has $\ell = 2\kappa + 1 - r - r'$ linearly independent solutions, but for solvability of the inhomogeneous equation it needs $\ell' = m - r - r'$ conditions and thus the index of the problem is equal to

$$\ell - \ell' = 2\kappa + 1 - m.$$

Let us assume now $\kappa < 0$. Since in this case the function $\varphi_0(z) = z^{-\kappa} \varphi(z)$ is holomorphic in G^+, then (1.68) is the Schwartz condition for the holomorphic function $\varphi_0(z)$, so we obtain

$$z^{-\kappa} \varphi(z) = \frac{1}{i} \int_\Gamma \tilde{L}(z, t) \, h_0(t) \, dt + iC, \tag{1.84}$$

provided the solvability conditions

$$\frac{1}{i} \int_\Gamma h_0(t) \, \Phi_j(t) \, dt = 0, \quad j = 1, \ldots, m \tag{1.85}$$

are fulfilled. This condition may be written in terms of ρ as

$$\operatorname{Re} \iint_{G^+} \rho(z) \, \Phi_j^*(z) \, dx \, dy = 0, \quad j = 1, \ldots, m. \tag{1.86}$$

Moreover, since $\varphi(z)$ must be continuous at the origin $z = 0$, from (1,84) we obtain also $-2\kappa - 1$ real conditions for ρ:

$$-\frac{1}{\pi} \iint_{G^+} \zeta^{-\kappa-1} \rho(\zeta) \, dG_\zeta^+ + \frac{1}{2\pi i} \int_\Gamma \Big[\iint_{G^+} \tilde{\ell}(t, \zeta) \, \zeta^{-\kappa} \rho(\zeta) \, dG_\zeta^+$$

$$+ \iint_{G^+} \tilde{L}(z, \zeta) \, \bar{\zeta}^{-\kappa} \overline{\rho(\zeta)} \, dG_\zeta^+ \Big] \frac{dt}{t} + iC = 0,$$

$$-\frac{1}{\pi} \iint_{G^+} \zeta^{-\kappa-1-j} \rho(\zeta) \, dG_\zeta^+ + \frac{1}{2\pi i} \int_\Gamma \Big[\iint_{G^+} \tilde{\ell}(t, \zeta) \, \zeta^{-\kappa} \rho(\zeta) \, dG_\zeta^+$$

$$+\iint\limits_{G^+} \tilde{L}(t,\zeta)\,\bar{\zeta}^{-\kappa}\,\overline{\rho(\zeta)}\,dG_\zeta^+ + \overline{T_{G^+}(\zeta^{-\kappa}\rho(\zeta))}\Big]\frac{dt}{t^{j+1}} = 0,$$

$$j = 1, \ldots, \kappa - 1. \tag{1.87}$$

In case the conditions (1.86), (1.87) are fulfilled, from (1.84) we find the holomorphic function $\varphi(z)$ as some linear continuous operator over ρ:

$$\varphi(z) = \iint\limits_{G^+}\left[\tilde{\ell}_\kappa(z,\zeta)\,\zeta^{-\kappa}\rho(\zeta) - \tilde{I}_\kappa(z,\bar{\zeta})\,\bar{\zeta}^{-\kappa}\,\overline{\rho(\zeta)}\right]dG_\zeta^+, \tag{1.88}$$

where

$$\tilde{\ell}_\kappa(z,\zeta) = \frac{1}{2\pi i}\int_\Gamma \frac{\tilde{\ell}(z,t)\,dt}{t^{-\kappa}(t-\zeta)},$$

$$\tilde{I}_\kappa(z,\bar{\zeta}) = \frac{1}{2\pi^2 i}\int_\Gamma \frac{dt}{t^{-\kappa}(\bar{t}-\bar{\zeta})(t-z)}.$$

Differentiating (1.67) we obtain

$$w_{\bar{z}} = \rho, \quad w_z = S_{G^+}\rho + \varphi'(z), \tag{1.72}$$

where

$$\varphi'(z) = -\iint\limits_{G^+} I_\kappa(z,\bar{\zeta})\,\bar{\zeta}^{-\kappa}\,\overline{\rho(\zeta)}\,dG_\zeta^+$$

$$+\iint\limits_{G^+} \ell_k(z,\zeta)\,\zeta^{-\kappa}\rho(\zeta)\,dG_\zeta^+, \tag{1.89}$$

and where $I_\kappa(z,\bar{\zeta})$ is the singular kernel:

$$I_\kappa(z,\bar{\zeta}) = \frac{1}{2\pi^2 i}\int_\Gamma \frac{dt}{t^{-\kappa}(\bar{t}-\bar{\zeta})(t-z)^2} \tag{1.90}$$

and $\ell_\kappa(z, \zeta)$ is the regular kernel, which is analytic in both variables:

$$\ell_\kappa(z, \zeta) = \frac{1}{2\pi i}\int_\Gamma \frac{\ell(z,t)\,dt}{t^{-\kappa}(t-\zeta)}.$$

Substituting (1.67) and (1.72) into (1.6) we obtain

$$\rho - K_-^{(\kappa)}\rho = g_1, \tag{1.91}$$

where $K_-^{(\kappa)}$ is a linear continuous operator in $L_p(G^+)$: $K_-^{(\kappa)} = K_\kappa^0 + K_\kappa^1$ with principal part:

$$K_\kappa^0 \equiv q_1\Big(S_{G^+}\rho - \iint_{G^+} I_\kappa(z,\bar\zeta)\bar\zeta^{-\kappa}\,\overline{\rho(\zeta)}\,dG_\zeta^+\Big)$$

$$+\, q_2\Big(\overline{S_{G^+}\rho} - \overline{\iint_{G^+} I_\kappa(z,\zeta)\zeta^{-\kappa}\,\overline{\rho(\zeta)}\,dG_\zeta^+}\Big)$$

and regular part

$$K_\kappa^1\rho \equiv q_1 \iint_{G^+} \ell_k(z,\zeta)\zeta^{-\kappa}\rho(\zeta)\,dG_\zeta^+ + q_2\overline{\iint_{G^+} \ell_k(z,\tau)\zeta^{-\kappa}\rho(\zeta)\,dG_\zeta^+}$$

$$+\, a_1\Big[\iint_{G^+} \tilde I_\kappa(z,\bar\zeta)\,\bar\zeta^{-\kappa}\,\overline{\rho(\zeta)}\,dG_\zeta^+ - \iint_{G^+} \tilde\ell_k,\zeta)\,\zeta^{-\kappa}\rho(\zeta)\,dG_\zeta^+\Big]$$

$$+\, b_1\Big[\overline{\iint_{G^+} \tilde I_k(z,\zeta)\zeta^{-\kappa}\,\overline{\rho(\zeta)}\,dG_\zeta^+ - \iint_{G^+} \tilde\ell_k(z,\zeta)\zeta^{-\kappa}\rho(\zeta)\,dG_\zeta^+}.$$

Consider the following singular operator:

$$S'_{G^+}\rho \equiv \frac{\partial T_1\rho}{\partial z} + \mathring{K}_\kappa\rho,$$

where

$$\mathring{K}_\kappa\rho \equiv \iint_{G^+} I_k(z,\bar\zeta)\bar\zeta^{-\kappa}\,\overline{\rho(\zeta)}\,dG_\zeta^+ .$$

Then for $\rho \in C_0^\infty(G^+)$ we obtain as above

$$\| S'_{G^+}\rho \|^2_{L_2(\overline{G^+})} = \| \rho \|^2_{L_2(\overline{G^+})} + \| \overset{\circ}{K}_\kappa \rho \|^2_{L_2(\overline{G^+})} - \sum_{k=1}^{\infty} | C_k |^2.$$

Since for $z \in G^+$ and $\tau \in G^-$

$$\frac{1}{\tau - z} = \sum_{k=1}^{\infty} C_k^0 \tilde{\omega}_k(z),$$

where $\{\omega_k(z)\}$ is a complete orthonormal system of holomorphic functions in G^+, then we obtain

$$\tilde{I}_\kappa(z, \bar{\zeta}) = \sum_{k=1}^{\infty} \tilde{\omega}_k(z) \overline{\omega_k^*(\zeta)},$$

because we may represent by Green's formula applied to the domain G^-:

$$I_\kappa(z, \bar{\zeta}) = -\frac{1}{\pi} \iint_{G^-} \frac{dG_\tau^-}{\tau^{-\kappa}(\bar{\tau} - \bar{\zeta})^2(\tau - z)}.$$

Hence

$$K_k^0 \bar{\rho} = \sum_{k=1}^{\infty} \Big(\overline{\iint_{G^+} \omega_k^*(\zeta) \zeta^{-\kappa} \rho(\zeta) dG_\zeta^-} \Big) \omega_k(z) = \sum_{k=1}^{\infty} A_k \omega_k(z)$$

and we have

$$\| S'_{G^+}\rho \|^2_{L_2(\overline{G^+})} = \| \rho \|^2_{L_2(\overline{G^+})} + \sum_{k=1}^{\infty} (| A_k |^2 - | C_k |^2).$$

In case G^+ is the unit disc we have for $|z| < 1$, $|\tau| > 1$,

$$\frac{1}{\tau - z} = \sum_{k=1}^{\infty} \frac{z^k}{\tau^{k+1}}$$

and

$$K'_\kappa \bar{\rho} = \frac{1}{\pi} \iint_{|\zeta|<1} \frac{\bar{\zeta}^{-2\kappa}\overline{\rho(\zeta)}\, dG^+_\zeta}{(1-\bar{\zeta} z)^2}$$

$$= \sum_{j=1}^{\infty} \Big(\frac{j}{\pi} \iint_{|\zeta|<1} \bar{\zeta}^{-2\kappa+j-1}\, \overline{\rho(\zeta)}\, dG^+_\zeta \Big) z^{j-1} = \sum_{j=1}^{\infty} \sqrt{(\tfrac{j}{\pi})}\, C_{-2\kappa+j}\, z^{j-1},$$

so

$$\| K'_\kappa \rho \|^2_{L_2} = \sum_{j=1}^{\infty} | C_{-2\kappa+j} |^2.$$

Hence in the case of a disc for $\rho \in C_0^\infty$

$$\| S'_{G^+} \rho \|^2_{L_2(\overline{G^+})} = \| \rho \|^2_{L_2(\overline{G^+})} - \sum_{j=1}^{-2\kappa} | C_j |^2,$$

and the equality $\| S'_{G^+} \rho \|^2_{L_2} = \| \rho \|^2_{L_2}$ is reached here if we take ρ such that

$$\iint_{|\zeta|<1} \rho(\zeta)\, \zeta^j\, d\xi\, d\eta = 0, \quad j = 1, 2, \dots .$$

Consequently $\| S'_{G^+} \|_{L_2(\overline{G^+})} = 1$. Therefore we may apply Fredholm's theorem to the equation (1.91), because the operator $q_1 S'_{G^+} \rho + q_2 \overline{S'_{G^+}}$ is contractive in $L_p(G^+)$. If $\rho^0 \in L_p(G^+)$ is a solution of the homogeneous equation corresponding to (1.91), then the function

$$\mathring{w}(z) = T_{G^+} \rho^0 + \iint_{G^+} \Big[\tilde{l}_\kappa(z,\zeta)\, \zeta^{-\kappa} \rho^0(\zeta) - \tilde{l}_\kappa(z,\zeta)\, \bar{\zeta}^{-\kappa}\, \overline{\rho(\zeta)} \Big] dG^+_\zeta \qquad (1.92)$$

will be a solution of the homogeneous problem (1.6′), (1.66). Then by means of the similarity principle $w = \Phi(W)e^s$, representing ρ^0 as a sum $\rho^0_0 + \rho^0_k$, where ρ^0_0 is the element of the space $L_p(G^+)$, satisfying the conditions (1.86), (1.87) and $\rho^0_k = c_0 \bar{z}^{k-1} + c_1 \bar{z}^{k-2} + \dots$, $(c_0 = \bar{c}_0)$, $k = -\kappa > 0$, is the polynomial which uniquely determines by (1.87). Then boundary condition (1.66) can be written as

$$\operatorname{Re} z^k \varphi = \operatorname{Re} \rho^0_k .$$

Then by the same reason as above we prove that $\rho_k^0 \equiv 0$ and $\varphi \equiv 0$, so the right-hand side of (1.92) will be equal to zero identically. Thus we obtain $\rho^0 \equiv 0$ by differentiating with respect to $\bar{z}$. This means that the equation (1.91) is uniquely solvable. In this case the homogeneous (RH) problem has no non-trivial solution $\ell = 0$ and the inhomogeneous (RH) problem is solvable if and only if $\ell' = m - 2\kappa - 1$ conditions of the form

$$\operatorname{Re} \iint_{G^+} g_1 v_j \, dx\, dy = 0, \quad j = 1,..,\ell'$$

are fulfilled, and these follow from (1.86), (1.87) if we substitute there $\rho = g_1 + \tilde{R} g_1$, obtained from (1.91). Summarizing, we have the following:

Theorem 1.3. *If $\lambda \neq 0$ on Γ, then the RH problem (1.66) for the general equation (1.6) has the Noetherian property: the homogeneous problem ($g_1 \equiv 0$) may have only a finite number of nontrivial solutions, and for solvability of the inhomogeneous problem it is necessary and sufficient that a finite number of conditions*

$$\operatorname{Re} \iint_{G^+} g_1 v_j \, dx\, dy = 0, \quad j = 1,..,\ell'$$

are fulfilled. The index of the RH problem is equal to

$$\ell - \ell' = 2\kappa + 1 - m$$

with $\kappa = \operatorname{Ind}_\Gamma \lambda$. In case $\kappa < 0$ the homogeneous problem has no nontrivial solutions.

1.2 Second-order elliptic systems in the plane

1.2.1. Introduction

We consider now a linear second-order system of two real equations with two real unknowns:

$$a_{11}(x,y)\frac{\partial^2 u_1}{\partial x^2} + a_{12}(x,y)\frac{\partial^2 u_2}{\partial x^2} + b_{11}(x,y)\frac{\partial^2 u_1}{\partial x \partial y} + b_{12}(x,y)\frac{\partial^2 u_2}{\partial x \partial y}$$

$$+ c_{11}(x, y)\frac{\partial^2 u_1}{\partial y^2} + c_{12}(x, y)\frac{\partial^2 u_2}{\partial y^2}$$

$$+ A_{11}(x, y)\frac{\partial u_1}{\partial x} + A_{12}(x, y)\frac{\partial u_2}{\partial x} + B_{11}(x, y)\frac{\partial u_1}{\partial y} + B_{12}(x, y)\frac{\partial u_2}{\partial y}$$

$$+ C_{11}(x, y)u_1 + C_{12}(x, y)u_2 = F_1(x, y)$$

$$a_{21}(x, y)\frac{\partial^2 u_1}{\partial x^2} + a_{22}(x, y)\frac{\partial^2 u_2}{\partial x^2} + b_{21}(x, y)\frac{\partial^2 u_1}{\partial x \partial y} + b_{22}(x, y)\frac{\partial^2 u_2}{\partial x \partial y}$$

$$+ c_{21}(x, y)\frac{\partial^2 u_1}{\partial y^2} + c_{22}(x, y)\frac{\partial^2 u_2}{\partial y^2}$$

$$+ A_{21}(x, y)\frac{\partial u_1}{\partial x} + A_{22}(x, y)\frac{\partial u_2}{\partial x} + B_{21}(x, y)\frac{\partial u_1}{\partial y} + B_{22}(x, y)\frac{\partial u_2}{\partial y}$$

$$+ C_{21}(x, y)u_1 + C_{22}(x, y)u_2 = F_2(x, y).$$

These two equations may be put in the single complex form

$$\tilde{a}(z)w_{\bar{z}^2} + \tilde{b}(z)w_{\bar{z}z} + \tilde{c}(z)\bar{w}_{z^2} + \tilde{\alpha}(z)\bar{w}_{\bar{z}^2} + \tilde{\beta}(z)\bar{w}_{\bar{z}z} + \tilde{\gamma}(z)\bar{w}_{z^2}$$

$$+ \tilde{A}(z)w_{\bar{z}} + \tilde{B}(z)w_z + \tilde{C}(z)\omega\bar{w}_z + \tilde{D}(z)\bar{w}_{\bar{z}} + \tilde{E}(z)w + \tilde{F}(z)\bar{w} = \tilde{G}(z). \tag{1.93}$$

The characteristic form of (1.93) is

$$\chi(z, \zeta) = |\,\tilde{a}(z)\zeta^2 + \tilde{b}(z)\,|\zeta|^2 + \tilde{c}(z)\bar{\zeta}^2\,|^2 - |\,\tilde{\alpha}(z)\zeta^2 + \tilde{\beta}(z)\,|\zeta|^2 + \tilde{\gamma}(z)\bar{\zeta}^2\,| \tag{1.94}$$

with $\zeta = \xi_1 + i\xi_2$.

The ellipticity of (1.93) at the point z means the definiteness of the form $\chi(z, \zeta)$ at z.

It is rather difficult to give any satisfactory canonical form for the general elliptic equation (1.93). Nevertheless in the case of constant coefficients in the principal part the following canonical forms hold for the second-order elliptic system (see [44]), which in terms of (1.93) can be written with the following coefficients:

$$\tilde{a} = (\lambda - k)(k + 1),\ \tilde{b} = -(\lambda - k)(k - 1),\ \tilde{c} = 0,$$

$$\tilde{\alpha} = (\lambda + k)(k - 1), \ \tilde{\beta} = -(\lambda + k)(k + 1), \ 0 < k \le 1.$$

The following second-order equation with special coefficients

$$\left(a_1 \frac{\partial}{\partial \bar{z}} + c_1 \frac{\partial}{\partial z}\right)\left(a_2 \frac{\partial w}{\partial \bar{z}} + c_2 \frac{\partial w}{\partial z} + d_2 \frac{\partial \bar{w}}{\partial \bar{z}} + b_2 \frac{\partial \bar{w}}{\partial z}\right)$$

$$+\left(d_1 \frac{\partial}{\partial z} + b_1 \frac{\partial}{\partial z}\right)\left(\bar{d}_2 \frac{\partial w}{\partial z} + \bar{b}_2 \frac{\partial w}{\partial \bar{z}} + \bar{a}_2 \frac{\partial \bar{w}}{\partial z} + c_2 \frac{\partial \bar{w}}{\partial \bar{z}}\right) + \cdots = \tilde{f} \tag{1.95}$$

is obtained as a product of two first-order equations like (1.1), and has characteristic form

$$\chi = (|a_1\zeta + c_1\bar{\zeta}|^2 - |d_1\zeta + b_1\bar{\zeta}|^2)\,(|a_2\zeta + c_2\bar{\zeta}|^2 - |d_2\zeta + b_2\bar{\zeta}|^2),$$

which is a product of two elliptic quadratic forms. If we denote

$$v = a_2 w_{\bar{z}} + c_2 w_z + d_2 \bar{w}_{\bar{z}} + b_2 \bar{w}_z \tag{1.96}$$

then in case the coefficients $A = |a_1|^2 - |b_1|^2$, $B = |d_1|^2 - |c_1|^2$, $\alpha = a_1 \bar{c}_1 - \bar{b}_1 d_1$, $\beta = a_1 \bar{d}_1 - \bar{b}_1 c_1$ satisfy the condition (a) of p. 2 the equation (1.95) may be written as

$$v_{\bar{z}} - q_1 v_z - q_2 \overline{v_z} + \cdots = \tilde{f}_1 \tag{1.97}$$

with q_1, q_2 satisfying the condition

$$|q_1| + |q_2| < 1.$$

By means of linear transformation (1.12) in this case the equation (1.97) reduces to (1.18). If the coefficients on the right-hand side of (1.96) satisfy, say, the condition (a) with

$$A = |a_2|^2 - |b_2|^2, \ B = |d_2|^2 - |c_2|^2, \ \alpha = a_2 \bar{c}_2 - \bar{b}_2 d_2, \ \beta = a_2 \bar{d}_2 - \bar{b}_2 c_2,$$

then we obtain

$$w_{\bar{z}} - \tilde{q}_1 w_z - \tilde{q}_2 \overline{w_z} + \cdots = \bar{a}_2 v - b_2 \bar{v} \tag{1.98}$$

with

$$|\tilde{q}_1| + |\tilde{q}_2| < 1.$$

By means of linear transformation over

$$w : \psi = \frac{1 - |\tilde{q}_1|^2 + |\tilde{q}_2|^2 \sqrt{\tilde{\sigma}}}{2(1 - |\tilde{q}_1|^2)} w - \frac{\tilde{q}_2}{1 - |\tilde{q}_1|^2} \bar{w}$$

we reduce the left-hand side in (1.98) to the Beltrami operator $B\psi \equiv \psi_{\bar{z}} - \tilde{q}\psi_z$ with $|\tilde{q}| < 1$ and then we obtain the following expression for v:

$$v(z) = \tilde{a}B\psi + \tilde{b}\,\overline{B\psi} \tag{1.99}$$

with

$$|\tilde{a}| \neq |\tilde{b}|.$$

Now taking into account (1.98) we obtain the following equation for ψ:

$$\left(\frac{\partial}{\partial \bar{z}} - q\frac{\partial}{\partial z}\right)\left(\tilde{A}\,B\psi + \tilde{B}\,\overline{B\psi}\right) + \cdots = \tilde{F}_0, \tag{1.100}$$

where

$$\tilde{A} = \frac{1 - |q_1|^2 + |q_2|^2 + \sqrt{\sigma}}{2(1 - (q_1|^2)} \tilde{a} - \frac{q_2}{1 - |q_1|^2} \bar{\tilde{b}},$$

$$\tilde{B} = \frac{1 - |q_1|^2 + |q_2|^2 + \sqrt{\sigma}}{2(1 - (q_1|^2)} \tilde{b} - \frac{q_2}{1 - |q_1|^2} \bar{\tilde{a}}.$$

Equation (1.100) is called the canonical form for equation (1.95). The characteristic form of (1.100) is

$$|\zeta - q\bar{\zeta}|^2 |\zeta - \tilde{q}\bar{\zeta}|^2$$

apart from a non-zero function multiple, because it is

$$(|\tilde{A}|^2 - |\tilde{B}|^2)|\zeta - q\bar{\zeta}|^2 |\zeta - \tilde{q}\bar{\zeta}|^2,$$

$$|\tilde{A}|^2 - |\tilde{B}|^2 = \frac{1 - |q_1|^2 + |q_2|^2 + \sqrt{\sigma}}{4(1 - (q_1|^2)} (|\tilde{a}|^2 - |\tilde{b}|^2) \neq 0.$$

We used above a special linear transformation over an unknown function.

Lemma 1.3. *The linear transformation*

$$w = Av + B\bar{v}, \quad |A|^2 - |B|^2 \neq 0$$

reduces the general equation (1.93) to the same kind of equation with the same characteristic form apart from a non-zero multiple.

Indeed it is enough to note that for $v(z)$ we obtain the equation

$$a^* v_{\bar{z}^2} + b^* v_{\bar{z}z} + c^* v_{z^2} + \alpha^* \bar{v}_{\bar{z}z} + \beta^* \bar{v}_{\bar{z}z} + \gamma^* \bar{v}_{z^2} + \cdots = f^*,$$

with

$$a^* = \tilde{a}A + \tilde{\alpha}B, \ b^* = \tilde{b}A + \tilde{\beta}\bar{B}, \ c^* = \tilde{c}A + \tilde{\gamma}\bar{B},$$

$$\alpha^* = \tilde{a}B + \tilde{\alpha}\bar{A}, \ \beta^* = \tilde{b}B + \tilde{\beta}\bar{A}, \ \gamma^* = \tilde{c}B + \tilde{\gamma}\bar{A}$$

and so

$$\chi^*_{(z,\zeta)} = |a^*\zeta^2 + b^*|\zeta|^2 + \gamma^*\bar{\zeta}^2|^2 - |\alpha^*\zeta^2 + \beta^*|\zeta|^2 + \gamma^*\bar{\zeta}^2|^2$$

$$= (|A|^2 - |B|^2) \cdot \chi(z,\zeta).$$

We also note that (1.93) may be written as

$$a^0 w_{xx} + b^0 w_{xy} + c^0 w_{yy} + \alpha^0 \bar{w}_{xx} + \beta^0 \bar{w}_{xy} + \gamma^0 \bar{w}_{yy} + \cdots = g^0$$

with characteristic form

$$\chi = |a^0\xi_1^2 + b^0\xi_1\xi_2 + c^0\xi_2^2|^2 - |\alpha^0\xi_1^2 + \beta^0\xi_1\xi_2 + \gamma^0\xi_2^2|^2.$$

1.2.2. Conjugation problems for second-order elliptic equations

In this section we state the conjugation problems for the elliptic equation (1.93) given on plane $\mathbb{C}$. We assume that

$$|\tilde{b}|^2 - |\tilde{\beta}|^2 \neq 0$$

on $\mathbb{C}\backslash\Gamma$. The eliminating $\bar{w}_{\bar{z}z}$ we reduce (1.93) to the following form, in terms of the function $u = w$:

$$u_{\bar{z}z} + a_1(z)u_z{}^2 + a_2(z)\overline{u_z}{}^2 + b_1(z)u_{\bar{z}}{}^2 + b_2(z)\overline{u_{\bar{z}}}{}^2$$

$$+ A_1(z)u_z + A_2(z)\overline{u_z} + B_1(z)u_{\bar{z}} + B_2(z)\overline{u_{\bar{z}}} + c_1(z)u + c_2(z)\bar{u} = F(z). \quad (1.101)$$

Also we assume that $a_k(z)$, $b_k(z)$ are bounded measurable functions in $\mathbb{C}\backslash\Gamma$, vanishing at infinity, and $A_k(z)$, $B_k(z)$, $C_k(z, F(z)$ are given functions belonging $L_{p,2}(\mathbb{C})$, $p > 2$. The first conjugation problem (I) is to find a solution to equation (1.101) bounded in $\mathbb{C}$ with piecewise Hölder continuous first-order derivatives, vanishing at infinity and with Laplacian in $L_{p,2}(\mathbb{C})$, $p > 2$, and satisfying the boundary conditions

$$u_z^+(t) - G(t)u_z^-(t) = 0, \quad u_{\bar{z}}^+(t) - H(t)\, u_{\bar{z}}^-(t) = 0 \quad (1.102)$$

with given Hölder continuous functions $G(t) \neq 0$, $H(t) \neq 0$, $t \in \Gamma$. The second conjugation problem (II) is to find a piecewise Hölder continuous solution to (1.101) with piecewise Hölder continuous derivative u_z vanishing at infinity with Laplacian in $L_{p,2}(\mathbb{C})$, $p > 2$, and satisfying the boundary conditions

$$u_z^+(t) - G(t)u_z^-(t) = 0, \quad u^+(t) - \frac{1}{\overline{G(t)}}\, u^-(t) = 0 \quad (1.103)$$

with given Hölder continuous function $G(t) \neq 0$, $t \in \Gamma$. The third conjugation problem (III) is to find a piecewise Hölder continuous solution to (1.101) with piecewise continuous derivative $u_{\bar{z}}$ vanishing at infinity, with Laplacian in $L_{p,2}(\mathbb{C})$, $p > 2$ and satisfying the boundary conditions

$$u_{\bar{z}}^+(t) - - \overline{G(t)}\, u_{\bar{z}}^-(t) = 0, \quad u^+(t) - \frac{1}{G(t)} u^-(t) = 0. \quad (1.104)$$

Since by conjugation and by transformation the third problem reduces to the second, we need investigate only (I) and (II).

Let us start with problem (I). Denote Laplacian $u_{\bar{z}z}$ by ρ. Then for unknown function $v = u_z$ we obtain the Riemann problem:

$$v_{\bar{z}} = \rho \quad (1.105)$$

in $\mathbb{C} - \Gamma$ and

$$v^+ - Gv^- = 0 \quad (1.106)$$

on Γ, and for an unknown function $w = \overline{u_{\bar z}}$ we obtain the Riemann problem

$$w_{\bar z} = \bar\rho \tag{1.107}$$

in $\mathbb{C}\setminus\Gamma$ and

$$w^+ - \bar H w^- = 0 \tag{1.108}$$

on Γ.

Let

$$\kappa_1 = \mathrm{Ind}_\Gamma G = \frac{1}{2\pi i}\int_\Gamma d\log G(t),\quad \kappa_2 = \mathrm{Ind}_\Gamma H = \frac{1}{2\pi i}\int_\Gamma d\log H(t).$$

There are four possibilities:

(a) $\kappa_1 \geq 0,\ \kappa_2 > 0$ (b) $\kappa_1 \geq 0,\ \kappa_2 \leq 0$

(c) $\kappa_1 < 0,\ \kappa_2 > 0$ (d) $\kappa_1 < 0,\ \kappa_2 \leq 0.$

By means of Lemma 1.1, in case (a),

$$\begin{aligned} v(z) &= T\rho + X_1(z)\,(T\theta_1\rho - \theta_1(z)T\rho) + X_1(z)P_{\kappa_1-1}(z) \\ &= T_0^{(1)}\rho + \sum_{j=0}^{\kappa_1-1} a_j\, z^j X_1(z), \end{aligned} \tag{1.109}$$

$$\theta_1(z) = \frac{1}{2\pi i}\int_\Gamma \frac{(1-G(t)\,dt}{X_1^+(t)\,(t-z)}$$

and

$$w(z) = -\frac{X_2(z)}{\pi}\iint_{\mathbb{C}} \frac{\overline{\rho(\zeta)}\,d\xi\,d\eta}{X_2(\zeta)\,(\zeta-z)}, \tag{1.110}$$

provided the conditions

$$\iint_{\mathbb{C}} \overline{\rho(z)}\,\frac{z^j}{X_2(z)}\,dx\,dy = 0,\quad j\ \ldots,\ \kappa_2-1 \tag{1.111}$$

are fulfilled, where $X_1(z)$ is the piecewise holomorphic canonical solution for the problem (1.106) and $X_2(z)$ is the same for (1.108); in case (b) we have formula (1.109) and formula

$$w(z) = T\bar{\rho} + X_2(z)\,(T\theta_2\bar{\rho} - \theta_2(z)T\bar{\rho}) + X_2(z)P_{\kappa_2-1}(z)$$
$$+ T_0^{(2)}\bar{\rho} + \sum_{j=0}^{\kappa_2-1} b_j\, z^j X_2(z), \tag{1.112}$$

in case (c) we have formula (1.110) with complementary condition (1.111) and formula

$$v(z) = -\frac{X_1(z)}{\pi} \iint_{\mathbb{C}} \frac{\rho(\zeta)d\xi d\eta}{X_1(\zeta)(\zeta - z)} \tag{1.113}$$

with complementary conditions

$$\iint_{\mathbb{C}} \rho(z)\frac{z^j}{X_1(z)}\,dx\,dy = 0,\;\; j = 0, ..., \kappa_1 - 1 \tag{1.114}$$

and finally in case (d) we have formula (1.112) and formula (1.113) with complementary conditions (1.114).

The function $u(z)$ is then expressed in terms of v and w by the curvilinear integral[2]

$$u(z) = \int_{z_0}^{z} v\,dz + \bar{w}\,d\bar{z} + \hat{C} \tag{1.115}$$

with arbitrary complex constant $\hat{C}$.

From (1.115) we obtain

$$u_z = v,\;\; u_{\bar{z}} = \bar{w},$$

$$u_{z\bar{z}} = v_{\bar{z}} = \bar{w}_z = \rho,\;\; u_{z^2} = v_z,\;\; u_{\bar{z}^2} = \bar{w}_{\bar{z}}. \tag{1.116}$$

Substituting (1.115) and (1.116) into the equation (1.101) we reduce problem (I) to four singular integral equations, corresponding to four cases with complementary conditions, except case (b) where there are no complementary conditions. For instance in case (a) the equation is

[2] Since $v_{\bar{z}} - \bar{w}_z = u_{\bar{z}z} - u_{\bar{z}z} \equiv 0$, the curvilinear integral does not depend on the path connecting z_0 and z.

$$\rho - K_a\rho = F + \sum_{j=0}^{\kappa_1 - 1} a_j F_j^{(1)} + \bar{a}_j F_j^{(2)}, \tag{1.117}$$

where K_a is a linear continuous operator with principal singular part

$$K_a\rho = a_1 S\rho + a_2 \overline{S\rho} + b_1 \bar{X}_2 \overline{S\left(\frac{\bar{\rho}}{X_2}\right)} + b_2 X_2 S\frac{\bar{\rho}}{X_2},$$

which is a contractive operator in case

$$|a_1| + |a_2| + |b_1| + |b_2| < 1. \tag{1.118}$$

Hence in this case the equation (1.117) is reducible to the Fredholm equation. This equation is equivalent to the problem (I) in case (a) provided its solution satisfies the conditions (1.111). Thus in this case we have the following result:

Theorem 1.4. *If principal coefficients of (1.101) satisfy the inequality (1.118), and $G(t) \neq 0$, $H(t) \neq 0$ on Γ, then the problem* (I) *is Noetherian: the homogeneous problem* (I) *may have only finite number k of nontrivial solutions and the inhomogeneous problem* (I) *is solvable if its right-hand side F satisfies a finite number k' conditions*

$$\iint_{\mathbb{C}} F(z)v_j(z)\,dx\,dy = 0, \quad j = 1, \ldots, k'.$$

The index of the problem is equal to

$$k - k' = 2(\kappa_1 + \kappa_2),$$

if linearly independence holds over the field of real numbers.

The same result is obtained for other cases.

Corollary 1.2. *The problem* (I), *under the assumptions of Theorem 1.4, has Fredholm's property, i.e. has index equal to zero if $H = 1/G$.*
Now we turn to problem (II).

Theorem 1.5. *If (1.118) holds and $G \neq 0$ on Γ, then problem* (II) *has Fredholm's property: the corresponding homogeneous problem may have only a finite number k of linearly independent nontrivial solutions and the inhomogeneous problem* (II) *is solvable if and only if its right-hand side satisfies the same number k of conditions*

$$\iint_{\mathbb{C}} F(z)v_j(z)\,dx\,dy = 0, \quad j = 1, ..., k,$$

i.e. the index of the problem (II) *is equal to zero.*

Proof. If we denote by ρ the Laplacian $u_{\bar{z}z}$, then for unknown

$$v = u_z \tag{1.119}$$

we obtain the Riemann problem (1.105), (1.106).

Hence by Lemma 1.1, in case $\kappa \geq 0$ we obtain

$$v(z) = T\rho + X(z)\,[\,T\theta\rho - \theta(z)\,T\rho\,] + \sum_{j=0}^{\kappa-1} a_j\, z^j X(z)$$

$$= T_0\rho + \sum_{j=0}^{\kappa-1} a_j\, \varphi_j(z), \tag{1.120}$$

and in case $\kappa < 0$ we have

$$v(z) = -\frac{X(z)}{\pi} \iint_{\mathbb{C}} \frac{\rho(\zeta)\,d\xi\,d\eta}{X(\zeta)(\zeta - z)} = X(z)T\frac{\rho}{X} \tag{1.121}$$

with complementary conditions

$$\iint_{\mathbb{C}} \rho(z)\frac{z^j}{X(z)}\,dx\,dy = 0, \quad j = 0, ..., -\kappa - 1. \tag{1.122}$$

Thus in case $\kappa \geq 0$ we obtain the Riemann problem for unknown function $w(z) = \overline{u(z)}$:

$$w_{\bar{z}} = \overline{T_0\,\rho} + \sum_{j=0}^{\kappa-1} \overline{a_j}\;\overline{\varphi_j(z)} \tag{1.123}$$

in $\mathbb{C} \setminus \Gamma$ and

$$w^+ - \frac{1}{G}w^- = 0 \tag{1.124}$$

on Γ and in case $\kappa < 0$:

$$w_{\bar{z}} = \overline{X(z)}\,\overline{T\left(\frac{\rho}{X}\right)} \tag{1.125}$$

in $\mathbb{C} \setminus \Gamma$ and

$$w^+ - \frac{1}{G} w^- = 0 \tag{1.126}$$

on Γ with complementary conditions (1.122).

Since the index of $1/G$ is $-\kappa \leq 0$, then problem (1.123), (1.124) is solvable if and only if

$$\sum_{j=0}^{\kappa-1} \left(\iint_{\mathbb{C}} \varphi_k(z)\,\overline{\varphi_j(z)}\,dx\,dy \right) \overline{a_j} = -\iint_{\mathbb{C}} \overline{T_0\,\rho}\,\varphi_k\,dx\,dy, \tag{1.127}$$

because $\varphi_k = z^k X(z)$ are the piecewise holomorphic solutions of the problem $\varphi^+ - G\varphi^- = 0$, which is adjoint to (1.124). But

$$\det\left(\iint_{\mathbb{C}} \varphi_k(z)\,\overline{\varphi_j(z)}\,dx\,dy \right) \neq 0$$

because it is the Gramian of linearly independent functions $\{\varphi_j\}$, and thus all constant a_j are determined uniquely through constants

$$\iint_{\mathbb{C}} \bar{T}\varphi_k\,\bar{\rho}\,dx\,dy$$

and then we obtain from (1.123), (1.124):

$$\overline{u(z)} = w(z) = -\frac{1}{\pi X(z)} \iint_{\mathbb{C}} \frac{X(\zeta)\overline{T_0\,\rho}}{\zeta - z}\,d\xi\,d\eta - \sum_{j=0}^{\kappa-1} \left(\frac{1}{\pi X(z)} \iint_{\mathbb{C}} \frac{X(\zeta)\overline{\varphi_j(\zeta)}}{\zeta - z} \right) \overline{a_j}$$

$$= \frac{1}{X(z)} \{ T(X\overline{T_0\,\rho}) + \sum_{j=0}^{\kappa-1} T(X\Phi_j) \}. \tag{1.128}$$

Analogously from (1.125), (1.126) we obtain, for $\kappa < 0$,

$$\overline{u(z)} = w(z) = T\left(\bar{X}\,\overline{T\left(\frac{\rho}{X}\right)}\right) + \frac{1}{X(z)}\left[T\theta\,\bar{X}\,\overline{T\left(\frac{\rho}{X}\right)} - \theta(z)\,T\left(\bar{X}\,\overline{T\left(\frac{\rho}{X}\right)}\right) \right]$$

$$+\sum_{j=0}^{\kappa-1} a_j \frac{z^j}{X(z)} \tag{1.129}$$

with complementary conditions (1.122), because the canonical solution for the problem (1.124) coincides with $1/X(z)$.

From (1.128) by differentiation and integration by parts we obtain

$$u_{\bar{z}z} = \rho,\ u_{z^2} = S\rho + X(z)\left[S\theta\rho - \theta(z)S\rho\right] + X'(z)\left[T\theta\rho - \theta(z)T\rho\right]$$
$$-X(z)\theta'(z)T\rho + \sum_{j=0}^{\kappa-1} a_j\,\varphi_j',$$

$$u_{\bar{z}} = \frac{1}{X}\,S(\bar{X}\,T_0\rho) + \sum_{j=0}^{\kappa-1} \frac{1}{X} S(\bar{X}\,\varphi_j)a_j,$$

$$u_{\bar{z}^2} = -\frac{1}{X}\bar{S}(\bar{X}\rho) - \sum_{j=0}^{\kappa-1} \frac{1}{X}\, S(\overline{X'}\,\varphi_j)a_j + \cdots . \tag{1.130}$$

Substituting (1.128) and (1.130) with (a) obtained form (1.127) into (1.101) we obtain the following integral equation for $\rho \in L_{p,2}(\mathbb{C})$.

$$\rho - L_+\rho = \tilde{F}^*, \tag{1.131}$$

where L_+ is linear continuous operator in $L_{p,2}(\mathbb{C})$ with principal part $\overset{0}{L}_+$:

$$\overset{0}{L}_+\rho \equiv a_1\,S\rho + a_2\,\overline{S\,\rho} - \frac{b_1}{X}\bar{S}(\bar{X}\rho) - \frac{b_2}{X}S(X\bar{\rho}). \tag{1.132}$$

This operator is contractive in $L_{p,2}(\mathbb{C})$ with some $p > 2$ provided inequality (1.118) holds. Therefore multiplying both sides of (1.132) by the invertible operator $(I - \overset{0}{L}_+)^{-1}$ we reduce (1.131) to Fredholm's equation.

Now from (1.129) we obtain

$$u_{\bar{z}z} = \rho,\ u_{z^2} = X\,S\frac{\rho}{X} + \cdots,$$

$$u_{\bar{z}} = \bar{S}\,(X\,T\,(\frac{\rho}{X})) + \cdots + \sum_{j=0}^{\kappa-1} a_j\,\bar{\psi}_j, \tag{1.133}$$

$$u_{\bar{z}^2} = -\bar{S}\rho + \cdots + \sum_{j=0}^{\kappa-1} a_j \overline{\psi'_j}$$

and then substituting into (1.101) we obtain the following integral equation for $\rho \in L_{p,2}(\mathbb{C})$:

$$\rho - L - \rho = F^* + \sum_{j=1}^{2\kappa} \alpha_j \tilde{F}_j, \tag{1.134}$$

which is equivalent to problem (II) in case $\kappa > 0$ if complementary conditions (1.222) are fulfilled, where L_- is a linear continuous operator in $L_{p,2}(\mathbb{C})$ with

principal singular part L_-^0:

$$L_-^0\rho \equiv a_1 X S(\rho/X) + a_2 \bar{X} \overline{S(\rho/X)} - b_1 \bar{S}\rho - b_2 S\bar{\rho},$$

which is contractive in $L_{p,2}(\mathbb{C})$ in case inequality (1.118) holds. Thus we may apply Fredholm's theorem to equation (1.134) as well as to (1.131). Now by the same arguments we applied above we conclude that the following theorem is true:

Theorem 1.5'. *If the inequality (1.118) holds and* $G \neq 0$ *on* Γ, *then problem* (II) *has Fredholm's property: the inhomogeneous problem* (II) *is solvable if and only if its right-hand side* F *satisfies the finite number* k *of conditions*

$$\iint_{\mathbb{C}} F(z)\, v_j(z)\, dx\, dy = 0, \quad j = 1,\dots,k$$

and the corresponding homogeneous problem (II) may have only an equal finite number k *of linearly independent nontrivial solutions.*

1.2.3. The Dirichlet and Neumann problems

(a) The simplest boundary-value problem for the second-order elliptic equation is to find a solution $u(z)$ of the equation (1.93) in a bounded domain G^+ continuous in closure $\bar{G}^+ = G^+ + T$ with Laplacian in $L_p(G^+)$ satisfying the Dirichlet condition

$$u(t) = h(t), \quad t \in \Gamma$$

for a given function $h(t)$ continuous on Γ.

This problem comes also from elasticity as a displacement boundary-value problem: to find in G^+ a solution of the Navier displacement equations of equilibrium (in isotropic elasticity):

$$\begin{pmatrix} \lambda+2\mu & 0 \\ 0 & \mu \end{pmatrix} \frac{\partial^2 u}{\partial x^2} + \begin{pmatrix} 0 & \lambda+\mu \\ \lambda+\mu & 0 \end{pmatrix} \frac{\partial^2 u}{\partial x \partial y} + \begin{pmatrix} \mu & 0 \\ 0 & \lambda+2\mu \end{pmatrix} \frac{\partial^2 u}{\partial y^2} = \begin{pmatrix} f_1 \\ f_2 \end{pmatrix} \tag{1.135}$$

with displacement $u = (u_1, u_2)$ specified at all points of the boundary Γ of domain G^+.

Equation (1.135) may be written in canonical complex form (1.93) with $k = 1$ as in anisotropic elasticity (for an orthotropic body), when Navier equations for plane strain may also be written in form (1.93) with $0 < k < 1$.

From elasticity comes also another boundary-value problem: to find in G^+ a solution of the Navier equation of equilibrium with the normal stresses specified at the boundary Γ. This is called the stress boundary-value problem. Using stress-strain relations the stress boundary conditions may be written in terms of displacement as two equalities on Γ among the first derivatives of displacement.

The simplest boundary-value problem with given first-derivative relation on Γ is the Neumann problem: to find a solution of the equation (1.93) in G^+ with the first derivatives continuous in the closure $\bar{G}^+$ and with the Laplacian in $L_p(G^+)$ satisfying the boundary condition

$$\left.\frac{\partial u}{\partial \nu}\right|_{\Gamma} = h(t), \quad t \in \Gamma,$$

where $\partial/\partial\nu$ denotes the directional derivative along the unit vector ν normal to the boundary. Both the Dirichlet and Neumann problems as well as the stress boundary-value problems are completely solved, but only for the case when the equation (1.101) is strongly elliptic (see [50], [11]).

Following example (see [25])

$$u_{\bar{z}^2} = F(z) \tag{1.136}$$

in $|z| < 1$ and

$$u(t) = h(t)$$

on Γ shows that the homogeneous Dirichlet problem ($F = k \equiv 0$) has infinitely many linearly independent nontrivial solutions $u_k(z) = z^k(1 - |z|^2)$, $k = 0,1,...,$ and the inhomogeneous problem is solvable if and only if its right-hand side satisfies infinitely many conditions. Note that (1.136) is not strongly elliptic but merely elliptic. This means that if the general equation (1.93) is elliptic in G^+ the Dirichlet problem is not the natural boundary-value problem in the bounded domain G^+ because it is ill posed in general. It can be shown that the same holds for the Neumann problem. In the next section we state boundary-value problems which seem to be more natural for general equations.

(b) In this section we restrict ourselves to the Dirichlet problem for inhomogeneous equation (1.101) with homogeneous boundary condition

$$u(t) = 0 \tag{1.137}$$

on Γ.

Then we can represent $u(z)$ uniquely by means of Green operator (see [51]):

$$u(z) = G\rho \equiv -\frac{2}{\pi}\iint_{G^+} g(z,\zeta)\,\rho(\zeta)\,dG^+_\zeta, \tag{1.138}$$

where $g(z, \zeta)$ is the Green function of the Dirichlet problem for the Laplace equation in G^+ and $\rho \in L_p(G^+), p > 1,$ is an unknown density. Differentiating (1.138) we obtain (see [25])

$$u_z = T_1\rho = -\frac{1}{\pi}\iint_{G^+}\frac{\rho(\zeta)dG^+_\zeta}{\zeta - z} - \iint_{G^+} k_{G^+}(z,\bar\zeta)\rho(\zeta)\,dG^+_\zeta, \tag{1.139}$$

$$u_{\bar z} = \bar T_1\rho = -\frac{1}{\pi}\iint_{G^+}\frac{\rho(\zeta)dG^+_\zeta}{\bar\zeta - \bar z} - \iint_{G^+}\overline{k_{G^+}(z,\bar\zeta)}\,\rho(\zeta)\,dG^+_\zeta,$$

$$k_{G^+}(z,\bar\zeta) = \frac{1}{2\pi^2 i}\int_\Gamma \frac{d\bar t}{(\bar t - \bar\zeta)(t - z)} - \frac{1}{\pi}\iint_{G^+}\frac{\ell(z,\tau)\,dG^+_\zeta}{\bar\tau - \bar\zeta}, \tag{1.140}$$

$$u_{\bar z z} = \rho, \quad u_{z^2} = S_1\rho = S_{G^+}\rho - \iint_{G^+} K_0(z,\bar\zeta)\,\rho(\zeta)\,dG^+_\zeta,$$

$$u_{\bar z^2} = \bar S_1\rho = \bar S_{G^+}\rho - \iint_{G^+}\overline{K_0(z,\bar\zeta)}\,\rho(\zeta)\,dG^+_\zeta,$$

where $S_{G^+}, \bar S_{G^+}$ are finite Hilbert transformations, which are singular integral operators to be understood in the Cauchy principal value sense:

$$S_{G^+}\rho = -\frac{1}{\pi}\iint_{G^+}\frac{\rho(\zeta)dG_\zeta^+}{(\zeta - z)^2} = -\frac{1}{\pi}\lim_{\varepsilon\to 0}\iint_{G^+\setminus(|\zeta - z|<\varepsilon)} \tag{1.141}$$

$$\bar{S}_{G^+}\rho = -\frac{1}{\pi}\iint_{G^+}\frac{\rho(\zeta)dG_\zeta^+}{(\bar{\zeta} - \bar{z})^2} = -\frac{1}{\pi}\lim_{\varepsilon\to 0}\iint_{G^+\setminus(|\zeta - z|<\varepsilon)}$$

$$K_0(z,\bar{\zeta}) = \frac{1}{2\pi^2 i}\int_\Gamma \frac{d\bar{t}}{(\bar{t}-\bar{\zeta})(t-z)^2} - \frac{1}{\pi}\iint_{G^+}\frac{\ell_z'(z,z)\,dG_\tau^+}{\bar{\tau}-\bar{\zeta}}, \tag{1.142}$$

where $\ell(z, \zeta) = (2/\pi)\, h_{z\zeta}(z, \zeta)$ is holomorphic in $G^+ \times G^+$ and continuous in $\bar{G}^+ \times \bar{G}^+$, $h(z, \zeta)$ is the regular part of the Green function. Substituting (1.138)-(1.141) into (1.101) we obtain for the unknown function $\rho \in L_p(G^+)$ the integral equation

$$\rho + P\rho = F(z), \tag{1.143}$$

where P is a linear continuous operator in $L_p(G^+)$

$$P\rho \equiv a_1 S_1\rho + a_2\overline{S_1\rho} + b_1\bar{S}_1\rho + b_2 S_1\bar{\rho}$$

$$+ A_1 T_1\rho + A_2\overline{T_1\rho} + B_1\bar{T}_1\rho + B_2 T_1\bar{\rho} + C_1 G\rho + G_2\overline{G\rho}. \tag{1.144}$$

Let us consider now the homogeneous adjoint Dirichlet problem:

$$v_{\bar{z}z} + (\bar{a}_1 v)_{\bar{z}^2} + (a_2\bar{v})_{\bar{z}^2} + (\bar{b}_1 v)_{z^2} + (b_2\bar{v})_{z^2} - (\bar{A}_1 v)_{\bar{z}}$$

$$- (A_2\bar{v})_{\bar{z}} - (\bar{B}_1 v)_z - (B_2\bar{v})_z + \bar{C}_1 v + C_2\bar{v} = 0 \tag{1.145}$$

in G^+ and

$$v(t) = 0 \tag{1.146}$$

on Γ.

Transferring all terms except first in (1.145) from the left-hand side to the right-hand side we see that a solution of the adjoint problem represent as

$$v(z) = \frac{2}{\pi}\iint_{G^+} g(z,\zeta)\left\{(\overline{a_1(\zeta)}\,v)_{\bar{\zeta}^2} + (a_2(\zeta)\,\bar{v})_{\bar{\zeta}^2} + (\overline{b_1(\zeta)}v)_{\zeta^2}\right.$$

$$+ (b_2(\zeta)\bar{v})_{\zeta^2} - (\overline{A_1(\zeta)}\, v)_{\bar{\zeta}} - (A_2(\zeta)\,\bar{v})_{\bar{\zeta}} - \overline{B_1(\zeta)}\, v)_{\zeta} - (B_2(\zeta)\,\bar{v})_{\zeta}$$

$$+ \overline{C_1(\zeta)}\, v(\zeta) + C_2(\zeta)\, \overline{v(\zeta)} \Big\}\, dG^+_{\zeta}. \tag{1.147}$$

Integrating by parts and taking into account that $g(z, \zeta) = 0$ for $z \in G^+$ and $\zeta \in \Gamma$, we obtain for instance

$$\frac{2}{\pi} \iint_{G^+} g(z, \zeta)\, (\overline{a_1(\zeta)}\, v)_{\bar{\zeta}^2}\, dG^+_{\zeta} = -\frac{2}{\pi} \iint_{G^+} \frac{\partial g(z,\zeta)}{\partial \bar{\zeta}} (\overline{a_1(\zeta)}\, v)_{\bar{\zeta}}\, dG^+_{\zeta}.$$

Again integrating the right-hand side of this relation by parts and taking into account (1.146), we obtain

$$\frac{2}{\pi} \iint_{G^+} \frac{\partial g(z,\zeta)}{\partial \bar{\zeta}} (\overline{a_1(\zeta)}\, v)_{\zeta}\, dG^+_{\zeta} = -\frac{2}{\pi} \lim_{\varepsilon \to 0} \iint_{G^+ - (|\zeta - z| < \varepsilon)} \frac{\partial g(z,\zeta)}{\partial \bar{\zeta}} (\overline{a_1(\zeta)}\, v)_{\zeta}\, d\xi\, d\eta$$

$$= \lim_{\varepsilon \to 0} \frac{2}{\pi} \iint_{G^+ - (|\zeta - z| < \varepsilon)} \frac{\partial^2 g(z,\zeta)}{\partial \bar{\zeta}^2} \overline{a_1(\zeta)} v(\zeta)\, d\xi d\eta + \lim_{\varepsilon \to 0} \frac{1}{\pi i} \int_{|\zeta - z| = \varepsilon} \frac{\partial g(z,\zeta)}{\partial \bar{\zeta}} \overline{a_1(\zeta)} v(\zeta) d\xi$$

$$= \frac{2}{\pi} \iint_{G^+} \frac{\partial^2 g(z,\zeta)}{\partial \bar{\zeta}^2} \overline{a_1(\zeta)} v(\zeta)\, dG^+_{\zeta},$$

because

$$\lim_{\varepsilon \to 0} \frac{1}{\pi i} \int_{|\zeta - z| = \varepsilon} \frac{\partial g(z,\zeta)}{\partial \bar{\zeta}} \overline{a_1(\zeta)} v(\zeta) d\zeta = \lim_{\varepsilon \to 0} \frac{1}{2\pi i} \int_{|\zeta - z| = \varepsilon} \left(\frac{1}{\bar{\zeta} - \bar{z}} + O(\varepsilon) \right) \overline{a_1(\zeta)} v(\zeta)\, d\zeta$$

$$= 0.$$

Analogously

$$\frac{2}{\pi} \iint_{G^+} g(z, \zeta)\, (\overline{b_1(\zeta)} \bar{v})_{\zeta^2}\, dG^+ = \frac{2}{\pi} \iint_{G^+} \frac{\partial^2 g(z,\zeta)}{\partial \zeta^2} \overline{b_1(\zeta)} v(\zeta)\, dG^+_{\zeta}.$$

Also integrating the other terms by parts we conclude that (1.147) reduces to the following integral equation

$$v(z) - \frac{2}{\pi}\iint_{G^+}\Big\{\frac{\partial^2 g(z,\zeta)}{\partial\bar{\zeta}^2}\Big[\overline{a_1(\zeta)v(\zeta) + a_2(\zeta)\overline{v(\zeta)}}\Big] + \frac{\partial^2 g(z,\zeta)}{\partial\zeta^2}\Big[\overline{b_1(\zeta)v(\zeta) + b_2(\zeta)\overline{v(\zeta)}}\Big]$$

$$+ \frac{\partial g(z,\zeta)}{\partial\zeta}\Big[\overline{A_1(\zeta)v(\zeta) + A_2(\zeta)\overline{v(\zeta)}}\Big] + \frac{\partial g(z,\zeta)}{\partial\zeta}\Big[\overline{B_1(\zeta)v(Z) + B_2(\zeta)\overline{v(\zeta)}}\Big]$$

$$+ g(z,\zeta)\Big[\overline{C_1(\zeta)v(\zeta) + C_2(\zeta)\overline{v(\zeta)}}\Big]\Big\}\, dG^+_\zeta = 0. \tag{1.147'}$$

Since the operator P in (1.144) is

$$P\rho \equiv -\frac{2}{\pi}\iint_{G^+}\Big\{a_1(z)\frac{\partial^2 g(z,\zeta)}{\partial z^2}\rho(\zeta) + a_2(z)\frac{\partial^2 g(z,\zeta)}{\partial\bar{z}^2}\overline{\rho(\zeta)}$$

$$+ b_1(z)\frac{\partial^2 g(z,\zeta)}{\partial\bar{z}^2}\rho(\zeta) + b_2(z)\frac{\partial^2 g(z,\zeta)}{\partial z^2}\overline{\rho(\zeta)} + A_1(z)\frac{\partial g(z,\zeta)}{\partial z}\rho(\zeta)$$

$$+ A_2(z)\frac{\partial g(z,\zeta)}{\partial\bar{z}}\overline{\rho(\zeta)} + B_1(z)\frac{\partial g(z,\zeta)}{\partial\bar{z}}\rho(\zeta) + B_2(z)\frac{\partial g(z,\zeta)}{\partial z}\overline{\rho(\zeta)}$$

$$+ g(z,\zeta)\Big[C_1(z)\,\rho(\zeta) + C_2(z)\,\overline{\rho(\zeta)}\Big]\Big\}\, dG^+_\zeta, \tag{1.144'}$$

the homogeneous equation (1.147') is an adjoint to (1.143):

$$\nu + P^*\nu = 0, \tag{1.143'}$$

where P^* is an integral operator adjoint to operator P with respect to the metric

$$(\rho, \nu) = \operatorname{Re}\iint_{G^+}\rho\,\bar{\nu}\,dx\,dy.$$

It is easy to see that

$$P\rho = P_0\rho + P_1\rho,$$

where P_1 is an operator compact in $L_p(G^+)$ and P_0 is a principal singular part:

$$P_0\rho \equiv a_1\,S_1\rho + a_2\,\overline{S_1\,\rho} + b_1\,\overline{S}_1\rho + b_2\,S_1\,\bar{\rho}. \tag{1.148}$$

Now we represent the operator S_1 as

$$S_1\rho = \iint\limits_{G^+} \Big[\ell(z,\zeta) - \frac{1}{\pi(\zeta - z)^2} \Big] \rho(\zeta)\, dG^+_\zeta - \iint\limits_{G^+} \big[K_0(z,\bar\zeta) + \ell(z,\zeta) \big] \rho(\zeta)\, dG^+_\zeta \quad (1.149)$$

and denote by $\tau(z)$ the harmonic function in G^+ with $t'(s)$ as its boundary value on Γ:

$$\tau(z) = \frac{1}{2\pi} \int_\Gamma \frac{\partial g(z,t)}{\partial \nu_t}\, t'(s)\, ds = \frac{1}{2\pi} \int_\Gamma \frac{\partial g(z,t)}{\partial \nu_t}\, dt. \quad (1.150)$$

Evidently $|\tau(z)| \le 1$ for $z \in G^+$, because $|t'(s)| \equiv 1$. Taking into account the integral representation of the Bergman's kernel function $K(z, \bar\zeta)$ we can write

$$K_0(z,\bar\zeta) = \frac{1}{2\pi^2 i} \int_\Gamma \frac{\overline{t'^2(s)}\, dt}{(\bar t - \bar\zeta)(t - z)^2} - \frac{1}{\pi} \iint\limits_{G^+} \frac{\ell'_z(z,\tau)\, dG^+_\tau}{\bar\tau - \bar\zeta}$$

$$= \overline{\tau^2(\zeta)}\, K(z,\bar\zeta) + \frac{1}{2\pi^2 i} \int \frac{\overline{t'^2(s)} - \overline{\tau^2(\zeta)}}{(\bar t - \bar\zeta)(t - z)^2}\, dt$$

$$- \iint\limits_{G^+} \Big[\bar\tau^2(\zeta)\, \ell(\sigma, z)\, \overline{\ell(\sigma,\zeta)} + \frac{\ell'_z(z,\sigma)}{\pi(\bar\sigma - \bar\zeta)} \Big] dG^+_\sigma.$$

Hence the operator S_1 represents as

$$S_1\rho = \hat S_1\rho + K_1\rho, \quad (1.151)$$

where

$$\hat S_1\rho \equiv \iint\limits_{G^+} \Big[\ell(z,\zeta) - \frac{1}{\pi(\zeta - z)^2} \Big] \rho(\zeta)\, dG^+_\zeta - \iint\limits_{G^+} K(z,\bar\zeta)\, \overline{\tau^2(\zeta)}\, \rho(\zeta)\, dG^+_\zeta \quad (1.152)$$

and K_1 is an operator, compact in $L_p(G^+)$, because

$$\left| \frac{\overline{t'^2(s)} - \overline{\tau^2(\zeta)}}{\bar t - \bar\zeta} \right| < \frac{\text{const}}{|t - \zeta|^\varepsilon}, \quad 0 < \varepsilon < 1.$$

Consequently the equation (1.143) may be written as

$$\rho + P^0\rho + P^1\rho = F(z), \tag{1.143'}$$

where P^1 is an operator compact in $L_p(G^+)$ and

$$P^0_\rho \equiv a_1\,\hat{S}_1\rho + a_2\,\overline{\hat{S}_1\,\rho} + b_1\,\hat{S}_1\bar{\rho} + b_2\,\overline{\hat{S}_1\,\rho}\,. \tag{1.152}$$

Lemma 1.4. *The operator* $\hat{S}_1$ *is a linear continuous in* $L_p(G^+)$ *with* L_2*-norm, not greater than one.*

Proof. Since

$$\overline{\hat{S}_1\,\rho} = \frac{\partial \hat{T}_1\,\rho}{\partial z} - k\,(\bar{\tau}^2\rho), \tag{1.153}$$

where

$$\hat{T}_1\rho \equiv \iint_{G^+} \Big[\tilde{\ell}(z,\zeta) - \frac{1}{\pi(\zeta - z)}\Big]\,\rho(\zeta)\,dG^+_\zeta,$$

$$K\rho \equiv \iint_{G^+} K(z,\bar{\zeta})\,\rho(\zeta)\,dG^+_\zeta\,,$$

then for $\rho \in C_0^\infty(G^+)$, by means of the Green formula, we obtain

$$\|\hat{S}_1\rho\|^2_{L_2(\bar{G}^+)} = \iint_{G^+} \frac{\partial \hat{T}_1\,\rho}{\partial z}\,\overline{\frac{\partial \hat{T}_1\,\rho}{\partial \bar{z}}}\,dG^+_z - \iint_{G^+} \frac{\partial \hat{T}_1\,\rho}{\partial z}\,\overline{K(\bar{\tau}^2_\rho)}\,dG^+_\zeta$$

$$-\overline{\frac{\partial \hat{T}_1\,\rho}{\partial z}}\,K(\bar{\tau}^2\rho)\,dG^+_\zeta + \|K(\bar{\tau}^2\rho)\|^2_{L_2(\bar{G}^+)} = \iint_{G^+} \frac{\partial}{\partial \bar{z}}\Big(\frac{\partial \hat{T}_1\,\rho}{\partial z}\,\overline{\hat{T}_1\,\rho}\Big)\,dG^+_z$$

$$-\iint_{G^+} \frac{\partial^2 \hat{T}_1\,\rho}{\partial \bar{z}\,\partial z}\,\overline{\hat{T}_1\,\rho}\,dG^+_z - 2\,\mathrm{Re}\iint_{G^+} \frac{\partial \hat{T}_1\,\rho}{\partial z}\,\overline{K(\bar{\tau}^2\rho)}\,dG^+_z + \|K(\bar{\tau}^2\rho)\|^2_{L_2(\bar{G}^+)}$$

$$= \|K(\bar{\tau}^2\rho)\|^2_{L_2(\bar{G}^+)} - \iint_{G^+} \frac{\partial \rho}{\partial z}\,\overline{\hat{T}_1\,\rho}\,dG^+_z + \frac{1}{2i}\int_\Gamma \frac{\partial \hat{T}_1\,\rho}{\partial t}\,\overline{\hat{T}_1\,\rho}\,dt$$

$$-2\,\mathrm{Re}\iint\limits_{G^+}\frac{\partial}{\partial z}\left[\hat{T}_1\rho\,\overline{K(\bar{\tau}^2\rho)}\,\mathrm{d}G_z^+\right]=\|\rho\|^2_{L_2(\overline{G}^+)}+\|K(\bar{\tau}^2\rho)\|^2_{L_2(\overline{G}^+)}$$

$$+\iint\limits_{G^+}\overline{\rho(\zeta)}\iint\limits_{G^+}\rho(z)\Big\{-\frac{1}{2\pi^2 i}\int_\Gamma\frac{\mathrm{d}t}{(\bar{t}-\bar{\zeta})(t-z)^2}+\frac{1}{2i}\int_\Gamma \ell(t,z)\,\overline{\tilde{\ell}(t,z)}\,\mathrm{d}t$$

$$-\frac{1}{2\pi i}\int_\Gamma\frac{\overline{\tilde{\ell}(t,\zeta)}\,\mathrm{d}t}{(t-z)^2}+\frac{1}{2\sigma i}\int_\Gamma\frac{\ell(t,z)\,\mathrm{d}t}{\bar{t}-\bar{\zeta}}\Big\}\,\mathrm{d}G_\zeta^+\,\mathrm{d}G_z^+$$

$$+\mathrm{Re}\,\frac{1}{i}\int_\Gamma\hat{T}_1\rho\,\overline{K(\bar{\tau}^2\rho)}\,\mathrm{d}\bar{t}.$$

But as we have seen above

$$\frac{1}{2\pi i}\int_\Gamma\frac{\overline{\tilde{\ell}(t,\zeta)}\,\mathrm{d}t}{(t-z)^2}=\frac{1}{\pi}\iint\limits_{G^+}\frac{\overline{\ell(\tau,\zeta)}\,\mathrm{d}G_\tau^+}{\tau-z},$$

$$\frac{1}{2\pi i}\int_\Gamma\frac{\ell(t,z)\,\mathrm{d}t}{\bar{t}-\bar{\zeta}}=-\frac{1}{\pi}\iint\limits_{G^+}\frac{\ell(\tau,z)\,\mathrm{d}G_\tau^+}{(\bar{\tau}-\bar{\zeta})^2}$$

and

$$\frac{1}{\pi}\iint\limits_{G^+}\frac{\ell(\tau,\zeta)\,\mathrm{d}G_\tau^+}{(\bar{\tau}-\bar{z})^2}=\iint\limits_{G^+}\overline{\ell(\tau,z)}\,\ell(t,\zeta)\,\mathrm{d}G_\tau^+,$$

i.e.

$$\frac{1}{\pi}\iint\limits_{G^+}\frac{\overline{\ell(\tau,\zeta)}\,\mathrm{d}G_\tau^+}{(\tau-z)^2}=\iint\limits_{G^+}\overline{\ell(\tau,\zeta)}\,\ell(\tau,z)\,G_\tau^+,$$

$$-\frac{1}{2i}\int_\Gamma\tilde{\ell}(t,\zeta)\,K(z,\bar{t})\,\mathrm{d}\bar{t}=\iint\limits_{G^+}\ell(\tau,\zeta)\,K(z,\bar{\tau})\,\mathrm{d}G_\tau^+,$$

$$-\frac{1}{2i}\int_\Gamma\frac{K(z,\bar{t})\,\mathrm{d}\bar{t}}{t-\zeta}=-\iint\limits_{G^+}\frac{K(z,\bar{\tau})\,\mathrm{d}G_\tau^+}{(\tau-\zeta)^2}$$

then it follows that

$$\|\hat{S}_1\rho\|^2_{L_2} = \|\rho\|^2_{L_2} + \|K(\bar{\tau}^2\rho)\|^2_{L_2} - \iint_{G^+} \overline{\rho(\zeta)} \iint_{G^+} \rho(z)\,K(z,\bar{\zeta})\,\mathrm{d}G^+_\zeta\,\mathrm{d}G^+_z$$

$$+\operatorname{Re}\iint_{G^+} \rho(\zeta) \iint_{G^+} \overline{\tau^2(z)}\,\rho(z)\left\{\frac{1}{\mathrm{i}}\int_\Gamma \tilde{\ell}(t,\zeta)\,K(z,\bar{t})\,\mathrm{d}\bar{t} + \frac{1}{\mathrm{i}}\int_\Gamma \frac{K(z,\bar{t})\,\mathrm{d}\bar{t}}{t-\zeta}\right\}\mathrm{d}G^+_\zeta\,\mathrm{d}G^+_z$$

$$= \|\rho\|^2_{L_2} - \iint_{G^+} \overline{\rho(\zeta)} \iint_{G^+} \rho(z)\,K(z,\bar{\zeta})\,\mathrm{d}G^+_\zeta\,\mathrm{d}G^+_z + \|K(\bar{\tau}^2\rho)\|^2_{L_2}$$

$$= \|\rho\|^2_{L_2} + \|K(\bar{\tau}^2\rho)\|^2_{L_2} - \sum_{k=1}^{\infty} |a_k|^2,$$

where

$$a_k = \iint_{G^+} \rho(z)\,\omega_k(z)\,\mathrm{d}G^+_z.$$

But since

$$K(\bar{\tau}^2\rho) = \iint_{G^+} K(z,\bar{\zeta})\,\bar{\tau}^2(\zeta)\,\rho(\zeta)\,\mathrm{d}G^+_\zeta = \sum_{k=1}^{\infty} c_k\,\omega_k(z),$$

where

$$c_k = \iint_{G^+} \bar{\tau}^2(\zeta)\,\rho(\zeta)\,\overline{\omega_k(\zeta)}\,\mathrm{d}G^+_\zeta\,,$$

then we obtain

$$\|\hat{S}_1\rho\|^2_{L_2} = \|\rho\|^2_{L_2} - \sum_{k=1}^{\infty} (|a_k|^2 - |c_k|^2). \tag{1.154}$$

Since $|\tau(z)| \le 1$ for $z \in \bar{G}^+$, then it is evident that $|c_k| \le |a_k|$ and

$$\|\hat{S}_1\rho\|^2_{L_2} \le \|\rho\|^2_{L_2}, \text{ i.e. } \|\hat{S}_1\|_{L_2} \le 1.$$

Since L_p-norm of the operator $\hat{S}_1$ is continuous in p, then by the Riesz–Torin theorem $\| \hat{S}_1 \|_{L_p} < 1$ for some $p: 0 < 2 - p < \varepsilon$. Hence in case (1.118) holds the operator $I + P^0$ will be invertible the equation (1.143) reduces to the Fredholm's equation. The homogeneous adjoint equation (1.143) also equivalent to Fredholm's equation. Thus we have proved the following:

Theorem 1.6. *If the coefficients of the principal part of equation (1.101) are bounded measurable functions satisfying the inequality (1.118) and its other coefficients and right-hand side are in* $L_p(\bar{G}^+)$ *with* $p > 2$, *then the Dirichlet problem (1.101), (1.137) has the Fredholm property: the inhomogeneous problem is solvable if and only if its right-hand side satisfies the finite number* k *of conditions*

$$\operatorname{Re} \iint_{G^+} F(z)\, \overline{v_j(z)}\, dx\, dy = 0, \quad j \; 1,\dots,k, \tag{1.155}$$

where $v_1,\dots,v_k$ *are nontrivial solutions of an adjoint homogeneous problem (1.145), (1.146) and the corresponding homogeneous Dirichlet problem* $(F \equiv 0)$, *may have only the same finite number* k *of nontrivial solutions.*

(c) If we apply this theorem to the equation

$$a(z)u_{\bar{z}z} + b(z)\bar{u}_{\bar{z}2} + a_0(z)u_z + b_0(z)\overline{u_z} = F(z) \tag{1.156}$$

with $a(z) \neq 0$ in $\bar{G}^+$ it gives a well-known result, because in this case inequality (1.118) is

$$|\, b(z)\,| < |\, a(z)\,| \tag{1.157}$$

which implies the strong ellipticity of (1.156).

We show that the Dirichlet problem (1.137) for (1.156) has the Fredholm property also in case $|\, b(z)\,| > |\, a(z)\,|$.

Let us suppose that $a(z) \neq 0$, $z \in \bar{G}^+$ and the equation is just elliptic, i.e.

$$|\, a(z)\,| \neq |\, b(z)\,|$$

in $\bar{G}^+$. Then a solution of the Dirichlet problem (1.137) for (1.156) takes the form

$$u(z) = -\frac{2}{\pi} \iint_{G^+} g(z,\zeta)\, F_1(\zeta)\, dG_\zeta^+, \tag{1.158}$$

where

$$F_1(z) = F(z) - \frac{b(z)}{a(z)}\bar{u}_{\bar{z}2} - \frac{a_0(z)}{a(z)}u_z - \frac{b_0(z)}{a(z)}\bar{u}_{\bar{z}}.$$

Integrating by parts on the right-hand side of (1.158) and taking into account that $g(z,\zeta) = 0$ for $z \in G^+$, $\zeta \in \Gamma$, we obtain

$$u(z) = -\frac{2}{\pi}\iint_{G^+}\left(\frac{b(\zeta)}{a(\zeta)}g(z,\zeta)\right)_{\bar{\zeta}}\bar{u}_{\bar{\zeta}}\,dG_\zeta^+$$

$$+\frac{2}{\pi}\iint_{G^+}\frac{g(z,\zeta)}{a(\zeta)}(a_0(\zeta)u_\zeta + b_0(\zeta)\,\bar{u}_{\bar{\zeta}})\,dG_\zeta^+ + F_2(z), \tag{1.159}$$

where

$$F_2(z) = -\frac{2}{\pi}\iint_{G^+} g(z,\zeta)\,F(\zeta)\,dG_\zeta^+.$$

Differentiating (1.159) with respect to z we obtain the following integral equation with respect to unknown $\omega(z) = u_z \in L_p(G^+)$:

$$\omega(z) + \frac{b(z)}{a(z)}\overline{\omega(z)} - \iint_{G^+}\frac{b(\zeta)}{a(\zeta)}K(z,\bar{\zeta})\,\overline{\omega(\zeta)}\,dG_\zeta^+$$

$$+\tilde{K}\omega = F(z), \tag{1.160}$$

where $\tilde{K}$ is an integral operator compact in L_p and $K(z,\bar{\zeta})$ is the Bergman kernel of the domain G^+, which is expressed in terms of derivatives of Green's function

$$K(z,\bar{\zeta}) = -\frac{2}{\pi}\frac{\partial^2 g(z,\zeta)}{\partial z\,\partial\bar{\zeta}}.$$

The integral equation (1.160) is singular, because of the strong singularity of the kernel $K(z,\bar{\zeta})$ at the boundary points. Nevertheless it can be shown (see [25]) that this equation has the Fredholm property. Here we do all calculations for the case when a, b, are constant and the domain G^+ is the unit disc $|z| < 1$, i.e. let us consider the equation

$$\omega(z) + \lambda\,\overline{\omega(z)} - \lambda K\bar{\omega} = g(z) \tag{1.161}$$

in disc $|z| < 1$, where

$$K\bar{\omega} = \iint_{|\zeta|<1} \frac{\overline{\omega(\zeta)}\, d\xi\, d\eta}{(1-\bar{\zeta} z)^2}, \quad \lambda = b/a.$$

From (1.161) it follows that

$$\omega(z) = \frac{1}{1-|\lambda|^2}\left(\varphi(z) - \lambda\,\overline{\varphi(z)} + g(z) - \lambda\,\overline{g(z)}\right), \tag{1.162}$$

because the function

$$\varphi(z) = K\bar{\omega} \equiv \iint_{G^+} \frac{\overline{\omega(\zeta)}\, d\xi\, d\eta}{(1-\bar{\zeta} z)^2}$$

is holomorphic in the disc $|z| < 1$ for any $\omega \in L_p, p > 1$. Substituting (1.162) into (1.161) we obtain the integral equation for the holomorphic function $\varphi(z)$:

$$\varphi - \lambda K\bar{\varphi} = \lambda\,(Kg - \lambda K\bar{g}), \tag{1.163}$$

because of the reproducing property

$$K\varphi \equiv \varphi.$$

Hence by Green's formula and the residue theorem for any holomorphic function $\varphi(z)$,

$$K\bar{\varphi} = \frac{1}{\pi} \iint_{|\zeta|<1} \frac{\overline{\varphi(\zeta)}\, d\xi\, d\eta}{(1-\bar{\zeta} z)^2} = \frac{\partial}{\partial z}\left(\frac{z}{\pi} \iint_{|\zeta|<1} \frac{\overline{\varphi(\zeta)}\, d\xi\, d\eta}{1-\bar{\zeta} z}\right)$$

$$= \frac{\partial}{\partial z}\left(\frac{z}{\pi} \iint_{|\zeta|<1} \frac{(\overline{\varphi(\zeta)}\cdot\zeta)_\zeta}{1-\bar{\zeta} z}\, d\xi\, d\eta\cdot\right) = \frac{\partial}{\partial z}\left(\frac{z}{\pi} \iint_{|\zeta|<1} \frac{\partial}{\partial \zeta}\left(\frac{\overline{\varphi(\zeta)}\cdot\zeta}{1-\bar{\zeta} z}\right) d\xi\, d\eta\right)$$

$$= \frac{\partial}{\partial z}\left(-\frac{z}{2\pi i} \int_{|\zeta|=1} \frac{\overline{\varphi(\zeta)}\,\zeta\, d\bar{\zeta}}{1-\bar{\zeta} z}\right) = \frac{\partial}{\partial z}\left(z\,\overline{\frac{1}{2\pi i} \int_{|\zeta|=1} \frac{\varphi(\zeta)\, d\zeta}{\zeta(1-\zeta\bar{z})}}\right)$$

$$\frac{\partial}{\partial z}\left(\overline{\varphi(0)}\cdot z\right) = \overline{\varphi(0)}$$

and (1.163) becomes

$$\varphi(z) = \lambda\,\overline{\varphi(0)} + \lambda\,(Kg - \lambda K\bar{g}). \tag{1.164}$$

Since $|\lambda| \neq 1$ and

$$(K\bar{g} - \bar{\lambda} Kg)_{z=0} = \frac{1}{\pi} \iint_{|\zeta|<1} \overline{g(\zeta)}\, d\xi\, d\eta - \frac{\bar{\lambda}}{\pi} \iint_{|\zeta|<1} g(\zeta)\, d\xi\, d\eta,$$

then

$$\varphi(0) = \frac{1}{\pi} \iint_{|\zeta|<1} g(\zeta)\, d\xi\, d\eta.$$

Thus the solution of (1.161) is

$$\omega(z) = \frac{\lambda}{\pi(1-|\lambda|^2)} \iint_{|\zeta|<1} \overline{g(\zeta)} - \bar{\lambda}\, g(\zeta)\, d\xi\, d\eta + \frac{g(z) - \lambda \overline{g(z)}}{1-|\lambda|^2}$$

$$+ \frac{\lambda}{1-|\lambda|^2}(Kg - \lambda K\bar{g}) + \frac{|\lambda|^2}{1-|\lambda|^2}(\bar{\lambda}\bar{K}g - \overline{Kg}),$$

where

$$\overline{Kg} = \frac{1}{\pi} \iint_{|\zeta|<1} \frac{g(\zeta)\, d\xi\, d\eta}{(1-\zeta\bar{z})^2},$$

i.e. (1.161) is uniquely solvable and a general equation reduces to Fredholm's equation as we desired.

1.2.4. The problem A_n for the inhomogeneous Laplace equation

(a) In the next section we are going to investigate the following general boundary-value problem (A_n): find a solution $u(z)$ to equation (1.101) continuous in the closure $\bar{G}^+$ with its first derivatives, with Laplacian in $L_p(G^+), p > 2$ and satisfying the boundary conditions

$$\mathrm{Re}\,[\,\overline{t'(s)}\,\overline{\lambda(t)}\,u(t)\,] = 0, \quad \mathrm{Re}\,[\,\overline{\lambda(t)}\,u_z(t)\,] = 0$$

on Γ.

In this section we restrict ourselves to the particular case of (1.101): first we consider this problem for the inhomogeneous Laplace equation

$$u_{z\bar{z}} = f(z). \tag{1.165}$$

A general solution of (1.165) is

$$u(z) = \varphi(z) + \overline{\psi(z)} - \frac{2}{\pi} \iint_{G^+} g(z, \zeta) f(\zeta)\, dG_\zeta^+, \tag{1.166}$$

where φ, ψ are arbitrary functions holomorphic in G^+, $g(z, \zeta)$ is the Green function of the domain G^+.

Since

$$g(z, \zeta) = \log \frac{1}{|\zeta - z|} + h(z, \zeta),$$

where $h(z, \zeta)$ is a regular harmonic function, having $\ln |z - \zeta|$ as boundary value on Γ, then we have

$$u_z = \varphi'(z) - \frac{2}{\pi} \iint_{G^+} h_z(z, \zeta) f(\zeta)\, dG_\zeta^+ - \frac{1}{\pi} \iint_{G^+} \frac{f(\zeta)\, d G_\zeta^+}{\zeta - z}$$

$$= \varphi_0(z) + T_{G^+} f, \tag{1.167}$$

φ_0 is holomorphic in G^+:

$$\varphi_0(z) = \varphi'(z) - \frac{2}{\pi} \iint_{G^+} h_z(z, \zeta) f(\zeta)\, dG_\zeta^+ .$$

Substituting (1.167) into the second condition (A_n) we get the following Riemann-Hilbert boundary condition for the holomorphic function:

$$\mathrm{Re}\, [\, \overline{\lambda(t)}\, \varphi_0(t)\,] = -\, \mathrm{Re}\, [\, \lambda(t)\, T_{G^+} f\,] = h_0(t) \tag{1.168}$$

on Γ.

This problem is solvable if and only if

$$\int_\Gamma h_0(t)\, \overline{\lambda(t)}\, \psi_j(t)\, dt = 0, \quad j = 1, \ldots, \ell', \tag{1.169}$$

where ψ_j are holomorphic solutions of the homogeneous problem adjoint to (1.168):

$$\mathrm{Re}\, [\, t'(s)\, \overline{\lambda(t)}\ \psi(t)\,] = 0 \tag{1.170}$$

on Γ.

In case (1.169) holds, a general solution of the problem (1.168) takes the form

$$\varphi_0(z) = Rh_0 + \sum_{j=1}^{\ell} c_j \varphi_0^{(j)}(z),$$

where R is some linear continuous integral operator on Γ and $\{\varphi_0^{(j)}\}$ is a complete system of linearly independent nontrivial solutions of the homogeneous problem corresponding to (1.168):

$$\operatorname{Re}[\lambda(t)\,\varphi_0(t)] = 0. \tag{1.168'}$$

By (1.167) we obtain

$$\varphi'(z) = Rh_0 + \frac{2}{\pi}\iint_{G^+} h_z(z,\zeta) f(\zeta)\, dG_\zeta^+ + \sum_{j=1}^{\ell} c_j \varphi_0^{(j)}(z). \tag{1.171}$$

Substituting (1.166) into the first condition (A_n) we obtain the following Riemann-Hilbert boundary condition for the holomorphic function ψ:

$$\operatorname{Re}[\, t'(s)\,\overline{\lambda(t)}\;\psi(t)\,] = -\operatorname{Re}[\, t'(s)\,\overline{\lambda(t)}\;\overline{\varphi(t)}] = h^0(t) \tag{1.172}$$

on Γ.

The problem (1.172) is solvable if and only if

$$\int_\Gamma h^0(t)\,\overline{t'(s)}\,\lambda(t)\,\varphi_0^{(j)}(t)\,dt = 0, \quad j = 1,\ldots,\ell, \tag{1.173}$$

where $\varphi_0^{(j)}$ are solutions of the homogeneous problem, adjoint to (1.172), which is just problem (1.169).

Let $\chi^{(j)}(t) = \operatorname{Im}[\,\lambda(t)\,\varphi_0^{(j)}(t)\,]$ on Γ.Then from (1.168) we have $\lambda(t)\varphi_0^{(j)} = i\chi$ and conditions (1.173) are

$$\int_\Gamma \chi^{(j)}(t)\, h^0(t\,ds = 0, \quad j = 1,\ldots,\ell$$

or

$$\operatorname{Re}\int_\Gamma \chi^{(j)}(t)\,\overline{\lambda(t)}\;\overline{\varphi(t)}]\,dt = 0, \quad j = 1,\ldots,\ell.$$

Using Green's formula we have then

$$\operatorname{Re}\iint_{G^+} \varphi_0^{(j)}(z)\,\overline{\varphi'(z)}\,dG_z^+ = \operatorname{Re}\frac{1}{2i}\int_\Gamma \varphi_0^{(j)}(t)\,\overline{\varphi(t)}\,dt$$

$$= \operatorname{Re}\frac{1}{2i}\int_\Gamma \lambda(t)\varphi_0^{(j)}(t)\,\overline{\lambda(t)}\;\overline{\varphi(t)}\,dt = \operatorname{Re}\frac{1}{2i}\int_\Gamma i\chi^{(j)}(t)\,\overline{\lambda(t)}\;\overline{\varphi(t)}\,dt = 0.$$

Substituting here instead φ' its expression from (1.171) we obtain

$$\sum_{i=1}^{\ell} C_i \operatorname{Re} \iint_{G^+} \varphi_0^{(j)}(z)\,\overline{\varphi_0^{(j)}(z)}\,dG_z^+ = -\operatorname{Re} \iint_{G^+} R_1 f\,\varphi_0^{(j)}(z)\,dG_z^+, \tag{1.174}$$

where

$$R_1 f = R h_0 + \frac{2}{\pi} \iint_{G^+} h_z(z,\zeta) f(\zeta)\,dG_\zeta^+.$$

The equalities (1.174) form an inhomogeneous algebraic system of equations for constants $C_1,\dots,C_\ell$, which is uniquely solvable because its determinant is the Gramian of linearly independent holomorphic functions $\{\varphi_0^{(j)}\}$.

Hence the problem (1.172) is solvable and we may represent its solution as

$$\psi(z) = R^0 h^0 + \sum_{i=1}^{\ell'} D_i \psi^{(i)}(z), \tag{1.175}$$

where R^0 is some linear continuous integral operator on Γ and D_i are arbitrary real constants. This means that for solvability of the problem (A_n) for an inhomogeneous Laplace equation it needs ℓ' conditions and the same number ℓ' of nontrivial solution as the corresponding homogeneous problem (A_n), (1.165) $(f \equiv 0)$, i.e. the index of the problem (A_n), (1.165) is equal zero.

Theorem 1.7. *The problem (A_n) for the inhomogeneous Laplace equation (1.165) has Fredholm's property: it has index equal zero.*

(b) To investigate problem (A_n) for the general equation (1.101) we need a representation formula for a solution of the problem (A_n) for the inhomogeneous Laplace equation. We assume that $z = 0$ is an inner point of G^+ and denote $\kappa = \kappa_0 + \kappa_1 + \dots + \kappa_m$,

$$\kappa_k = \frac{1}{2\pi i}\int_{\Gamma_k} d\log\lambda(t), \quad k = 0,\dots,m.$$

Then without loss of generality we may take instead $\overline{\lambda(t)}$ just $t^{-\kappa}$ (see [58]), i.e. the boundary conditions (A_n) we may take as

$$\mathrm{Re}\,[\,t'(s)\,t^{-\kappa}\,u(t)\,] = 0,\ \ \mathrm{Re}\,[\,t^{-\kappa}\,u_z(t)\,] = 0 \qquad (A_n)$$

on Γ.

First we assume $\kappa \geq 0$. Then the condition (1.168) becomes

$$\mathrm{Re}\,[\,t^{-\kappa}\,\varphi_0(t)\,] = h_0(t) = -\frac{1}{2\,t^{\kappa}}\,T_{G^+}f - \frac{1}{2\,\bar{t}^{\kappa}}\,\overline{T_{G^+}f}\,. \qquad (1.168')$$

Hence in case

$$\int_\Gamma h_0(t)\,\bar{t}^{-\kappa}\,\psi_j(t)\mathrm{d} = 0,\ \ j = 1,\ldots,m-r$$

for solutions ψ_j of the adjoint problem

$$\mathrm{Re}\,[\,t'(s)\,t^{\kappa}\psi(t)\,] = 0,$$

we may take φ_0 as in Section 1.1.4 to be of the form (see formula (1.71'):

$$\varphi_0(z) = -\,z^{\kappa}\iint_{G^+}\Big(\tilde{K}(z,\zeta) - \sum_{j=0}^{\kappa-1}\frac{\partial^j\tilde{K}(z,\zeta)}{\partial\zeta^j}\Big|_{\zeta=0}\frac{\zeta^j}{j!}\Big)\frac{\overline{f(\zeta)}}{\zeta^{\kappa}}\mathrm{d}G_\zeta^+$$

$$+\,z^{\kappa}\iint_{G^+}\Big\{\Big(\frac{1}{2\pi i}\int_\Gamma\frac{\tilde{\ell}(z,t)\,\mathrm{d}t}{t^{\kappa}(t-\zeta)}\Big)f(\zeta) + \Big(\frac{1}{2\pi i}\int_\Gamma\frac{\tilde{\ell}(z,t)\,\mathrm{d}t}{\bar{t}^{\kappa}(\bar{t}-\bar{\zeta})}\Big)\overline{f(\zeta)}\;\mathrm{d}G_\zeta^+$$

$$+\sum_{j=0}^{2\kappa+1-r} C_j\,\varphi^{(j)}(z). \qquad (1.176)$$

Now from (1.172):

$$\mathrm{Re}\,[\,t'(s)\,t^{\kappa}\,\psi(t)\,] = h^0(t) = -\,\mathrm{Re}[\,t'(s)\,t^{\kappa}\,\overline{\varphi(t)}\,] \qquad (1.172)$$

on Γ we can derive the representation formula, provided its right-hand side $h^0(t)$ satisfies the conditions (1.173):

$$\int_\Gamma h^0(t)\,\overline{t'(s)}\,\bar{t}^{\kappa}\,\varphi_j(t)\,\mathrm{d}t = 0,\ \ j = 1,\ldots,\ell \qquad (1.177)$$

where φ_j are a solution of the homogeneous problem

$$\operatorname{Re}[t^{-\kappa}\varphi(t)] = 0. \tag{1.168'}$$

The conditions (1.177) are

$$\sum_{j=1}^{2\kappa+1-r} C_j \operatorname{Re} \iint_{G^+} \varphi_i(z)\overline{\varphi_j(z)}\,dG_z^+ = -\frac{2}{\pi}\iint_{G^+}\varphi_i(z)\,dG_z^+ \iint_{G^+} h_z(z,\zeta) f(\zeta)\,dG_\zeta^+$$

$$+ \operatorname{Re}\frac{1}{i}\,\varphi_i(z)\,\bar{z}^\kappa \int_\Gamma \tilde{L}(z,t)\,h^0(t)\,dt. \tag{1.177'}$$

Since the function $z^\kappa \psi(z)$ is holomorphic in G^+, then from (1.172′) it follows that (see [25]):

$$z^\kappa\psi(z) = \frac{1}{\pi i}\int_\Gamma \frac{h^0(t)\,ds}{t-z} + \frac{1}{i}\int_\Gamma Y_\ell(z, t\, h^0(t)\,ds, \tag{1.178}$$

where

$$Y_\ell(z,t) = \int_0^t \ell(z, t(\,dt.$$

But from

$$z^\kappa\psi(z) = \frac{z^\kappa}{\pi i}\int_\Gamma \frac{h^0(t)\,dt}{t^\kappa(t-z)} + \frac{z^\kappa}{i}\int_\Gamma\left(\frac{1}{2\pi i}\int_\Gamma \frac{Y_\ell(\tau,t)}{\tau^\kappa(\tau-z)}\right) h^0(t)\,ds$$

$$+\sum_{j=0}^{\kappa-1}\frac{z^j}{\pi i}\int_\Gamma\left(\frac{1}{t^{j+1}} + \frac{1}{i}\int_\Gamma \frac{Y_\ell(\tau,t)\,d\tau}{\tau^{j+1}}\right) h^0(t)\,ds$$

it follows that in order for $\psi(z)$ to be continuous at $z = 0$ it is necessary and sufficient that

$$\int_\Gamma \frac{h^0(t)\,dt}{t^{j+1}} + \frac{1}{i}\int_\Gamma h^0(t)\,ds\int_\Gamma \frac{h_z(\tau,t)\,d\tau}{\tau^{j+1}} = 0,\ j = 0,\dots,\kappa-1,$$

then we obtain

$$\psi(z) = \frac{1}{\pi i}\int_\Gamma \frac{h^0(t)\,ds}{t^\kappa(t-z)} + \frac{1}{\pi i}\int_\Gamma\left(\frac{1}{\pi i}\int_\Gamma \frac{Y_\ell(\tau,t)\,d\tau}{\tau^\kappa(\tau-z)}\right) h^0(t)\,ds,$$

or since

$$h^0(t) = -\frac{t'(s)\,t^{\kappa}}{2}\,\overline{\varphi(t)} - \frac{\overline{t'(s)}\,\bar{t}^{\kappa}}{2}\,\varphi(t),$$

then we obtain

$$\psi(z) = -\frac{1}{2\pi i}\int_{\Gamma}\frac{\overline{\varphi(t)}\,dt}{t-z} - \frac{1}{2\pi i}\int_{\Gamma}\frac{\bar{t}^{\kappa}\varphi(t)\,d\bar{t}}{t^{\kappa}(t-z)}$$

$$-\frac{1}{\pi i}\int_{\Gamma} t^{\kappa} h_1(z,t)\,\overline{\varphi(t)}\,dt - \frac{1}{\pi i}\int_{\Gamma}\bar{t}^{\kappa} h_1(z,t)\,\varphi(t)\,d\bar{t}, \tag{1.179}$$

where

$$h_1(z,t) = \frac{1}{2\pi i}\int_{\Gamma}\frac{Y_l(\tau,t)\,d\tau}{\tau^{\kappa}(\tau-z)}.$$

Thus the holomorphic functions φ and ψ and so by (1.166) the solution $u(z)$ of the problem (A_n), (1.165) are expressed in terms of the right-hand side $f(z)$ provided $f(z)$ satisfies the solvability conditions. Let us turn to the case $\kappa < 0$. Since in this case the function $z^{-\kappa}\varphi_0(z)$ is holomorphic in G^+, then in case the solvability conditions

$$\frac{1}{i}\int_{\Gamma} h_0(t)\,\varphi_j(t)\,dt = 0,\ j = 1,..,m \tag{1.180}$$

are fulfilled, a solution of the Schwartz problem (1.168′) takes the form (see [25]):

$$z^{-\kappa}\varphi_0(z) = \frac{1}{i}\int_{\Gamma}\tilde{L}(z,t)\,h_0(t)\,ds + iC. \tag{1.181}$$

The conditions (1.180) are

$$\mathrm{Re}\iint_{G^+} f(\zeta)\,\tilde{\varphi}_j(\zeta)\,dG_{\zeta}^{+} = 0,\ j = 1,\ldots,m, \tag{1.180′}$$

where

$$\tilde{\varphi}_j(\zeta) = \frac{1}{2\pi i}\int_{\Gamma}\frac{\bar{t}^{\kappa}\Phi_j(t)\,dt}{t-\zeta},$$

$\Phi_j(z) = 2\partial u_j/\partial z$ being a harmonic measure of contour Γ_j at the point $z \in G$ with respect to domain G^+. Moreover, using continuity of the function $\varphi_0(z)$ at $z = 0$

from (1.181) we obtain also $-2\kappa - 1$ real solvability conditions

$$-\frac{1}{\pi}\iint_{G^+} \zeta^{-\kappa-1} f(\zeta)\,dG_\zeta^+ + \frac{1}{2\pi i}\int_\Gamma\Big(\iint_{G^+}\tilde{\ell}(t,\zeta)\,\zeta^{-\kappa} f(\zeta)\,dG_\zeta^+$$

$$+\iint_{G^+}\tilde{L}(z,\zeta)\,\bar{\zeta}^{-\kappa}\overline{f(\zeta)}\,dG_\zeta^+\Big)\frac{dt}{t} + iC = 0,$$

$$-\iint_{G^+}\zeta^{-\kappa-j-1} f(\zeta)\,dG_\zeta^+ + \frac{1}{2\pi i}\int_\Gamma\Big(\iint_{G^+}\tilde{\ell}(t,\zeta)\,\zeta^{-\kappa} f(\zeta)\,dG_\zeta^+$$

$$+\iint_{G^+}\tilde{L}(z,\zeta)\,\bar{\zeta}^{-\kappa}\overline{f(\zeta)}\,dG_\zeta^+ + \overline{T_{G^+}(\zeta^{-\kappa} f(\zeta))}\Big)\frac{dt}{t^{j+1}} = 0,$$

$$j = 1,\dots, -\kappa-1. \qquad (1.180'')$$

In this case we find φ_0 from (1.181) and then

$$\varphi'(z) = \varphi_0(z) + \frac{2}{\pi}\iint_{G^+} h_z(z,\zeta)\,f(\zeta)\,dG_\zeta^+$$

$$= \frac{2}{\pi}\iint_{G^+} h_z(z,\zeta)\,f(\zeta)\,dG_\zeta^+$$

$$+\iint_{G^+}\big[\tilde{\ell}_\kappa(z,\zeta)\,\zeta^{-\kappa} f(\zeta) - I_\kappa(z,\bar{\zeta})\,\bar{\zeta}^{-\kappa}\overline{f(\zeta)}\,\big]dG_\zeta^+$$

$$= \iint_{G^+}\Big(\frac{1}{2\pi^2 i}\int_\Gamma\frac{d\bar{t}}{(\bar{t}-\bar{\zeta})(t-z)}\Big)f(\zeta)\,dG_\zeta^+ + \iint_{G^+}\big[\tilde{\ell}_\kappa(z,\zeta)\,\zeta^{-\kappa} f(\zeta)$$

$$-I_\kappa(z,\bar{\zeta})\,\bar{\zeta}^{-\kappa}\overline{f(\zeta)}\,\big]dG_\zeta^+ + \iint_{G^+}\ell(\tau,z)\,\bar{T}_{G^+}f\,dG_\zeta^+, \qquad (1.182)$$

where

$$\tilde{\ell}_\kappa(z,\zeta) = \frac{1}{2\pi i}\int_\Gamma\frac{\tilde{\ell}(t,\zeta)\,dt}{t^{-\kappa}(t-\zeta)}$$

$$I_\kappa(z,\zeta) = \frac{1}{2\pi^2 i} \int_\Gamma \frac{dt}{t^{-\kappa}(\bar{t}-\bar{\zeta})(t-z)}. \tag{1.183}$$

We shall seek the holomorphic function $\psi(z)$ in the form $\psi = z^{-\kappa}\psi_0(z)$. Then for ψ_0 we obtain the Riemann-Hilbert condition

$$\operatorname{Re}[\,t'(s)\,\psi_0(t)\,] = h^0(t)$$

on Γ. The particular solution of this problem, in case the solvability conditions are fulfilled, takes the form

$$\psi_0(z) = \frac{1}{\pi i}\int_\Gamma \frac{h^0(t)\,ds}{t-z} + \frac{1}{i}\int_\Gamma Y_\ell(z,t)\,h^0(t)\,ds.$$

Hence

$$\psi(z) = \frac{z^{-\kappa}}{\pi i}\int_\Gamma \frac{h^0(t)\,ds}{t-z} + \frac{z^{-\kappa}}{i}\int_\Gamma \frac{\varphi(t)\,d\bar{t}}{t^{-\kappa}(t-z)} + \dots + \sum_{j=1}^{-\kappa} c_j\,\psi^{(j)}$$

$$= -\frac{z^{\kappa}}{2\pi i}\int_\Gamma \frac{\overline{\varphi(t)}\,dt}{t^{\kappa'}(t-z)} - \frac{z^{\kappa'}}{2\pi i}\int_\Gamma \frac{\varphi(t)\,d\bar{t}}{t^{\kappa}(t-z)} + \dots + \sum_{j=1}^{\kappa'} c_j\,\psi^{(j)}, \tag{1.184}$$

where we have dropped lower parts and put $\kappa' = -\kappa > 0$.

Thus in case $\kappa < 0$ we have expressed by (1.182) the function φ' and by (1.184) the function ψ in terms of $f(z)$, provided the solvability $-2\kappa - 1 + m$ conditions (1.180′), (1.180″) are fulfilled.

(c) In the special case when the domain G^+ is the unit disc $|z| < 1$, we obtain a more explicit representation formula. If $\kappa \geq 0$, then the Riemann-Hilbert problem (1.168) is solvable without any condition, so we obtain

$$\varphi_0(z) = -\frac{z^{2\kappa+1}}{\pi}\iint_{|\zeta|<1} \frac{\overline{f(\zeta)}\,d\xi\,d\eta}{1-\bar{\zeta}z} + \sum_{j=1}^{2\kappa+1} c_j\,\varphi^{(j)} \tag{1.185′}$$

and

$$\varphi'(z) = -\frac{1}{\pi}\iint_{|\zeta|<1} \frac{\zeta f(\zeta)\,d\xi\,d\eta}{1-\bar{\zeta}z} - \frac{z^{2\kappa+1}}{\pi}\iint_{|\zeta|<1} \frac{\overline{f(\zeta)}\,d\xi\,d\eta}{1-\bar{\zeta}z} + c_j\,\varphi^{(j)} \tag{1.185}$$

The Riemann-Hilbert problem (1.172′) for the holomorphic function ψ is

$$\mathrm{Re}\,[\,\mathrm{i}t^{\kappa+1}\,\psi(t)\,] = h^0(t) = \frac{t^{\kappa+1}}{2\mathrm{i}}\,\overline{\varphi(t)} - \frac{\bar{t}^{\kappa+1}}{2\mathrm{i}}\varphi(t),$$

so it has a solution if and only if

$$\int_\Gamma \frac{h^0(t)}{t^{j+1}}\mathrm{d}t = 0, \quad j = 0,\dots,\kappa$$

and

$$\psi(z) = -\frac{1}{2\pi\mathrm{i}}\int_\Gamma \frac{\overline{\varphi(t)}\,\mathrm{d}t}{t-z} + \frac{1}{2\pi\mathrm{i}}\int_\Gamma \frac{\bar{t}^{\,2\kappa+2}\varphi(t)\,\mathrm{d}\bar{t}}{t-z}. \tag{1.186}$$

If $\kappa < 0$, then the problem (1.168′) has solution if and only if

$$\int_\Gamma \frac{h_0(t)\,\mathrm{d}t}{t^{j+1}} = 0, \quad j = 0,\dots,\kappa' - 1$$

and for $\varphi_0(z)$ we have an expression

$$\varphi_0(z) = -\frac{1}{\pi}\iint_{|\zeta|<1} \frac{\bar{\zeta}^{-2\kappa-1}\overline{f(\zeta)}}{1-\bar{\zeta}z}\mathrm{d}\xi\,\mathrm{d}\eta$$

and so

$$\varphi'(z) = -\frac{1}{\pi}\iint_{|\zeta|<1} \frac{\bar{\zeta}f(\zeta)\,\mathrm{d}\xi\,\mathrm{d}\eta}{1-\bar{\zeta}z} - \frac{1}{\pi}\iint_{|\zeta|<1} \frac{\bar{\zeta}^{-2\kappa-1}\overline{f(\zeta)}\,\mathrm{d}\xi\,\mathrm{d}\eta}{1-\bar{\zeta}z}. \tag{1.187}$$

In case $\kappa < 0$ the problem (1.172) for a holomorphic function is the Riemann-Hilbert problem with nonnegative index $-\kappa - 1$ and so is solvable without any condition:

$$\psi(z) = z^{-2\kappa-2}\,\varphi(z) - \frac{z^{-\kappa-1}}{2\pi\mathrm{i}}\int_\Gamma \frac{\overline{\varphi(t)}\,\mathrm{d}t}{t^{-\kappa-1}(t-z)} + \sum_{j=1}^{-2\kappa-1} c_j\,\psi^{(j)}. \tag{1.188}$$

1.2.5. The problem (A_n) for the general elliptic equation

In the previous section we derived the formula for a solution of the problem (A_n) for an inhomogeneous Laplacian. Now we use this formula to investigate (A_n) for the general equation (1.101). If $u(z)$ is the solution of (A_n), then denote by $f(z)$ the Laplacian $u_{\bar{z}z}$ and obtain the above representations: the function $u(z)$ is expressed by (1.166) in terms of the holomorphic functions φ, ψ which are expressed in terms of $f(z)$ by (1.176), (1.182) and by (1.179), (1.184) (or by (1.185), (1.186) and by (1.187), (1.188) in the case of the unit disc).

(a) Let us begin with the unit disc $|z|<1$. In case $\kappa>0$, by differentiation with respect to z and integration by parts we obtain from (1.184)

$$\psi'(z) = -\frac{1}{2\pi i}\int_{|t|=1}\frac{\overline{\varphi(t)\,dt}}{\overline{(t-z)}^2} + \frac{1}{2\pi i}\int_{|t|=1}\frac{\bar{t}^{2\kappa+2}\varphi(t)}{(t-z)^2}\,dt$$

$$= -\frac{1}{2\pi i}\int_{|t|=1}\frac{\overline{\varphi'(t)}\,d\bar{t}}{t-z} + \frac{1}{2\pi i}\int_{|t|=1}\frac{\varphi'(t)\,dt}{t^{2\kappa+2}(t-z)} = \frac{1}{2\pi i}\int_{|t|=1}\frac{\varphi'(t)\,dt}{t^{2\kappa+2}(t-z)},$$

because by Cauchy's theorem

$$-\frac{1}{2\pi i}\int_{|t|=1}\frac{\overline{\varphi'(t)}\,d\bar{t}}{t-z} = \overline{\frac{1}{2\pi i}\int_{|t|=1}\frac{t\varphi'(t)\,dt}{1-\bar{z}t}} = 0.$$

Substituting (1.185) into the last relation we obtain

$$\psi'(z) = -\frac{1}{\pi}\iint_{|\zeta|<1}\Big(\frac{1}{2\pi i}\int_{|t|=1}\frac{dt}{t^{2\kappa+2}(1-t\zeta)(t-z)}\Big)\zeta f(\zeta)\,d\xi\,d\eta$$

$$-\frac{1}{\pi}\iint_{|\zeta|<1}\Big(\frac{1}{2\pi i}\int_{|t|=1}\frac{dt}{t(1-t\zeta)(t-z)}\Big)\overline{f(\zeta)}\,d\xi\,d\eta + \sum_{j=1}^{2\kappa+1} c_j\psi^{(j)}_{(z)}$$

$$= -\frac{1}{\pi}\iint_{|\zeta|<1}\frac{\bar{\zeta}^{2\kappa+3}f(\zeta)}{1-\bar{\zeta}z}\,d\xi\,d\eta - \frac{1}{\pi}\iint_{|\zeta|<1}\frac{\zeta\overline{f(\zeta)}\,d\xi\,d\eta}{1-\bar{\zeta}z} + \sum_{j=1}^{2\kappa+1} c_j\psi^{(j)}_{(z)}, \tag{1.189}$$

because by the residue theorem

$$\frac{1}{2\pi i}\Big(\int_{|t|=1}\!-\int_{\Gamma}\frac{dt}{t^{2\kappa+2}(1-t\bar{\zeta})(t-z)}\Big) = \frac{1}{z^{2\kappa+2}}\Big(\frac{1}{2\pi i}\int_{|t|=1}\frac{dt}{(1-t\bar{\zeta})(t-z)}$$

$$-\sum_{j=0}^{2\kappa+1}\frac{z^j}{2\pi i}\int_{\Gamma}\frac{dt}{(1-t\bar{\zeta})^{j+1}} = \frac{1}{z^{2\kappa+2}}\Big(\frac{1}{1-\bar{\zeta}z}-\sum_{j=0}^{2\kappa+1}(\bar{\zeta}z)^j\Big)$$

$$= \frac{1}{z^{2\kappa+2}}\Big(\frac{1}{1-\bar{\zeta}z}-\frac{1-(\bar{\zeta}z)^{2\kappa+2}}{1-\bar{\zeta}z}\Big) = \frac{\bar{\zeta}^{2\kappa+2}}{1-\bar{\zeta}z},$$

$$\frac{1}{2\pi i}\int_{|t|=1}\frac{dt}{t(1-t\bar{\zeta})(t-z)}=\frac{1}{z}\Big(\frac{1}{2\pi i}\int_{|t|=1}\frac{dt}{(1-t\bar{\zeta})(t-z)}$$

$$-\frac{1}{2\pi i}\int_{|t|=1}\frac{dt}{t(1-t\bar{\zeta})}\Big)=\frac{1}{z}\Big(\frac{1}{1-\bar{\zeta}z}-1\Big)=\frac{\bar{\zeta}}{1-\bar{\zeta}z}.$$

Differentiating (1.189) we obtain

$$\psi''(z)=-\frac{1}{\pi}\iint_{|\zeta|<1}\frac{\bar{\zeta}^{2(\kappa+2)}f(\zeta)}{(1-\bar{\zeta}z)^2}d\xi\,d\eta-\frac{1}{\pi}\iint_{|\zeta|<1}\frac{\bar{\zeta}^2\overline{f(\zeta)}\,d\xi\,d\eta}{(1-\bar{\zeta}z)^2}$$

$$+\sum_{j=1}^{2\kappa+1}c_j\,\psi_{(j)}(z). \qquad (1.190)$$

From (1.185′) by differentiation also we have

$$\varphi_0'(z)=-\frac{z^{2\kappa+1}}{\pi}\iint_{|\zeta|<1}\frac{\bar{\zeta}\,\overline{f(\zeta)}\,d\xi\,d\eta}{(1-\bar{\zeta}z)^2}-\frac{-(2\kappa+1)z^{2\kappa}}{\pi}\iint_{|\zeta|<1}\frac{\overline{f(\zeta)}\,d\xi\,d\eta}{1-\bar{\zeta}z}$$

$$+\sum_{j=1}^{2\kappa+1}c_j\,\varphi_{(j)}'(z). \qquad (1.191)$$

Hence by (1.166) we obtain the following formulae for derivatives of $u(z)$ through the unknown density $f(z)$:

$$u_{\bar{z}}=-\frac{1}{\pi}\iint_{|\zeta|<1}\frac{f(\zeta)\,d\xi\,d\eta}{\bar{\zeta}-\bar{z}}-\frac{1}{\pi}\iint_{|\zeta|<1}\frac{\zeta^{2\kappa+3}}{1-\zeta\bar{z}}\overline{f(\zeta)}\,d\xi\,d\eta$$

$$+\sum_{j=1}^{2\kappa+1}c_j\,\overline{\psi(j)}=\bar{T}^{(1)}f+\sum_{j=1}^{2\kappa+1}c_j\,\overline{\psi(j)}, \qquad (1.192)$$

$$u_z=-\frac{1}{\pi}\iint_{|\zeta|<1}\frac{f(\zeta)\,d\xi\,d\eta}{\zeta-z}-\frac{z^{2\kappa+1}}{\pi}\iint_{|\zeta|<1}\frac{\overline{f(\zeta)}\,d\xi\,d\eta}{1-\bar{\zeta}z}+\sum_{j=1}^{2\kappa+1}c_j\,\varphi^{(j)} \qquad (1.193)$$

$$=T_{(1)}f+\sum_{j=1}^{2\kappa+1}c_j\,\varphi^{(j)}, \qquad (1.194)$$

$$u_{\bar{z}z}=f(z),$$

$$u_{\bar{z}^2} = -\frac{1}{\pi}\iint_{|\zeta|<1}\frac{f(z)\,d\xi\,d\eta}{(\zeta-z)^2} - \frac{1}{\pi}\iint_{|\zeta|<1}\frac{\zeta^{2(\kappa+2)}\overline{f(\zeta)}}{(1-\zeta\bar{z})^2}\overline{f(\zeta)}\,d\xi\,d\eta$$

$$+\sum_{j=1}^{2\kappa+1} c_j\overline{\psi(j)} = S^{(1)}f + \sum_{j=1}^{2\kappa+1} c_j\overline{\psi(j)}, \tag{1.195}$$

$$u_{z^2} = -\frac{1}{\pi}\iint_{|\zeta|<1}\frac{f(z)\,d\xi\,d\eta}{(\zeta-z)^2} - \frac{z^{2\kappa+1}}{\pi}\iint_{|\zeta|<1}\frac{\zeta\,\overline{f(\zeta)}\,d\xi\,d\eta}{(1-\bar{\zeta}z)^2}$$

$$-(2\kappa+1)\frac{z^{2\kappa}}{\pi}\iint_{|\zeta|<1}\frac{\overline{f(\zeta)}}{1-\zeta\bar{z}} + \sum_{j=1}^{2\kappa+1} c_j\varphi^{(j)}$$

$$= S_{(1)}f + T^{(0)}\bar{f} + \sum_{j=1}^{2\kappa+1} c_j\varphi^{(j)}(z). \tag{1.196}$$

Now substituting (1.166) and (1.192)-(1.196) into (1.101) we obtain the following integral equation for unknown density:

$$f(z) + Q_\kappa^+ f = F(z) + \sum_{j=1}^{2\kappa+1} c_j F_j(z), \tag{1.197}$$

where Q_κ^+ is a linear continuous integral operator on $L_p(|z| \le 1)$ such that $Q_\kappa^+ = Q_\kappa^0 + \tilde{Q}_\kappa$ with $\tilde{Q}_\kappa$ a linear integral operator, compact in L_p,

$$Q_\kappa^0 f = a_1 S_{(1)}f + a_2\overline{S_{(1)}f} + b_1 S^{(1)}f + b_2\overline{S^{(1)}f} \tag{1.198}$$

and $F_j(z)$ are given functions.

If $\kappa < 0$, then by means of (1.187) we obtain from (1.188)

$$\psi'(z) = z^{-2\kappa-2}\varphi'(z) - \frac{z^{-\kappa-1}}{2\pi i}\int_{|t|=1}\frac{\overline{\varphi'(t)}\,d\bar{t}}{t^{-\kappa-1}(t-z)} - 2(\kappa+1)z^{-2\kappa-3}\varphi(z)$$

$$+\frac{(\kappa+1)z^{-\kappa-2}}{2\pi i}\int_{|t|=1}\frac{\overline{\varphi'(t)}\,d\bar{t}}{t^{-\kappa-1}(t-z)} + \sum_{j=1}^{-2\kappa-1} c_j\tilde{\psi}^{(j)}$$

$$= -\frac{z^{-2\kappa-2}}{\pi}\iint_{|\zeta|<1}\frac{\zeta f(\zeta)\,d\xi\,d\eta}{1-\bar{\zeta}z} - \frac{z^{-2\kappa-2}}{\pi}\iint_{|\zeta|<1}\frac{\zeta^{-2\kappa-1}\overline{f(\zeta)}}{1-\bar{\zeta}z}\,d\xi\,d\eta$$

$$+\frac{z^{-\kappa-1}}{\pi}\iint_{|\zeta|<1}\Big(\frac{1}{2\pi i}\int_{|t|=1}\frac{d\bar t}{\bar t^{-\kappa-1}(1-\bar t\zeta)(t-z)}\Big)\zeta\overline{f(\zeta)}\,d\xi\,d\eta$$

$$+\frac{z^{-\kappa-1}}{\pi}\iint_{|\zeta|<1}\Big(\frac{1}{2\pi i}\int_{|t|=1}\frac{d\bar t}{\bar t^{-\kappa-1}(1-\bar t\zeta)(t-z)}\Big)\zeta^{-2\kappa-1}f(\zeta)\,d\xi\,d\eta$$

$$+\cdots+\sum_{j=1}^{-2\kappa-1}c_j\tilde\psi^{(j)}$$

$$=-\frac{z^{-2\kappa-2}}{\pi}\iint_{|\zeta|<1}\frac{\zeta f(\zeta)\,d\xi\,d\eta}{1-\bar\zeta z}-\frac{z^{-2\kappa-2}}{\pi}\iint_{|\zeta|<1}\frac{\bar\zeta^{-2\kappa-1}\overline{f(\zeta)}}{1-\bar\zeta z}\,d\xi\,d\eta \tag{1.199}$$

$$+\cdots+\sum_{j=1}^{-2\kappa-1}c_j\tilde\psi^{(j)},$$

because

$$\overline{\frac{1}{2\pi i}\int_{|t|=1}\frac{t^{-\kappa}\,dt}{(1-t\bar\zeta)(t-z)}}=0$$

for $|z|<1$, and in (1.199) we have dropped the lower terms.
Hence

$$\psi''(z)=-\frac{z^{-2\kappa-2}}{\pi}\iint_{|\zeta|<1}\frac{\zeta^2 f(\zeta)\,d\xi\,d\eta}{(1-\bar\zeta z)^2}-\frac{z^{-2\kappa-2}}{\pi}\iint_{|\zeta|<1}\frac{\bar\zeta^{-2\kappa}\overline{f(\zeta)}\,d\xi\,d\eta}{(1-\bar\zeta z)^2}$$

$$+\cdots+\sum_{j=1}^{-2\kappa-1}c_j\tilde\psi^{(j)} \tag{1.200}$$

and we obtain the following expressions for derivatives of unknown function in case $\kappa<0$:

$$u_z=-\frac{1}{\pi}\iint_{|\zeta|<1}\frac{f(\zeta)\,d\xi\,d\eta}{\zeta-z}-\frac{1}{\pi}\iint_{|\zeta|<1}\frac{\bar\zeta^{-2\kappa-1}}{1-\bar\zeta z}\overline{f(\zeta)}\,d\xi\,d\eta\equiv T^{(2)}f+\cdots \tag{1.201}$$

$$u_{\bar z}=-\frac{1}{\pi}\iint_{|\zeta|<1}\Big(\frac{1}{\bar\zeta-\bar z}-\frac{\zeta}{1-\zeta\bar z}\Big)f(\zeta)\,d\xi\,d\eta-\frac{\bar z^{-2\kappa-2}}{\pi}\iint_{|\zeta|<1}\frac{\zeta\overline{f(\zeta)}\,d\xi\,d\eta}{1-\zeta\bar z}$$

$$-\frac{\bar{z}^{-2\kappa-2}}{\pi}\iint_{|\zeta|<1}\frac{\zeta^{-2\kappa-1}f(\zeta)}{1-\zeta\bar{z}}d\xi\, d\eta+\cdots+\sum_{j=1}^{-2\kappa-1}c_j\tilde{\psi}^{(j)}$$

$$\equiv T_{(2)}f+\cdots+\sum_{j=1}^{-2\kappa-1}c_j\tilde{\psi}^{(j)}, \tag{1.202}$$

$$u_{\bar{z}z}=f, \tag{1.203}$$

$$u_{z^2}=-\frac{1}{\pi}\iint_{|\zeta|<1}\frac{f(\zeta)d\xi d\eta}{(\zeta-z)^2}-\frac{1}{\pi}\iint_{|\zeta|<1}\frac{\bar{\zeta}^{-2\kappa}\overline{f(\zeta)}}{(1-\bar{\zeta}z)^2}d\xi\, d\eta\equiv S^{(2)}f \tag{1.204}$$

$$\bar{u}_{z^2}=-\frac{1}{\pi}\iint_{|\zeta|<1}\frac{\overline{f(\zeta)}\, d\xi d\eta}{(\xi-z)^2}+\frac{1}{\pi}\iint_{|\zeta|<1}\frac{\zeta^2\overline{f(\zeta)}\, d\xi d\eta}{(1-\bar{\zeta}z)^2}-\frac{z^{-2\kappa-2}}{\pi}\iint_{|\zeta|<1}\frac{\bar{\zeta}^2 f(\zeta)d\xi d\eta}{(1-\bar{\zeta}z)^2}$$

$$-\frac{z^{-2\kappa-2}}{\pi}\iint_{|\zeta|<1}\frac{\bar{\zeta}^{-2\kappa}\overline{f(\zeta)}}{(1-\bar{\zeta}z)^2}d\xi\, d\eta+\sum_{j=1}^{-2\kappa+1}c_j\bar{\psi}^{(j)}(z). \tag{1.205}$$

Since $z^{-2\kappa-2}\zeta^{-2\bar{\kappa}}=\bar{\zeta}^2+((\bar{\zeta}z)^{-2\kappa-2}-1)\bar{\zeta}^2$, then

$$\frac{z^{-2\kappa-2}}{\pi}\iint_{|\zeta|<1}\frac{\bar{\zeta}^{-2\kappa}f(\zeta)}{(1-\bar{\zeta}z)^2}d\xi\, d\eta=\frac{1}{\pi}\iint_{|\zeta|<1}\frac{\bar{\zeta}^2\overline{f(\zeta)}\, d\xi d\eta}{(1-\bar{\zeta}z)^2}$$

$$-\frac{1}{\pi}\iint_{|\zeta|<1}\frac{(1-(\bar{\zeta}z)^{-2\kappa-2})\bar{\zeta}^2}{(1-\bar{\zeta}z)^2}\overline{f(\zeta)}\, d\xi\, d\eta=\frac{1}{\pi}\iint_{|\zeta|<1}\frac{\bar{\zeta}^2\overline{f(\zeta)}\, d\xi d\eta}{(1-\bar{\zeta}z)^2}$$

$$-\frac{1}{\pi}\iint_{|\zeta|<1}\frac{\bar{\zeta}^2(1+\bar{\zeta}z+\cdots+(\bar{\zeta}z)^{-2\kappa-3})}{1-\bar{\zeta}z}\overline{f(\zeta)}\, d\xi\, d\eta$$

$$=\frac{1}{\pi}\iint_{|\zeta|<1}\frac{\bar{\zeta}^2\overline{f(\zeta)}\, d\xi d\eta}{(1-\bar{\zeta}z)^2}+\cdots,$$

so the last equality can be rewritten as

$$u_{\bar{z}^2}=-\frac{1}{\pi}\iint_{|\zeta|<1}\frac{f(\zeta)d\xi d\eta}{(\zeta-\bar{z})^2}-\frac{\bar{z}^{-2\kappa-2}}{\pi}\iint_{|\zeta|<1}\frac{\zeta^2\overline{f(\zeta)}\, d\xi d\eta}{(1-\zeta\bar{z})^2}+\cdots+$$

$$+\sum_{j=1}^{-2\kappa+1} c_j \overline{\tilde{\Psi}^{(j)'}_{(z)}} = \bar{S}_{(2)} f + \cdots + \sum_{j=1}^{-2\kappa+1} c_j \overline{\tilde{\Psi}^{(j)'}_{(z)}} \,. \tag{1.205'}$$

In this case for density $f(z)$ we obtain the following integral equation:

$$f + Q_\kappa' f = F(z) + \sum_{j=1}^{-2\kappa+1} F^{(j)}(z) \tag{1.206}$$

with $Q_\kappa^- = Q_\kappa' + \tilde{Q}_\kappa^0$ with compact $\tilde{Q}_\kappa^0$ and

$$Q_\kappa' f \equiv a_1 S^{(2)} f + a_2 \overline{S^{(2)}} f + b_1 \bar{S}_{(2)} f + b_2 S_{(2)} \bar{f}, \tag{1.207}$$

where $F^{(j)}(z)$ are given functions.

Thus we have proved the following:

Lemma 1.5. *Let G^+ be the unit disc $|z|<1$. Then in case $\kappa \ge 0$ the problem (A_n), (1.101) is equivalent to integral equation (1.197) with $2\kappa+1$ complementary real conditions on $f(z)$, which are equalities*

$$\operatorname{Re} \iint_{|\zeta|<1} f(z)\, \overline{v_j(z)}\, dx\, dy = 0, \quad j ,..,2\kappa+1, \tag{1.208}$$

where $\{v_j(z)\}$ is a complete system of completely determined functions and, in case $\kappa<0$, this problem is equivalent to integral equation (1.206) with $-2\kappa+1$ complementary real conditions on f like (1.208).

To investigate the above integral equations we consider the following integral operator for $n, n' \ge 0$:

$$\Pi_{n,n'} f \equiv -\frac{1}{\pi} \iint_{|\zeta|<1} \frac{f(\zeta)\, d\xi\, d\eta}{(\zeta - z)^2} - \frac{z^{n'}}{\pi} \iint_{|\zeta|<1} \frac{\zeta^n\, \overline{f(\zeta)}\, d\xi\, d\eta}{(1-\bar{\zeta} z)^2}$$

$$= \frac{\partial Tf}{\partial z} + K_{n,n'} f, \tag{1.209}$$

where

$$Tf \equiv -\frac{1}{\pi} \iint_{|\zeta|<1} \frac{f(\zeta)\, d\xi\, d\eta}{(\zeta - z)}, \quad K_{n,n'} \equiv -\frac{z^{n'}}{\pi} \iint_{|\zeta|<1} \frac{\zeta^n f(\zeta)\, d\xi\, d\eta}{1-\bar{\zeta} z}.$$

Lemma 1.6. *The operator $\Pi_{n,n'}$ is linear continuous in $L_p, p>1$, and has L_2-*

norm equal to one.

Proof. Let $\rho \in C^\infty$ have compact support in the disc $|z| < 1$. Then by means of Green's formula we obtain

$$\|\Pi_{n,n'}\rho\|^2_{L_2} = (\Pi_{n,n'}\rho, \Pi_{n,n'}\rho) = \left(\frac{\partial T\rho}{\partial z}, \frac{\partial T\rho}{\partial z}\right)$$

$$+\left(\frac{\partial T\rho}{\partial z}, K_{n,n'}\rho\right) + \left(K_{n,n'}\rho, \frac{\partial T\rho}{\partial z}\right) + \left(K_{n,n'}\rho, K_{n,n'}\rho\right)$$

$$= \frac{1}{2\mathrm{i}} \int_{|t|=1} \overline{T\rho}\, \frac{\partial T\rho}{\partial t} \mathrm{d}t + 2\,\mathrm{Re} \iint_{|z|<1} \frac{\partial T\rho}{\partial t}\, \overline{K_{n,n'}\rho}\, \mathrm{d}x\,\mathrm{d}y$$

$$- \iint_{|z|<1} \overline{T\rho}\, \frac{\partial^2 T\rho}{\partial \bar{z}\,\partial z} \mathrm{d}x\,\mathrm{d}y + \|K_{n,n'}\rho\|^2_{L_2} = - \iint_{|z|<1} \frac{\partial \rho}{\partial z}\, \overline{T\rho}\, \mathrm{d}x\,\mathrm{d}y$$

$$+ \|K_{n,n'}\rho\|^2_{L_2} + \iint_{|\zeta|<1} \overline{\rho(\tau)} \iint_{|\tau|<1} \rho(\tau) \left(\frac{1}{2\pi^2 \mathrm{i}} \int_{|t|=1} \frac{\mathrm{d}t}{(\bar{\zeta} - \bar{\tau})(\tau - t)}\right) \mathrm{d}\xi\,\mathrm{d}\eta\,\mathrm{d}\xi'\,\mathrm{d}\eta'$$

$$+ \mathrm{Re} - \frac{1}{\mathrm{i}} \iint_{|\zeta|<1} \rho(\zeta) \iint_{|\tau|<1} \rho(\tau)\, \frac{\bar{t}^{\,n'}\,\mathrm{d}\bar{t}}{(\zeta - t)(1 - \zeta\bar{t})^2} \tau^n\, \mathrm{d}\xi\,\mathrm{d}\eta\,\mathrm{d}\xi'\,\mathrm{d}\eta'$$

$$= \|\rho\|^2_{L_2} + \|K_{n,n'}\rho\|^2_{L_2} - \frac{1}{\pi} \iint_{|\zeta|<1} \overline{\rho(\zeta)} \iint_{|\tau|<1} \rho(\tau)\, \frac{\mathrm{d}\xi\,\mathrm{d}\tau\,\mathrm{d}\xi'\,\mathrm{d}\eta'}{(1 - \bar{\zeta}\tau)^2},$$

because

$$\frac{1}{2\pi^2 \mathrm{i}} \int_{|t|=1} \frac{\mathrm{d}t}{(\bar{t} - \bar{\zeta})(t - \tau)^2} = \frac{\partial}{\partial \tau} \frac{1}{2\pi^2 \mathrm{i}} \int_{|t|=1} \frac{t\,\mathrm{d}t}{(1 - t\bar{\zeta})(t - \tau)}$$

$$= \frac{\partial}{\partial \tau}\left(\frac{\tau}{\pi(1 - \tau\bar{\zeta})}\right) = \frac{1}{\pi(1 - \tau\bar{\zeta})^2}, \int_{|t|=1} \frac{\bar{t}^{\,n'}\,\mathrm{d}t}{(t - \zeta)(1 - \tau\bar{t})^2}$$

$$= \int_{|t|=1} \frac{\bar{t}^{\,n'+1}\,\mathrm{d}\bar{t}}{(1 - \zeta\bar{t})(1 - \tau\bar{t})^2} = \overline{\int_{|t|=1} \frac{t^{n'+1}\,\mathrm{d}t}{(1 - \bar{\zeta}t)(1 - \bar{\tau}t)^2}} = \frac{\partial}{\partial \bar{\tau}} \int_{|t|=1} \frac{\bar{\tau}\,t^{n'+1}\,\mathrm{d}t}{(1 - \bar{\zeta}t)(1 - \bar{\zeta}t)}.$$

Hence by

$$\frac{1}{(1-\bar{\zeta}\tau)^2} = \frac{\partial}{\partial\tau}\,\frac{\tau}{1-\bar{\zeta}\tau} = \frac{\partial}{\partial\tau}\,\tau\sum_{k=0}^{\infty}\bar{\zeta}^k\tau^k = \sum_{k=0}^{\infty}(k+1)\,\bar{\zeta}^k\tau^k$$

we obtain

$$\|\Pi_{n,n'}\rho\|^2_{L_2} = \|\rho\|^2_{L_2} + \|K_{n,n'}\rho\|^2_{L_2} - \frac{1}{\pi}\sum_{k=0}^{\infty}(k+1)\,|\alpha_k|^2,$$

where

$$\alpha_k = \iint\limits_{|z|<1}\rho(z)\,z^k\,dx\,dy.$$

But

$$K_{n,n'}\rho = -\frac{z^{n'}}{\pi}\iint\limits_{|z|<1}\sum_{k=0}^{\infty}(k+1)\,\bar{\zeta}\,\zeta^{k+n}\,\overline{\rho(\zeta)}\,d\xi\,d\eta\,z^k$$

$$= -\frac{1}{\pi}\sum_{k=0}^{\infty}(k+1)\,\bar{\alpha}_k\,z^{k+n'}$$

and so

$$\|K_{n,n'}\rho\|^2 = \frac{1}{\pi}\sum_{k=0}^{\infty}\frac{(k+1)^2|\alpha_{n+k}|^2}{n'+k+1},$$

because of

$$\iint\limits_{|z|<1} z^k\bar{z}^\ell\,dx\,dy = \begin{cases}\dfrac{\pi}{k+1}, & \ell = k,\\[2ex] 0, & \ell \neq k.\end{cases}$$

Hence

$$\|\Pi_{n,n'}\rho\|^2_{L_2} = \|\rho\|^2_{L_2} - \frac{1}{\pi}\sum_{k=0}^{\infty}(k+1)\left(|\alpha_k|^2 - \frac{k+1}{n'+k+1}|\alpha_{n+k}|^2\right) \le \|\rho\|^2_{L_2},$$

because

$$|\alpha_{n+k}|^2 = |\iint\limits_{|z|<1} \overline{\rho(z)}\,\bar{z}^{n+k}\,dx\,dy|^2 \le \|\rho\|^2_{L_2} \cdot \iint\limits_{|z|<1} |z|^{2(n+k)}dx\,dy = \frac{\pi}{n+k+1}\|\rho\|^2_{L_2},$$

$$|\alpha_k|^2 \le \frac{\pi}{k+1}\|\rho\|^2_{L_2}, \quad \frac{k+1}{n'+k+1}|\alpha_{n+k}|^2 \le |\alpha_{n+k}|^2 \le \frac{\pi}{k+1}\|\rho\|^2_{L_2} = |\alpha_k|^2$$

$$|\alpha_k|^2 - \frac{k+1}{k+n'+1}|\alpha_{n+k}|^2 \ge 0.$$

If ρ satisfies the equalities $\iint\limits_{|z|<1} \rho(z) z^k\,dx\,dy = 0$, $k = 1,2,\ldots$, i.e. $\alpha_k = 0$, then it follows that $\|\Pi_{n,n'}\rho\|^2_{L_2} = \|\rho\|^2_{L_2}$. It means that $\|\Pi_{n,n'}\| = 1$.

Considering

$$\tilde{\Pi}_{n,n'}f \equiv -\frac{1}{\pi}\iint\limits_{|\zeta|<1} \frac{\overline{f(\zeta)}\,d\xi\,d\eta}{(\zeta - z)^2} - \frac{z^{n'}}{\pi}\iint\limits_{|\zeta|<1} \frac{\zeta^n f(\zeta)\,d\xi\,d\eta}{(1-\bar{\zeta}z)^2},$$

we state by the same argument the following.

Lemma 1.7. *The operator* $\tilde{\Pi}_{n,n'}$ *has* L_2*-norm equal to one.*

Combining the above results we have, in case G^+ is the unit disc, the following:

Theorem 1.8. *If the coefficients of the principal part of equation (1.101) are bounded measurable functions satisfying the inequality (1.118), and its other coefficients and right-hand side are in* $L_p(|z| \le 1)$ *with* $p > 2$, *then the problem* (A_n), *(1.101) has Fredholm's property, i.e. has index equal zero.*

(b) Turning to the case of general multiply connected domain G^+, in case $\kappa \ge 0$ we obtain from (1.179) by differentiation and by integration by parts

$$\psi'(z) = -\frac{1}{2\pi i}\int_\Gamma \frac{\bar{t}^\kappa \varphi'(t)\,d\bar{t}}{t^\kappa (t-z)} - \frac{1}{2\pi i}\int_\Gamma \frac{\overline{\varphi'(t)\,dt}}{t-z} + \cdots$$

$$= -\iint\limits_{G^+} I_{(0)}(z,\zeta) f(\zeta)\,dG^+_\zeta + \iint\limits_{G^+} I^{(0)}(z,\zeta)\overline{f(\zeta)}\,dG^+_\zeta \cdot \ldots, \qquad (1.210)$$

where

$$I_{(0)}(z,\zeta)=\frac{1}{4\pi^2 i}\int_\Gamma \frac{\bar t^{\kappa}\overline{\tau^2(t)}\,d\bar t}{t^{\kappa}(t-z)(\bar t-\bar\zeta)}+\frac{1}{2\pi^2 i}\int_\Gamma \frac{[\psi^{(\kappa)}(t)-\psi^{(\kappa)}(z)]\,d\bar t}{(\bar t-\bar\zeta)(t-z)}+\cdots,$$

$$I^{(0)}(z,\zeta)=\frac{1}{4\pi^2 i}\int_\Gamma \frac{d\bar t}{(\bar t-\bar\zeta)(t-z)}+\frac{1}{2\pi^2 i}\int_\Gamma \frac{[\varphi^{(\kappa)}(t)-\varphi^{(\kappa)}(z)]\,dt}{\bar t^{\kappa}(\bar t-\bar\zeta)(t-z)}+\cdots,$$

$$\varphi^{(\kappa)}(z)=\frac{1}{2\pi i}\int_\Gamma \frac{\bar t^{\kappa}\,d\bar t}{t-z},\quad \psi^{(\kappa)}(z)=\frac{1}{2\pi i}\int_\Gamma \frac{\bar t^{\kappa}\,dt}{t^{\kappa}(t-z)}.$$

Differentiating (1.210) we obtain

$$\psi''(z)=-\iint_{G^+} I_{(1)}(z,\zeta)\,f(\zeta)\,dG_\zeta^+ +\iint_{G^+} I^{(1)}(z,\zeta)\,\overline{f(\zeta)}\,dG_\zeta^+\cdot\ldots, \tag{1.211}$$

where

$$I_{(1)}(z,\zeta)=\frac{\partial I_{(0)}(z,\zeta)}{\partial z}=\frac{1}{4\pi^2 i}\int_\Gamma \frac{\bar t^{\kappa}\overline{\tau^2(t)}\,d\bar t}{t^{\kappa}(\bar t-\bar\zeta)(t-z)^2}$$

$$+\frac{1}{2\pi^2 i}\int_\Gamma \frac{[\psi^{(\kappa)}(t)-\psi^{(\kappa)}(z)]\,dt}{(\bar t-\bar\zeta)(t-z)^2}+\cdots,$$

$$I^{(1)}(z,\zeta)=\frac{\partial I_{(0)}(z,\zeta)}{\partial z}=\frac{1}{4\pi^2 i}\int_\Gamma \frac{d\bar t}{(\bar t-\bar\zeta)(t-z)^2}$$

$$+\frac{1}{2\pi^2 i}\int_\Gamma \frac{[\varphi^{(\kappa)}(t)-\varphi^{(\kappa)}(z)]\,dt}{\bar t^{\kappa}(\bar t-\bar\zeta)(t-z)^2}+\cdots.$$

Also in case $\kappa<0$ from (1.184) by the same way we obtain

$$\psi'(z)=-\frac{z^{\kappa'}}{2\pi i}\int_\Gamma \frac{\overline{\varphi'(t)}\,d\bar t}{t^{\kappa'}(t-z)}-\frac{z^{\kappa'}}{2\pi i}\int_\Gamma \frac{\varphi'(t)\,dt}{t^{\kappa'}(t-z)}+\cdots$$

$$=-z^{\kappa'}\iint_{G^+}\left(\frac{1}{2\pi i}\int_\Gamma \frac{k(t,\zeta)\,d\bar t}{t^{\kappa'}(t-z)}\right)f(\zeta)\,dG_\zeta^+$$

$$-z^{\kappa'}\iint_{G^+}\Big(\frac{1}{2\pi i}\int_\Gamma\frac{\tilde{\ell}_\kappa(t,\zeta)\,d\bar{t}}{t^{\kappa'}(t-z)}\Big)\zeta^{\kappa'}f(\zeta)\,dG_\zeta^+ + z^{\kappa'}\iint_{G^+}\Big(\frac{1}{2\pi i}\int_\Gamma\frac{I_\kappa(t,\bar{\zeta})\,d\bar{t}}{t^{\kappa'}(t-z)}\Big)\bar{\zeta}^{\kappa'}\overline{f(\zeta)}\,dG_\zeta^+$$

$$-z^{\kappa'}\iint_{G^+}\Big(\frac{1}{2\pi i}\int_\Gamma\frac{\ell(\tau,t)\,d\bar{t}}{t^{\kappa'}(t-z)}\Big)\bar{T}f\,dG_\zeta^+ - z^{\kappa'}\iint_{G^+}\Big(\frac{1}{2\pi i}\int_\Gamma\frac{\overline{k(t,\bar{\zeta})}\,d\bar{t}}{t^{\kappa'}(t-z)}\Big)\overline{f(\zeta)}\,dG_\zeta^+$$

$$-z^{\kappa'}\iint_{G^+}\Big(\frac{1}{2\pi i}\int_\Gamma\frac{\overline{\tilde{\ell}_\kappa(t,\zeta)\,dt}}{t^{\kappa'}(t-z)}\Big)\bar{\zeta}^{\kappa'}f(\zeta)\,dG_\zeta^+$$

$$+z^{\kappa'}\iint_{G^+}\Big(\frac{1}{2\pi i}\int_\Gamma\frac{\overline{I_\kappa(t,\zeta)}\,d\bar{t}}{t^{\kappa'}(t-z)}\Big)\zeta^{\kappa'}f(\zeta)\,dG_\zeta^+$$

$$-z^{\kappa'}\iint_{G^+}\Big(\frac{1}{2\pi i}\int_\Gamma\frac{\overline{\ell(\tau,t)}\,d\bar{t}}{t^{\kappa'}(t-z)}\Big)T\bar{f}\,dG_\zeta^+ + \cdots .$$

Since

$$\tilde{I}_{(0)}(z,\bar{\zeta}) = \frac{1}{2\pi^2 i}\int_\Gamma\frac{\overline{k(t,\bar{\zeta})}\,d\bar{t}}{t^{\kappa'}(t-z)} = \frac{1}{4\pi^2 i}\int_\Gamma\frac{\tau^2(t)\,d\bar{t}}{t^{\kappa'}(t-z)(t-\zeta)} + \cdots$$

$$= \frac{1}{2\pi(\zeta-z)}\Big(\iint_{G^+}\Big(\frac{1}{2\pi i}\int_\Gamma\frac{\tau^2(t)\,d\bar{t}}{t^{\kappa'}(t-\zeta)} - \frac{1}{2\pi i}\int_\Gamma\frac{\tau^2(t)\,d\bar{t}}{t^{\kappa'}(t-z)}\Big) + \cdots$$

$$= [\psi_{\kappa'}(\zeta) - \psi_{\kappa'}(z)]\,(2\pi(\zeta-z))^{-1},\quad \psi_{\kappa'}(z) = \frac{1}{2\pi i}\int_\Gamma\frac{\tau^2(t)\,d\bar{t}}{t^{\kappa'}(t-z)},$$

and

$$\tilde{I}^{(0)}(z,\zeta) = \frac{1}{2\pi i}\int_\Gamma\frac{\overline{I_\kappa(t,\bar{\zeta})}\,d\bar{t}}{t^{\kappa'}(t-z)} = \frac{1}{4\pi^2 i}\int_\Gamma\frac{d\bar{t}}{|t|^{2\kappa'}(t-z)(t-\zeta)} + \cdots$$

$$= \frac{1}{2\pi(\zeta-z)}\Big(\iint_{G^+}\Big(\frac{1}{2\pi i}\int_\Gamma\frac{d\bar{t}}{|t|^{2\kappa'}(t-\zeta)} - \frac{1}{2\pi i}\int_\Gamma\frac{d\bar{t}}{|t|^{2\kappa'}(t-z)}\Big) + \cdots$$

$$= \frac{\psi^{(\kappa)}(\zeta) - \psi^{(\kappa)}(z)}{2\pi(\zeta - z)} + \cdots, \quad \psi^{(\kappa)}(z) = \frac{1}{2\pi i}\int_\Gamma \frac{d\bar{t}}{|t|^{2\kappa'}(t-z)},$$

because as $z \to t \in \Gamma$,

$$I_\kappa(t,\bar{\zeta}) = \lim_{z\to t} I_\kappa(z,\bar{\zeta}) = \frac{1}{2\pi t^{\kappa'}(\bar{t}-\bar{\zeta})} + \frac{1}{2\pi^2 i}\int_\Gamma \frac{d\tau}{\tau^{\kappa'}(\bar{\tau}-\bar{\zeta})(\tau - t)}.$$

Thus the derivatives of the functions $\tilde{I}_{(0)}(z,\zeta)$ and $\tilde{I}^{(0)}(z,\zeta)$ have a weak singularity at $\zeta = z$ and

$$\underset{\sim}{I}_{(0)}(z,\bar{\zeta}) = \frac{1}{2\pi i}\int_\Gamma \frac{k(t,\bar{\zeta})\,d\bar{t}}{t^{\kappa'}(t-z)} = \frac{1}{4\pi^2 i}\int_\Gamma \frac{\overline{\tau^2(t)}\,d\bar{t}}{t^{\kappa'}(\bar{t}-\bar{\zeta})(t-z)} + \cdots$$

$$\tilde{I}^{(0)}(z,\bar{\zeta}) = \frac{1}{2\pi i}\int_\Gamma \frac{I_\kappa(t,\bar{\zeta})\,d\bar{t}}{t^{\kappa'}(t-z)} = \frac{1}{4\pi^2 i}\int_\Gamma \frac{d\bar{t}}{t^{2\kappa'}(\bar{t}-\bar{\zeta})(t-z)} + \cdots.$$

Thus, dropping lower terms, we obtain

$$\psi''(z) = -z^{\kappa'}\iint_{G^+}\Big(\frac{1}{4\pi^2 i}\int_\Gamma \frac{\overline{\tau^2(t)}\,dt}{t^{\kappa'}(\bar{t}-\bar{\zeta})(t-z)^2}\Big)f(\zeta)\,dG_\zeta^+$$

$$+ z^{\kappa'}\iint_{G^+}\Big(\frac{1}{4\pi^2 i}\int_\Gamma \frac{d\bar{t}}{t^{2\kappa'}(\bar{t}-\bar{\zeta})(t-z)^2}\Big)\bar{\zeta}^{\kappa'}\overline{f(\zeta)}\,dG_\zeta^+ + \cdots. \quad (1.212)$$

Consequently, in case $\kappa \ge 0$,

$$u_{\bar{z}z} = f(z),$$

$$u_{z^2} = S_{G^+}f - z^{2\kappa'}\frac{\partial}{\partial z}\iint_{G^+}\frac{1}{(\bar{\zeta}z)^\kappa}\Big(\tilde{K}(z,\bar{\zeta}) - \sum_{j=0}^{\kappa-1}\frac{\partial^j \tilde{K}(z,\bar{\zeta})}{\partial\bar{\zeta}^j}\Big|_{\bar{\zeta}=0}\frac{\bar{\zeta}^j}{j!}\Big)\overline{f(\zeta)}\,dG_\zeta^+$$

$$u_{\bar{z}^2} = \overline{S_{G^+}}f + \iint_{G^+}\Big(\frac{1}{2\pi^2 i}\int_\Gamma \frac{t^\kappa\tau^2(t)\,dt}{\bar{t}^\kappa(t-\zeta)(\bar{t}-\bar{z})^2}\Big)\overline{f(\zeta)}\,dG_\zeta^+ + \cdots, \quad (1.213)$$

and in case $\kappa < 0$,

$$u_{z^2} = S_{G^+}f - \iint_{G^+}\Big(\frac{1}{2\pi^2 i}\int_\Gamma \frac{dt}{t^{\kappa'}(\bar{t}-\bar{\zeta})(t-z)^2}\Big)\zeta^{\kappa'}\overline{f(\zeta)}\,dG_\zeta^+ + \cdots$$

$$u_{\bar{z}^2} = \overline{S}_{G^+}f + \iint_{G^+}\Big(\frac{1}{2\pi^2 i}\int_\Gamma \frac{dt}{(t-\zeta)(\bar{t}-\bar{z})^2}\Big)f(\zeta)\,dG_\zeta^+$$

$$- z^{\kappa'}\iint_{G^+}\Big(\frac{1}{2\pi^2 i}\int_\Gamma \frac{dt}{\bar{t}^{2\kappa'}(t-\zeta)(\bar{t}-\bar{z})^2}\Big)\zeta^{\kappa'}f(\zeta)\,dG_\zeta^+ \tag{1.214}$$

$$+ z^{\kappa'}\iint_{G^+}\Big(\frac{1}{2\pi^2 i}\int_\Gamma \frac{\tau^2(t)\,dt}{t^{\kappa'}(t-\zeta)(\bar{t}-\bar{z})^2}\Big)\overline{f(\zeta)}\,dG_\zeta^+ + \ldots,$$

where

$$S_{G^+}f = -\frac{1}{\pi}\iint_{G^+}\frac{f(\zeta)\,dG_\zeta^+}{(\zeta-z)^2},\quad \overline{S}_{G^+}f = -\frac{1}{\pi}\iint_{G^+}\frac{f(\zeta)\,dG_\zeta^+}{(\bar{\zeta}-\bar{z})^2}. \tag{1.215}$$

Substituting all these expressions into equation (1.101) we reduce the problem (A_n) to integral equations on f with complementary conditions, and as above we prove Theorem 1.8, in the general case.

Remark. The problem (A_n) in case $n = 0$ was stated and investigated in [25]. In this case as it turns out this problem has Fredholm's property not only for equations like (1.101), but also for some equations (1.93) in case $\tilde{c}(z) \equiv 0$. For equations (1.93) with $\tilde{a}(z) \neq 0$ by the same property possess the problem $(\bar{A}_0)$ with boundary conditions

$$\mathrm{Re}\,[\, t'(s)u(t)\,] = 0,\quad \mathrm{Re}\,[\, u_{\bar{z}}(t)\,] = 0$$

on Γ.

1.2.6 The problem B_n for the general second-order elliptic equation

In this section we consider the following problem (B_n): find a solution to (1.101) with the first order derivatives Hölder continuous in closure $\bar{G}^+$ with Laplacian in $L_p(\bar{G}^+)$, $p > 2$ satisfying the conditions:

$$\mathrm{Re}\,[\,\overline{G(t)}\,u_z(t)\,] = 0,\quad \mathrm{Re}\,[\,\overline{H(t)}\,u_{\bar{z}}\,] = 0,\quad t \in \Gamma \tag{B_n}$$

where $G(t) \neq 0$, $H(t) \neq 0$ are given functions Hölder continuous on Γ. Let $\kappa_1 = \mathrm{Ind}_\gamma G$, $\kappa_2 = \mathrm{Ind}_\gamma H$.

Theorem 1.9. *If the coefficients of the equation (1.101) in its principal part are bounded measurable in $\bar{G}^+$ functions, satisfying the inequality and other coefficients and the right-hand side are in $L_p(\bar{G}^+)$ with $p > 2$, then the problem (B_n) is Noetherian; the corresponding homogeneous problem ($f \equiv 0$) has only a finite number k of linearly independent nontrivial solutions, and the inhomogeneous problem (B_n) is solvable if and only if its right-hand side satisfies finite number k' conditions*

$$\mathrm{Re} \iint_{G^+} F(z)\, \overline{v_j(z)}\, dx\, dy\ 0 \quad j = 1, \ldots, k'.$$

The index of the problem (B_n) is even and equal to

$$k - k' = 2(\kappa_1 + \kappa_2 + 2 - m).$$

This theorem can be proved as above: first we consider this problem (B_n) for the inhomogeneous Laplace equation

$$u_{\bar{z}z} = \rho(z). \tag{1.216}$$

Integrating this equation we have (see (1.166):

$$u(z) = \varphi(z) + \overline{\psi(z)} - \frac{2}{\pi} \iint_{G^+} g(z, \zeta)\, \rho(\zeta)\, d\xi\, d\eta. \tag{1.217}$$

Substituting this expression into the boundary conditions (B_n) we obtain the Riemann-Hilbert boundary conditions

$$\mathrm{Re}\, \overline{G(t)}\, \varphi_0(t) = -\,\mathrm{Re}\, T_{G^+} \rho, \quad \mathrm{Re}\, \overline{H(t)}\, \psi_0(t) = -\,\mathrm{Re}\, \bar{T}_G \rho \tag{1.218}$$

for holomorphic functions

$$\psi_0(z) = \varphi'(z) - \frac{2}{\pi} \iint_{G^+} h_z(z, \zeta)\, \rho(\zeta)\, dG^+_\zeta, \quad \psi_0(z) = \psi'(z) - \frac{2}{\pi} \iint_{G^+} h_z(z, \zeta)\, \overline{\rho(\zeta)}\, dG^+_\zeta.$$

The assertions of Theorem 1.9 follow from those of the Riemann-Hilbert problems by means of the singular integral equations on the plane.

Corollary 1.3. *Under the assumptions of Theorem 1.9 the problem* (B_n) *has Fredholm's property in case* $H(t) = t'(s)t\,G(t)$. *Indeed in this case*

$$\kappa_2 = \mathrm{Ind}_\gamma \bar{H} = -\kappa_1 + m - 2.$$

We will give full calculations for the case when G^+ is the unit disc: $|z| < 1$. Without loss of generality we take $G(t) \equiv t^n$. Thus we consider the problem with boundary conditions

$$\mathrm{Re}\,[\,t^{-n}\,u_z^+(t)\,] = 0, \quad \mathrm{Re}\,[\,i\,t^{-(n+2)}\,u_{\bar{z}}^+(t)\,] = 0 \tag{B_n'}$$

on Γ.

Then for holomorphic functions $\varphi_0(z), \psi_0(z)$ we have boundary conditions

$$\mathrm{Re}\,[\,t^{-n}\varphi_0(t)\,] = -\,\mathrm{Re}\,[\,t^{-n}T_{G^+}\rho\,] = h_0(t),$$

$$\mathrm{Re}\,[\,i\,t^{(n+2)}\,\psi_0(t)\,] = -\,\mathrm{Re}\,i\,t^{(n+2)}\,T_{G^+}\rho\,] = h^0(t). \tag{1.218$'$}$$

Let $n \ge 0$. Then from the first condition (1.218′) we obtain

$$\varphi_0(z) = -\frac{z^{2n+1}}{\pi}\iint_{|\zeta|<1}\frac{\overline{\rho(\zeta)}\,d\xi\,d\eta}{1-\bar{\zeta}z} \tag{1.219}$$

$$+\,i\,c_0\,z^n + \sum_{j=0}^{n-1}\alpha_j(z^j - z^{2n-j}) + i\,\beta_j\,(z^j + z^{2n-j}),$$

where c_0, α_j, β_j are arbitrary real constants. Since in the case of the unit disc $h(z,\zeta) = \log|1-\bar{\zeta}z|$, then

$$\varphi'(z) = \varphi_0(z) + \frac{2}{\pi}\iint_{|\zeta|<1}\frac{\partial}{\partial z}\log|1-\bar{\zeta}z|\,\rho(\zeta)\,d\xi\,d\eta$$

$$= -\frac{1}{\pi}\iint_{|\zeta|<1}\frac{\bar{\zeta}\,\rho(\zeta)}{1-\bar{\zeta}z}\,d\xi\,d\eta - \frac{z^{2n+1}}{\pi}\iint_{|\zeta|<1}\frac{\overline{\rho(\zeta)}\,d\xi\,d\eta}{1-\bar{\zeta}z}$$

$$+\,i\,c_0\,z^n + \sum_{j=0}^{n-1}\alpha_j(z^j - z^{2n-j}) + i\,\beta_j\,(z^j + z^{2n-j})$$

$$= -\frac{1}{\pi}\iint_{|\zeta|<1}\sum_{j=0}^{\infty}\bar{\zeta}^{j+1}\rho(\zeta)\,d\xi\,d\eta\,z^j - \frac{1}{\pi}\iint_{|\zeta|<1}\sum_{j=0}^{\infty}\bar{\zeta}^j\,\overline{\rho(\zeta)}\,d\xi\,d\eta\,z^{2n+1+j}$$

$$+\mathrm{i}\, c_0 z^n + \sum_{j=0}^{n-1} \alpha_j (z^j - z^{2n-j}) + \mathrm{i}\, \beta_j (z^j + z^{2n-j}).$$

Hence

$$\varphi(z) = -\frac{1}{\pi} \iint_{|\zeta|<1} \sum_{j=0}^{\infty} \frac{(\bar{\zeta} z)^{j+1}}{j+1} \rho(\zeta)\, d\xi\, d\eta - \frac{1}{\pi} \iint_{|\zeta|<1} \sum_{j=0}^{\infty} \frac{\bar{\zeta}^j z^{2n+2-j}}{2(n+1)+j} \overline{\rho(\zeta)}\, d\xi\, d\eta$$

$$+ \frac{\mathrm{i}\, c_0 z^{n+1}}{n+1} + \sum_{j=0}^{n-1} \alpha_j \left(\frac{z^{j+1}}{j+1} - \frac{z^{2n+1-j}}{2n+1-j} \right) + \mathrm{i}\, \beta_j \left(\frac{z^{j+1}}{j+1} - + \frac{z^{2n+1+j}}{2n+1+j} \right) + a_0 + \mathrm{i}\, b_0. \tag{1.220}$$

From the second boundary condition (1.218′) we have

$$\mathrm{i}\, z^{n+2} \psi_0(z) = \frac{1}{\pi \mathrm{i}} \int_{|t|=1} \frac{h^0(t)\, dt}{t-z} - \frac{1}{2\pi \mathrm{i}} \int_{|t|=1} \frac{h^0(t)\, dt}{t}$$

$$= \frac{z^{n+2}}{\pi \mathrm{i}} \int_{|t|=1} \frac{h^0(t)\, dt}{t^{n+2}(t-z)} - \sum_{j=1}^{n+1} \frac{z^j}{\pi \mathrm{i}} \int_{|t|=1} \frac{h^0(t)\, dt}{t^{j+1}} - \frac{3}{2\pi \mathrm{i}} \int_{|t|=1} \frac{h^0(t)\, dt}{t}.$$

Hence $\psi_0(z)$ will be continuous at $z = 0$ if and only if

$$\int_{|t|=1} \frac{h^0(t)\, dt}{t^{j+1}} = 0, \; j = 0, \ldots, n+1,$$

i.e.

$$\iint_{|\zeta|<1} \left[\zeta^{n+1-j}\, \overline{\rho(\zeta)} + \bar{\zeta}^{n+1+j} \rho(\zeta) \right] d\xi\, d\eta = 0,$$

$$j = 0, \ldots, n+1. \tag{1.221}$$

In this case we obtain

$$\psi_0(z) = -\frac{1}{\pi} \int_{|t|=1} \frac{h^0(t)\, dt}{t^{n+2}(t-z)} = \frac{1}{2\pi \mathrm{i}} \int_{|t|=1} \frac{\overline{T_{G^+} \rho}\, dt}{t^{n+2}(t-z)}$$

$$= \frac{1}{\pi} \iint_{|\zeta|<1} \frac{\bar{\zeta}^{2n+3} \rho(\zeta)}{1 - \bar{\zeta} z}\, d\xi\, d\eta,$$

i.e.

$$\psi'(z) = -\frac{1}{\pi}\iint_{|\zeta|<1} \frac{\zeta\,\overline{\rho(\zeta)}\,d\xi\,d\eta}{1-\zeta\bar{z}} + \frac{1}{\pi}\iint_{|\zeta|<1} \frac{\bar{\zeta}^{2n+3}\rho(\zeta)}{1-\bar{\zeta}z}\,d\xi\,d\eta$$

$$-\frac{1}{\pi}\iint_{|\zeta|<1}\sum_{j=0}^{\infty} \bar{\zeta}^{j+1} z^j\,\overline{\rho(\zeta)}\,d\xi\,d\eta + \frac{1}{\pi}\iint_{|\zeta|<1}\sum_{j=0}^{\infty} \bar{\zeta}^{2n+3+j} z^j\rho(\zeta)\,d\xi\,d\eta$$

and

$$\psi(z) = -\frac{1}{\pi}\iint_{|\zeta|<1}\sum_{j=0}^{\infty}\frac{(\bar{\zeta}z)^{j+1}}{j+1}\,\overline{\rho(\zeta)}\,d\xi\,d\eta + \frac{1}{\pi}\iint_{|\zeta|<1}\sum_{j=0}^{\infty}\frac{\bar{\zeta}^{2n+3+j}z^{j+1}}{j+1}\,\rho(\zeta)\,d\xi\,d\eta$$

$$+\frac{a_0}{2} - \frac{i\,b_0}{2}. \tag{1.222}$$

Substituting (1.220) and (1.222) into (1.217) we obtain the solution of the problem (B_n') for the inhomogeneous Laplace equation (1.216) represented in the case $n \geq 0$ by the formula

$$u(z) = -\frac{2}{\pi}\iint_{|\zeta|<1} \mathrm{Re}\Big[\sum_{j=0}^{\infty}\frac{(\bar{\zeta}z)^{j+1}}{j+1}\Big]\rho(\zeta)\,d\xi\,d\eta$$

$$-\frac{1}{\pi}\iint_{|\zeta|<1}\sum_{j=0}^{\infty}\Big[\frac{\bar{\zeta}z^{2n+2+j}}{2(n+1)+j} - \frac{\zeta^{2n+3+j}\bar{z}^{j+1}}{(j+1)}\Big]\overline{\rho(\zeta)}\,d\xi\,d\eta$$

$$-\frac{2}{\pi}\iint_{|\zeta|<1}\ln\left|\frac{1-\bar{\zeta}z}{\zeta-z}\right|\rho(\zeta)\,d\xi\,d\eta + a_0 + i\,b_0$$

$$+\frac{i\,c_0 z^{n+1}}{n+1} + \sum_{j=0}^{n-1}\alpha_j\Big(\frac{z^{j+1}}{j+1} - \frac{z^{2n+1-j}}{2n+1-j}\Big) + i\,\beta_j\Big(\frac{z^{j+1}}{j+1} + \frac{z^{2n+1-j}}{2n+1-j}\Big), \tag{1.223}$$

provided the function $\rho(z)$ satisfies the conditions (1.221). According to (1.223) the homogeneous problem (B_n) has exactly $2n+3$ nontrivial linearly independent solutions, and separating the real and imaginary parts in (1.221) we have also $2n+3$ real solvability conditions, i.e. the index of the problem (B_n) for the inhomogeneous Laplace equation is zero.

Now let $n < 0$, so that the boundary conditions (1.218′) are

$$\operatorname{Re} t^{n'} \varphi_0(t) = h_0(t) = -\operatorname{Re} t^{n'} T\rho, \ \operatorname{Re}\left[\mathrm{i}\, t^{-n'+2} \psi_0(t)\right] = h^0(t) = -\operatorname{Re} \mathrm{i}\, t^{-n'+2} T\bar{\rho}\right]. \tag{1.218''}$$

If $n < -1$, i.e. $n' - 2 \geq 0$, then from the second condition we obtain

$$\psi_0(z) = \frac{z^{2n'-3}}{\pi} \iint_{|\zeta|<1} \frac{\rho(\zeta) d\xi d\eta}{1 - \bar{\zeta} z} - \frac{2}{\pi} \iint_{|\zeta|<1} \log \frac{1}{|\zeta - z|} \rho(\zeta)\, d\xi\, d\eta$$

$$+ \alpha_n' z^n + \sum_{j=0}^{n'-3} \beta_j' (z^j + z^{2n'-4-j}) - \mathrm{i}\, \alpha_j' (z^j - z^{2n'-4-j})$$

and in case the equalities

$$\iint_{|\zeta|<1} \zeta^{n'+1-j} \rho(\zeta)\, d\xi\, d\eta + \iint_{|\zeta|<1} \bar{\zeta}^{n'+1+j}\, \overline{\rho(\zeta)}\, d\xi\, d\eta = 0 \tag{1.224}$$

$$j = 0, \ldots, n' - 1,$$

are fulfilled we obtain from the first condition (1.218′)

$$\varphi_0(z) = -\frac{1}{\pi} \iint_{|\zeta|<1} \frac{\bar{\zeta}^{2n'-1}\, \overline{\rho(\zeta)}}{1 - \bar{\zeta} z}\, d\xi\, d\eta.$$

Hence

$$\psi(z) = -\frac{1}{\pi} \iint_{|\zeta|<1} \sum_{j=0}^{\infty} \frac{(\bar{\zeta} z)^{j+1}}{j+1}\, \overline{\rho(\zeta)}\, d\xi\, d\eta$$

$$+ \frac{1}{\pi} \iint_{|\zeta|<1} \sum_{j=0}^{\infty} \frac{\bar{\zeta}^j z^{2n'-2+j}}{2n'-2+j} \rho(\zeta)\, d\xi\, d\eta + \frac{c_0' z^{n'+1}}{n+1}$$

$$+ \sum_{j=0}^{n'-3} \beta_j' \left(\frac{z^{j+1}}{j+1} - \frac{z^{2n'-3+j}}{2n'-3+j} \right) - \mathrm{i}\, \alpha_j' \left(\frac{z^{j+1}}{j+1} + \frac{z^{2n'-3+j}}{2n'-3+j} \right) + \frac{a_0'}{2} - \frac{\mathrm{i}\, b_0'}{2}$$

$$\varphi(z) = -\frac{1}{\pi} \iint_{|\zeta|<1} \sum_{j=0}^{\infty} \frac{(\bar{\zeta} z)^{j+1}}{j+1} \rho(\zeta)\, d\xi\, d\eta$$

$$- \frac{1}{\pi} \iint_{|\zeta|<1} \sum_{j=0}^{\infty} \frac{\bar{\zeta}^{2n'-1+j} z^{j+1}}{j+1}\, \overline{\rho(\zeta)}\, d\xi\, d\eta + \frac{a_0'}{2} - \frac{\mathrm{i}\, b_0'}{2}$$

and consequently

$$-\frac{1}{\pi}\iint_{|\zeta|<1}\sum_{j=0}^{\infty}\left[\frac{\bar{\zeta}^{2n'-1+j}z^{j+1}}{j+1}-\frac{\zeta^{j}z^{2n'-2+j}}{2n'+j-2}\right]\overline{\rho(\zeta)}\,d\xi\,d\eta$$

$$+a_0'+ib_0'+\frac{c_0'\bar{z}^{n'+1}}{u'+1}\sum_{j=0}^{n'-3}\beta_j'\left(\frac{\bar{z}^{j+1}}{j+1}-\frac{\zeta^{j}\bar{z}^{2n'-3+j}}{2n'+3+j}\right)+i\alpha_j'\left(\frac{\bar{z}^{j+1}}{j+1}+\frac{\zeta^{j}\bar{z}^{2n'-3+j}}{2n'+3+j}\right),$$

(1.225)

provided the conditions (1.224) fulfilled. According to this formula the homogeneous problem has exactly $2n' - 1 = -2n - 1$ nontrivial solutions, and separating real and imaginary parts in (1.224) we obtain the same number $2n' - 1 = -2n - 1$ of solvability conditions, i.e. the index of the problem (B'_n) for the inhomogeneous Laplace equation in case $n < 0$ is also zero. Now we turn to the general equation (1.101). For this we calculate the derivatives of (1.223) and (1.225). First from (1.223) we have

$$u_{\bar{z}} = \bar{T}\rho+\frac{1}{\pi}\iint_{|\zeta|<1}\frac{\zeta^{2n+3}\overline{\rho(\zeta)}}{1-\zeta\bar{z}}\,d\xi\,d\eta, \tag{1.226}$$

$$u_z = T\rho-\frac{z^{2n+1}}{\pi}\iint_{|\zeta|<1}\frac{\overline{\rho(\zeta)}\,d\xi\,d\eta}{1-\bar{\zeta}z}+\sum_{j=0}^{2n}c_j\,\varphi_j(z), \tag{1.227}$$

$$u_{\bar{z}z} = \rho, \tag{1.228}$$

$$u_{z^2} = S\rho-\frac{z^{2n}}{\pi}\iint_{|\zeta|<1}\frac{\overline{\rho(\zeta)}\,d\xi\,d\eta}{(1-\bar{\zeta}z)^2}-\frac{2nz^{2n}}{\pi}\iint_{|\zeta|<1}\frac{\overline{\rho(\zeta)}\,d\xi\,d\eta}{1-\bar{\zeta}z} \tag{1.229}$$

$$+\sum_{j=0}^{2n}c_j\,\varphi_j'(z),$$

$$u_{\bar{z}^2} = \bar{S}\rho+\frac{1}{\pi}\iint_{|\zeta|<1}\frac{\zeta^{2(n+2)}\overline{\rho(\zeta)}}{(1-\zeta\bar{z})^2}\,d\xi\,d\eta. \tag{1.230}$$

From (1.225) we have

$$u_z = T\rho-\frac{1}{\pi}\iint_{|\zeta|<1}\frac{\bar{\zeta}^{2n'-1}\overline{\rho(\zeta)}\,d\xi\,d\eta}{(1-\bar{\zeta}z)} \tag{1.231}$$

$$u_z = T\rho - \frac{1}{\pi}\iint_{|\zeta|<1} \frac{\bar{\zeta}^{2n'-1}\overline{\rho(\zeta)}\,d\xi\,d\eta}{(1-\bar{\zeta}z)} \tag{1.231}$$

$$u_{\bar{z}} = \bar{T}\rho + \frac{\bar{z}^{2n'-3}}{\pi}\iint_{|\zeta|<1} \frac{\overline{\rho(\zeta)}\,d\xi\,d\eta}{1-\bar{\zeta}z} + \sum_{j=0}^{2n'-2} c_j'\,\overline{\tilde{\varphi}_j(z)}, \tag{1.232}$$

$$u_{\bar{z}z} = \rho \tag{1.233}$$

$$u_{z^2} = S\rho - \frac{1}{\pi}\iint_{|\zeta|<1} \frac{\bar{\zeta}^{2n'}\overline{\rho(\zeta)\,d\xi\,d\eta}}{(1-\bar{\zeta}z)^2}, \tag{1.234}$$

$$u_{\bar{z}^2} = \bar{S}\rho + \frac{\bar{z}^{2(n'-1)}}{\pi}\iint_{|\zeta|<1} \frac{\overline{\rho(\zeta)}\,d\xi\,d\eta}{(1-\zeta\bar{z})^2} + \sum_{j=0}^{2(n'-1)} c_j'\,\overline{\tilde{\varphi}_j'(z)}. \tag{1.235}$$

Substituting (1.223), (1.226)-(1.230) into (1.101) we obtain the following integral equation for an unknown density ρ in case $n \geq 0$:

$$\rho + K_+\rho = F + \sum_{j=0}^{2n} c_j \hat{F}_j \tag{1.236}$$

where $K_+ = \overset{0}{K}_+ + \tilde{K}_+$, $\tilde{K}_+$ is a linear integral operator, compact in L_p, and $\overset{0}{K}_+$ is a singular integral operator

$$\overset{0}{K}_+\rho \equiv a_1 S^{(n)}\rho + a_2\overline{S^{(n)}\rho} + b_1 S_{(n)}\rho + b_2\overline{S_{(n)}\rho}, \tag{1.237}$$

$$S^{(n)}\rho \equiv S\rho - \frac{z^{2n}}{\pi}\iint_{|\zeta|<1} \frac{\overline{\rho(\zeta)}\,d\xi\,d\eta}{(1-\bar{\zeta}z)^2}, \tag{1.238}$$

$$S_{(n)}\rho \equiv \bar{S}\rho + \frac{1}{\pi}\iint_{|\zeta|<1} \frac{\zeta^{2(n+2)}\overline{\rho(\zeta)}\,d\xi\,d\eta}{(1-\zeta\bar{z})^2}. \tag{1.239}$$

Substituting (1.225), (1.231)-(1.235) into (1.101), we obtain the following integral equation for an unknown density ρ in case $n < 0$:

$$\rho + K_-\rho = F + \sum_{j=0}^{2(n'-1)} c_j \tilde{F}_j, \tag{1.240}$$

where $K_- = K_-^0 + \tilde{K}_-$, $\tilde{K}_-$ is a linear integral operator, compact in L_p and K_-^0 is a singular integral operator

$$K_-^0\rho \equiv a_1 S'_{(n)}\rho + a_1 \overline{S'_{(n')}} + b_1 S_1^{(n')}\rho + \overline{S_1^{(n')}}\rho, \tag{1.241}$$

$$S'_{(n)}\rho \equiv S\rho - \frac{1}{\pi} \iint_{|\zeta|<1} \frac{\bar{\zeta}^{2n'}\overline{\rho(\zeta)}}{(1-\bar{\zeta}z)^2} d\xi\, d\eta, \tag{1.242}$$

$$S_1^{(n')}\rho \equiv \bar{S}\rho + \frac{\bar{z}^{2(n'-1)}}{\pi} \iint_{|\zeta|<1} \frac{\overline{\rho(\zeta)}\, d\xi\, d\eta}{(1-\zeta\bar{z})^2}. \tag{1.243}$$

The equation (1.236) is equivalent to the problem (B_n'), provided ρ satisfies the conditions (1.221) and the equation (1.240) is equivalent to the problem (B_n'), provided ρ satisfies the conditions (1.224).

As we have seen above the L_2-norm of the operators $S^{(n)}$, $S_1^{(n')}$, $S_{(n)}$, $S'_{(n)}$ are equal to one and the operators $I + K_+^0$, $I + K_-^0$ are invertible in L_p for some $p > 2$. Hence as above we conclude that the equations (1.236) and (1.240) possess Fredholm's property. This proves Corollary 1.3 in case G^+ is the disc $|z| < 1$.

2 Higher-dimensional linear elliptic systems

2.1 First-order systems

2.1.1 Introduction

In this chapter we try to achieve for higher dimensions work analogous to that in the previous chapter. We are not going to consider previous problems for general systems, because we cannot do this. We restrict ourselves to special model systems, which are high-dimensional analogues of the inhomogeneous Cauchy-Riemann (C.R.) equation,

$$w_{\bar{z}} = f(z), \tag{2.1}$$

the inhomogeneous Laplace (L) equation

$$w_{z\bar{z}} = f(z). \tag{2.2}$$

and the inhomogeneous Bitsadze equation

$$w_{\bar{z}^2} = f(z). \tag{2.3}$$

We consider also some overdetermined elliptic systems.
Let us begin with the first-order linear system of real equations

$$\sum_{i=1}^{n} A_i(x)u_{x_i} + B(x)u = F(x), \tag{2.4}$$

where $x = (x_1,...,x_n) \in \mathbb{R}^n$, $u = (u_1,...,u_m) \in \mathbb{R}^m$, $A_i(x)$ are given real $(\ell \times m)$-matrices and $F(x) = (F_1,...,F_\ell) \in \mathbb{R}$ is a given real vector. For the first systems considered in the previous chapter, $\ell = m = n = 2$. If $\ell = m$, i.e. the system (2.4) is just determined, then the ellipticity of (2.4) at point $x \in \mathbb{R}^n$ means that

$$\det \chi(x, \xi) = \det \sum_{i=1}^{n} A_i(x)\xi_i \neq 0 \tag{2.5}$$

for all $\xi = (\xi_1,...,\xi_n) \in \mathbb{R}^n \setminus 0$.

It can be shown (see [1]) that in this case there exists a maximum value of n for which a system can be elliptic. For instance the system (2.4) with $B \equiv 0$ cannot be elliptic if $\ell = 2$ and $n > 2$. For given n, in case $\ell = m = (2k_1 + 1)2^{k_2}$, $k_2 = k_3 + 4k_4$ with k_1, k_2, k_3, k_4 nonnegative integers such that $0 \le k_3 < 4$, it has been shown in [1] the maximal number N of matrices $A_i(x)$ such that $\det \Sigma A_i(x)\xi_i \neq 0$ for all real $\xi_i \neq 0$ is equal[1] to $N = 2^{k_3} + 8k_4$.

In case $\ell > m$, i.e. the system (2.4) is overdetermined, the ellipticity of (2.4) at point $x \in \mathbb{R}^n$ means that

$$\operatorname{rank} \chi(x, \xi) = \operatorname{rank} \sum_{i=1}^{n} A_i(x)\xi_i = m \tag{2.6}$$

for all $x \in \mathbb{R}^n \setminus 0$.

The first-order system in $\mathbb{R}^3$ defined by

$$\sum_{i,j=1}^{3} a_{ij}(x) \frac{\partial u_j}{\partial x_i} + (A(x), u) = f(x),$$

$$\operatorname{grad} u_0(x) + \operatorname{curl} u + [\, B(x) \times u \,] = F(x), \tag{2.7}$$

where A, B are given real (3×3) matrices, $F(x) = (F_1, F_2, F_3)$ is a given real vector and $a_{ij}(x)$, $f(x)$ are given real functions such that the quadratic form

$$\sum_{i,j=1}^{3} a_{ij}(x)\xi_i \xi_j \tag{2.8}$$

is determined at point $x \in \mathbb{R}^3$, is elliptic at the point $x \in \mathbb{R}^3$, because for (2.7):

$$\det \chi = -|\xi|^2 \sum_{i,j=1}^{3} a_{ij}(x)\xi_i \xi_j \neq 0, \ \forall \xi \in \mathbb{R}^3 \setminus 0.$$

In particular if $a_{ij} = \delta_{ij}$ is Kronecker and $A = B = 0$, (2.7) reduces to an important system:

$$\operatorname{div} u = f^0(x),$$

$$\operatorname{grad} u_0 + \operatorname{curl} u = F(x) \tag{2.7'}$$

---

1 It follows then there are no first-order elliptic systems of 6, 10, 14, 18 equations in more than two independent variables.

called the inhomogeneous Moisil–Theodorescu system, which is an analogue in $\mathbb{R}^3$ of the inhomogeneous Cauchy-Riemann equation. Another important example of an analogue of the inhomogeneous CR equation in $\mathbb{R}^4$ is

$$\begin{aligned}
&\frac{\partial u_1}{\partial x_1} - \frac{\partial u_2}{\partial x_2} + \frac{\partial u_3}{\partial x_3} + \frac{\partial u_4}{\partial x_4} = f_1,\\
&\frac{\partial u_2}{\partial x_1} + \frac{\partial u_1}{\partial x_2} - \frac{\partial u_3}{\partial x_4} + \frac{\partial u_4}{\partial x_3} = f_2,\\
&\frac{\partial u_3}{\partial x_1} - \frac{\partial u_1}{\partial x_3} + \frac{\partial u_2}{\partial x_4} + \frac{\partial u_4}{\partial x_2} = f_3,\\
&\frac{\partial u_4}{\partial x_1} - \frac{\partial u_1}{\partial x_4} - \frac{\partial u_2}{\partial x_3} - \frac{\partial u_3}{\partial x_2} = f_4.
\end{aligned} \tag{2.9}$$

The overdetermined system

$$\sum_{i,j=1}^{3} a_{ij}(x) \frac{\partial u_j}{\partial x_i} + (A, u) = f, \tag{2.10}$$

$$\operatorname{curl} u + [B \times u] = F$$

with determined quadratic form (2.8) is elliptic, because

$$\operatorname{rank}\begin{pmatrix} \chi_1 & \chi_2 & \chi_3 \\ 0 & -\xi_3 & \xi_2 \\ \xi_3 & 0 & -\xi_1 \\ -\xi_2 & \xi_1 & 0 \end{pmatrix} = 3, \quad \chi_j = \sum_{i=1}^{3} a_{ij}(x)\xi_j$$

and, in particular, the overdetermined Maxwell equations

$$\operatorname{div} u = f, \quad \operatorname{curl} u = F= \tag{2.11}$$

are elliptic for an electrostatic field.

Now let us introduce some second-order elliptic operators generated by the first-order one considered above. Let D_3 and D_4 be the first-order matrix differential operators determined by the left-hand sides of (2.7′) and (2.9):

$$D_3 \equiv \begin{pmatrix} 0 & \mathrm{div} \\ \mathrm{grad} & \mathrm{curl} \end{pmatrix}, \tag{2.12}$$

$$D_4 \equiv \begin{pmatrix} \frac{\partial}{\partial x_1} & -\frac{\partial}{\partial x_2} & \frac{\partial}{\partial x_3} & \frac{\partial}{\partial x_4} \\ \frac{\partial}{\partial x_2} & \frac{\partial}{\partial x_1} & -\frac{\partial}{\partial x_4} & \frac{\partial}{\partial x_3} \\ -\frac{\partial}{\partial x_3} & \frac{\partial}{\partial x_4} & \frac{\partial}{\partial x_1} & \frac{\partial}{\partial x_2} \\ -\frac{\partial}{\partial x_4} & -\frac{\partial}{\partial x_3} & -\frac{\partial}{\partial x_2} & \frac{\partial}{\partial x_1} \end{pmatrix} \tag{2.13}$$

and let D_3', D_4' be the corresponding transposed operators

$$D_3' \equiv \begin{pmatrix} 0 & \mathrm{div} \\ \mathrm{grad} & -\mathrm{curl} \end{pmatrix}, \tag{2.12′}$$

$$D_4' \equiv \begin{pmatrix} \frac{\partial}{\partial x_1} & \frac{\partial}{\partial x_2} & -\frac{\partial}{\partial x_3} & -\frac{\partial}{\partial x_4} \\ -\frac{\partial}{\partial x_2} & \frac{\partial}{\partial x_1} & \frac{\partial}{\partial x_4} & -\frac{\partial}{\partial x_3} \\ \frac{\partial}{\partial x_3} & -\frac{\partial}{\partial x_4} & \frac{\partial}{\partial x_1} & -\frac{\partial}{\partial x_2} \\ \frac{\partial}{\partial x_4} & \frac{\partial}{\partial x_3} & \frac{\partial}{\partial x_2} & \frac{\partial}{\partial x_1} \end{pmatrix} \tag{2.13′}$$

Then we obtain

$$D_3 D_3' \equiv I_3 \cdot \Delta_3, \quad D_4 D_4' \equiv I_4 \cdot \Delta_4, \tag{2.14}$$

where Δ_n is the Laplacian in $\mathbb{R}^n$ and

$$\Lambda_3 \equiv D_3 D_3 \equiv \mathrm{grad}\,\mathrm{div} + \mathrm{curl}\,\mathrm{curl}, \tag{2.15}$$

$$\Lambda_4 \equiv D_4 D_4 \equiv \begin{bmatrix} \frac{\partial^2}{\partial x_1^2} - \sum_{j=2}^{4} \frac{\partial^2}{\partial x_j^2}, & -2\frac{\partial^2}{\partial x_1 \partial x_2}, & 2\frac{\partial^2}{\partial x_1 \partial x_3}, & 2\frac{\partial^2}{\partial x_1 \partial x_4} \\ 2\frac{\partial^2}{\partial x_1 \partial x_2}, & \frac{\partial^2}{\partial x_j^2} - \sum_{j=2}^{4} \frac{\partial^2}{\partial x_j^2}, & -2\frac{\partial^2}{\partial x_1 \partial x_3}, & 2\frac{\partial^2}{\partial x_1 \partial x_4} \\ -2\frac{\partial^2}{\partial x_1 \partial x_3}, & 2\frac{\partial^2}{\partial x_1 \partial x_4}, & \frac{\partial^2}{\partial x_1^2} - \sum_{j=2}^{4} \frac{\partial^2}{\partial x_j^2}, & 2\frac{\partial^2}{\partial x_1 \partial x_2} \\ -2\frac{\partial^2}{\partial x_1 \partial x_4}, & -2\frac{\partial^2}{\partial x_1 \partial x_3}, & -2\frac{\partial^2}{\partial x_1 \partial x_2}, & \frac{\partial^2}{\partial x_1^2} - \sum_{j=2}^{4} \frac{\partial^2}{\partial x_j^2}. \end{bmatrix} \tag{2.16}$$

The operators Λ_3, Λ_4 are elliptic as a product of two elliptic operators, but they are not strongly elliptic in contrast to the Laplacians (2.14).

We also use the complex form of the equations (2.7′), (2.9). More precisely we may put the equations (2.7′) into complex form:

$$\frac{\partial u}{\partial x_1} + 2\frac{\partial v}{\partial z} = f, \quad -2\frac{\partial u}{\partial \bar{z}} + \frac{\partial v}{\partial x_1} = g, \tag{2.17}$$

where

$$u = u_1 + \mathrm{i}\, u, \quad v = u_2 + \mathrm{i}\, u_3, \quad f = f^0 + \mathrm{i}\, F_1, \quad g = F_2 + \mathrm{i}\, F_3,$$

$$\frac{\partial}{\partial z} = \frac{1}{2}\left(\frac{\partial}{\partial x_2} - \mathrm{i}\frac{\partial}{\partial x_3}\right), \quad \frac{\partial}{\partial \bar{z}} = \frac{1}{2}\left(\frac{\partial}{\partial x_2} + \mathrm{i}\frac{\partial}{\partial x_3}\right)$$

and also we may put (2.9) into complex form:

$$\frac{\partial u}{\partial \bar{z}_1} - \frac{\partial v}{\partial z_2} = f, \quad \frac{\partial u}{\partial \bar{z}_2} + \frac{\partial v}{\partial z_1} = g. \tag{2.18}$$

Let us introduce also the second-order elliptic complex operators generated by the left-hand sides (2.17), (2.18):

$$\bar{\partial}_x^2 \equiv \begin{pmatrix} \dfrac{\partial^2}{\partial x_1^2} - 4\dfrac{\partial^2}{\partial z\,\partial\bar{z}}, & 4\dfrac{\partial^2}{\partial x_1\,\partial z} \\ -4\dfrac{\partial^2}{\partial x_1\,\partial\bar{z}}, & \dfrac{\partial^2}{\partial x_1^2} - 4\dfrac{\partial^2}{\partial\bar{z}\,\partial z} \end{pmatrix} \tag{2.19}$$

$$\partial_{\bar{z}}^2 \equiv \begin{pmatrix} \dfrac{\partial^2}{\partial\bar{z}_1} - \dfrac{\partial^2}{\partial z_2\,\partial\bar{z}_2}, & -\dfrac{\partial^2}{\partial x_1\,\partial z_2} \\ \dfrac{\partial^2}{\partial x_1\,\partial\bar{z}_2}, & \dfrac{\partial^2}{\partial z_1^2} - \dfrac{\partial^2}{\partial\bar{z}_2\,\partial z_2} \end{pmatrix} \tag{2.20}$$

where $\bar{\partial}_x$, $\partial_{\bar{z}}$ are the first-order elliptic operators in $\mathbb{R}^3$ and $\mathbb{C}^2$:

$$\bar{\partial}_x \equiv \begin{pmatrix} \dfrac{\partial}{\partial x_1} & 2\dfrac{\partial}{\partial z} \\ -2\dfrac{\partial}{\partial\bar{z}} & \dfrac{\partial}{\partial x_1} \end{pmatrix}, \quad \partial_{\bar{z}} \equiv \begin{pmatrix} \dfrac{\partial}{\partial\bar{z}_1} & -\dfrac{\partial}{\partial z_2} \\ \dfrac{\partial}{\partial\bar{z}_2} & \dfrac{\partial}{\partial z_1} \end{pmatrix}, \tag{2.21}$$

$$\frac{\partial}{\partial\bar{z}_1} = \frac{1}{2}\left(\frac{\partial}{\partial x_1} + i\frac{\partial}{\partial x_2}\right), \quad \frac{\partial}{\partial\bar{z}_2} = \frac{1}{2}\left(\frac{\partial}{\partial x_3} + i\frac{\partial}{\partial x_4}\right)$$

$$\frac{\partial}{\partial z_1} \equiv \overline{\frac{\partial}{\partial\bar{z}_1}}, \quad \frac{\partial}{\partial z_2} = \overline{\frac{\partial}{\partial\bar{z}_2}}.$$

We also consider the following first-order elliptic operators:

$$\partial_x' \equiv \begin{pmatrix} \dfrac{\partial}{\partial x_1} & -2\dfrac{\partial}{\partial z} \\ 2\dfrac{\partial}{\partial\bar{z}} & \dfrac{\partial}{\partial x_1} \end{pmatrix}, \quad \partial_z' \equiv \begin{pmatrix} \dfrac{\partial}{\partial z_1} & \dfrac{\partial}{\partial z_2} \\ -\dfrac{\partial}{\partial\bar{z}_2} & \dfrac{\partial}{\partial\bar{z}_1} \end{pmatrix}. \tag{2.22}$$

2.1.2 Conjugation problems for first-order elliptic systems

In this section we state the conjugation problems like the Riemann problem for the inhomogeneous equation

$$\bar{\partial}_x u = f \tag{2.23}$$

in $\mathbb{R}^3$, with given complex-valued vector $f = (f_1, f_2)$, complex-valued unknown vector $u = (u_1, u_2)$ and the operator determined by (2.21). We also state the same problem for the inhomogeneous equation

$$\partial_{\bar{z}} u = f \tag{2.24}$$

with operator $\partial_{\bar{z}}$ determined by (2.21).

Let $\overset{n}{\Omega}$ be a finite collection of disjoint bounded n-dimensional bodies in $\mathbb{R}^n$ with smooth boundaries $\Gamma_1^{(n)}, \dots, \Gamma_m^{(n)}$, and let $\Gamma_0^{(n)}$ a closed surface consisting inside $\Gamma_k^{(n)}$ and let $\overset{n}{\Omega_0^-}$ be the exterior of Ω^+, bounded by $\overset{(n)}{\Gamma} = \bigcup_{j=0}^{m} \Gamma_j^{(n)}$. If $u(x)$ is a vector function defined on $\mathbb{R}^n \backslash \Gamma^{(n)}$, then for $\xi \in \Gamma^{(n)}$ we denote by $u^+(\xi)$ the limit of $u(x)$ (if it exists) as $x \to \xi$ from within $\overset{n}{\Omega^+}$ and by $u^-(\xi)$ the limit of $u(x)$ as $x \to \xi$ from within $\overset{n}{\Omega^-}$. If $u(x)$ is continuous in $\overset{(n)}{\Omega}{}^+ + \Gamma^{(n)} = \overline{\overset{(n)}{\Omega^+}}$ and in $\overset{(n)}{\Omega} + \Gamma^{(n)} = \overline{\overset{(n)}{\Omega}}{}^-$, then we say that $u(x)$ is piecewise continuous in $\mathbb{R}^n$. We assume that the right-hand side f in (2.23), (2.24) vanishes at infinity as $O(1/|x|^{1+\varepsilon})$.

Problem R_x. Find in $\overset{3}{\Omega^+} \cup \overset{3}{\Omega^-}$ a piecewise continuous solution $u = (u_1, u_2)$ of (2.23), vanishing at infinity and satisfying the jump condition

$$u^+ - G u^- = 0 \tag{2.25}$$

on $\Gamma^{(3)}$ with a Hölder continuous complex-valued (2×2) nondegenerating matrix G given on $\Gamma^{(n)}$.

Problem R_z. Find $\overset{4}{\Omega^+} \cup \overset{4}{\Omega^-}$, a piecewise continuous solution $u(z) = (u_1, u_2)$ of (2.24), vanishing at infinity and satisfying the jump condition

$$u^+ - G u = 0 \tag{2.26}$$

on $\Gamma^{(4)}$ with Hölder continuous complex-valued (2×2) nondegenerating matrix G given on $\Gamma^{(4)}$.

Here we consider only the case when the matrix

$$G = \begin{pmatrix} G_{11} & G_{12} \\ G_{21} & G_{22} \end{pmatrix}$$

is constant and not degenerate ($\det G \neq 0$).

If we introduce a new vector function by

$$w = \begin{cases} u & \text{in } \overset{k}{\Omega}{}^{+}, \\ Gu & \text{in } \overset{k}{\Omega}{}^{-}, \end{cases} \quad k = 3, 4, \tag{2.27}$$

then multiplying (2.23) from the left-hand side by ∂_x' and (2.24) by ∂_z' and taking into account (2.25), (2.26), we see that w will be the solution in $\mathbb{R}^k \backslash \Gamma^{(k)}$ of the inhomogeneous Laplace equation

$$\Delta w = F, \tag{2.28}$$

continuous on $\mathbb{R}^k$, where

$$F = \begin{cases} \partial_x' f & \text{in } \overset{3}{\Omega}{}^{+}, \\ G\partial_x' f & \text{in } \overset{3}{\Omega}{}^{-} \end{cases} \tag{2.29}$$

in the case of problem (R_x) and

$$F = \begin{cases} \partial_z' f & \text{in } \overset{4}{\Omega}{}^{+}, \\ G\partial_z' f & \text{in } \overset{4}{\Omega}{}^{-} \end{cases} \tag{2.30}$$

in the case of problem $(R_{\bar{z}})$.

Hence, in the case of problem (R_x),

$$w = -\frac{1}{|s_5|} \int_{\mathbb{R}^3} \frac{F(\xi)\,d\xi}{\sqrt{[(\xi_1 - x_1)^2 + |\zeta - z|^2]}}, \tag{2.31}$$

and in the case of problem $(R_{\bar{z}})$,

$$w = -\frac{1}{|s_4|}\int_{\mathbb{R}^4}\frac{F(\xi)\,d\xi}{|\zeta_1 - z_1|^2 + |\zeta_2 - z_2|^2}. \tag{2.32}$$

Taking into account the definitions (2.29), (2.30) and integrating by parts we obtain

$$w = \frac{1}{|s_3|}\int_{\Omega^{3+}} E^{(3)}(\xi - x) f(\xi)\,d\xi + \frac{G}{|s_3|}\int_{\Omega^{3-}} E^{(3)}(\xi - x) f(\xi)\,d\xi \tag{2.33}$$

in case (R_x) and

$$w = \frac{1}{|s_4|}\int_{\Omega^{4+}} E^{(4)}(\zeta - z) f(\xi)\,d\xi$$

$$+\frac{G}{|s_4|}\int_{\Omega^{4-}} E^{(4)}(\zeta - z) f(\xi)\,d\xi, \text{ in case } (\mathbb{R}_{\bar{z}}). \tag{2.34}$$

where

$$E^{(3)}(x) = \frac{1}{|x|^3}\begin{pmatrix} x_1 & -\bar{z} \\ z & x_1 \end{pmatrix},\quad E^{(4)}(z) = \frac{1}{|z|^4}\begin{pmatrix} z_1 & \bar{z}_2 \\ -z_2 & \bar{z}_1 \end{pmatrix}.$$

Now if we take into account the definition (2.27), then we obtain the following formula for the solution of problem (R_x):

$$u(x) = \begin{cases} \dfrac{1}{|s_3|}\displaystyle\int_{\Omega^{3+}} E^{(3)}(\xi - x) f(\zeta)\,d\xi + \dfrac{G}{|s_3|}\displaystyle\int_{\Omega^{3}} E^{(3)}(\xi - x) f(\xi)\,d\xi & \text{in } \Omega^{+} \\ \dfrac{G^{-1}}{|s_3|}\displaystyle\int_{\Omega^{3+}} E^{(3)}(\xi - x) f(\xi)\,d\xi + \dfrac{G}{|s_3|}\displaystyle\int_{\Omega^{3-}} E^{(3)}(\xi - x) f(\xi)\,d\xi & \text{in } \Omega^{-}. \end{cases} \tag{2.35}$$

and the following formula for the solution of problem $(R_{\bar{z}})$:

$$u(z) = \begin{cases} \frac{1}{|s_4|} \int\limits_{\overset{4}{\Omega}{}^+} E^{(4)}(\xi - z) f(\zeta)\, d\xi + \frac{G}{|s_4|} \int\limits_{\overset{4}{\Omega}{}^-} E^{(4)}(\xi - z) f(\xi)\, d\xi & \text{in } \overset{4}{\Omega}{}^+ \\ \frac{G^{-1}}{|s_4|} \int\limits_{\overset{4}{\Omega}{}^+} E^{(4)}(\xi - z) f(\xi)\, d\xi + \frac{G}{|s_4|} \int\limits_{\overset{4}{\Omega}{}^-} E^{(4)}(\xi - z) f(\xi) d\xi & \text{in } \overset{4}{\Omega}{}^- . \end{cases} \tag{2.36}$$

Thus we have proved the following:

Theorem 2.1. *If G is a nondegenerate constant matrix, then the conjugation problem (R_x) as well as the problem $(R_{\bar{z}})$ has a unique solution for any right-hand side belonging to* $L_p^{loc}(\mathbb{R}^k)$ with $p > k$ and vanishing at infinity as $O(1/|x|^{1+\varepsilon})$.

Let us consider more general equations

$$\bar{\partial}_x u + Au + B\bar{u} = f \tag{2.37}$$

in $\mathbb{R}^3 \backslash \Gamma^{(3)}$ and

$$\partial_{\bar{z}} u + Au + B\bar{u} = f \tag{2.38}$$

in $\mathbb{R}^4 \backslash \Gamma^{(4)}$ with complex-valued (2×2) matrix given in $\mathbb{R}^{(k)} \backslash \Gamma^{(k)}$ such that its elements, as well as the right-hand side f, are L_p^{loc} functions with $p > k$ vanishing at infinity as $O(1/|x|^{1+\varepsilon})$, $\varepsilon > 0$.

Then by means of formulas (2.35), (2.36) we see that the considered conjugation problem (2.26) with constant matrix may be reduced to Fredholm's integral equation

$$u(x) + \int\limits_{\mathbb{R}^k} K(x, \xi)\, u(\xi) + L(x, \xi)\, \overline{u(\xi)}\, d\xi = 0 \tag{2.39}$$

with kernels $K(x, \xi)$, $L(x, \xi)$ having weak singularity at $\xi = x$.

Thus we derive the following

Theorem 2.2. *If G is a nondegenerate constant matrix, then the conjugation problem (2.26) for general equations (2.37), (2.38) has Fredholm's property.*

2.1.3. Schwartz-type boundary-value problems

Let Ω be a bounded domain in $\mathbb{R}^3$ with smooth boundary Γ. We consider in Ω the inhomogeneous equation (2.23) and let us try to find a solution of this equation continuous in $\bar{\Omega} = \Omega + \Gamma$, with boundary condition

$$u_2 = 0 \tag{2.40}$$

on Γ.

Since it follows from (2.23) that u_2 is the solution of the inhomogeneous Laplace equation

$$\Delta u_2 = 2\frac{\partial f_1}{\partial \bar{z}} + \frac{\partial f_2}{\partial x_1}, \tag{2.41}$$

we obtain

$$u_2(x_1, z) = -\frac{1}{|s_3|}\int_\Omega G(x_1, z; \xi_1, \zeta)\,(2\frac{\partial f_1}{\partial \bar{\zeta}} + \frac{\partial f_2}{\partial \xi_1})\,d\xi,$$

where $G(x, \xi)$ is the Green's function of the domain Ω. Since the Green's function vanishes at the boundary, integration by parts gives

$$u_2(x_1, z) = \frac{2}{|s_3|}\int_\Omega G_{\bar{\zeta}}(x_1, z; \xi_1, \zeta)\, f_1(\xi)\,d\xi$$

$$+ \frac{1}{|s_3|}\int_\Omega G_{\xi_1}(x_1, z; \xi_1, \zeta)\, f_2(\xi)\,d\xi\,. \tag{2.42}$$

Hence by condition (2.40) the function u_2 is determined uniquely by the right-hand side f. Then (2.23) yields an overdetermined set of equations for u_1 :

$$\frac{\partial u_1}{\partial x_1} = -2\frac{\partial u_2}{\partial z} + f_1\,,$$

$$\frac{\partial u_2}{\partial \bar{z}} = \frac{1}{2}\frac{\partial u_2}{\partial x_1} - \frac{1}{2}f_2\,. \tag{2.43}$$

If we assume now that the cross of Ω by plane $x_1 = 0$ is a simply connected plane domain G with a smooth boundary which is a closed curve $\gamma = \partial G$ without cross points, then from the first equation (2.43) it follows that

$$u_1(x_1, z) = F^0(z) + \int_0^{x_1} [f_1(\sigma, z) - 2\frac{\partial u_2}{\partial z}(\sigma, z)]\,d\sigma, \tag{2.44}$$

where $F^0(z) \in C^1(G)$ is an arbitrary complex-valued function. But the second equation (2.43) means that F^0 is the solution of inhomogeneous Cauchy-Riemann equation:

$$F^0_{\bar{z}} = g(z), \tag{2.45}$$

where

$$g(z) = -\frac{1}{2} f_2(0, z) + \frac{1}{|s_3|} \int_\Omega [2\, G_{\bar{\zeta}_{x_1}}(0, z; \xi) f_1(\xi) + G_{\xi_1 x_1}(0, z; \xi) f_2(\xi)]\, d\xi. \tag{2.46}$$

Hence

$$F^0(z) = \varphi^0(z) + T_G\, g = \varphi^0(z) - \frac{1}{\pi} \iint_G \frac{g(\zeta)\, d G_\zeta}{\zeta - z}, \tag{2.47}$$

where $\varphi^0(z)$ is holomorphic in G and we obtain the following formula for u_1:

$$u_1(x_1, z) = \varphi^0(z) + \overset{(1)}{K} f, \tag{2.48}$$

where

$$\overset{(1)}{K} f = \frac{1}{2\pi} \iint_G \frac{f_2(0, \tau)\, d G_\tau}{\tau - z} + \int_\Omega \left[\overset{(1)}{K}_1(x, \xi) f_1(\xi) + \overset{(1)}{K}_2(x, \xi) f_2(\xi) \right] d\xi, \tag{2.49}$$

$$\overset{(1)}{K}_1(x, \xi) = -\frac{4}{|s_3|} \int_0^{x_1} G_{\bar{\xi}_z}(\sigma, z; \xi)\, d\sigma - \frac{1}{|s_3|} \iint_G \frac{G_{\bar{\xi}_{x_1}}(0, \tau; \xi)\, d G_\tau}{\tau - z},$$

$$\overset{(1)}{K}_2(x, \xi) = -\frac{2}{|s_1|} \int_0^{x_1} G_{\xi_1 z}(\sigma, z; \xi)\, d\sigma - \frac{1}{2|s_1|} \iint_G \frac{G_{\xi_1 x_1}(0, \tau; \xi)\, d G_\tau}{\tau - z}.$$

Hence the function u_1 is determined uniquely by the right-hand side f apart of arbitrary function $\varphi^0(z)$, holomorphic in plane domain G.

Thus a solution of inhomogeneous equation (2.23) with given condition (2.40) is uniquely determined by its right-hand side apart from an arbitrary function, holomorphic in G.

Analogously it can be shown that a solution of (2.23) with given $u_1 |_\Gamma = 0$ is uniquely determined by its right-hand side apart from an arbitrary function antiholomorphic in G.

This assertion gives an idea to state the following problem.

Problem $S^{(1)}$. Find in Ω a solution of (2.23), continuous in $\bar{\Omega}$, satisfying the boundary condition (2.40) on Γ and boundary condition:

$$\operatorname{Re} \bar{\lambda}\, u_1 = 0 \tag{2.50}$$

on γ with given Hölder continuous function $\lambda(t) \neq 0,\ t \in \gamma$.

Problem $S^{(2)}$. Find in Ω a solution of (2.23), continuous in $\bar{\Omega}$, satisfying the boundary condition:

$$u_1 = 0 \tag{2.51}$$

on Γ and the boundary condition

$$\operatorname{Re} \lambda u_2 = 0 \tag{2.52}$$

on γ with given Hölder continuous function $\lambda(t) \neq 0,\ t \in \gamma$. Let $\kappa = \operatorname{Ind}_\gamma \lambda$ where

$$\operatorname{Ind}_\gamma \lambda = \frac{1}{2\pi i} \int_\gamma d \log \lambda(t).$$

Theorem 2.3. *If $\kappa \geq 0$, then the problem $S^{(1)}$ as well as the problem $S^{(2)}$ is solvable for any right-hand side $f(x) \in L_p(\Omega)$, $p > 3$ and the corresponding homogeneous problem ($f \equiv 0$) has exactly $2\kappa + 1$ nontrivial solutions, but if $\kappa < 0$, then the inhomogeneous problem $S^{(1)}$ and $S^{(2)}$ are solvable if and only if the right-hand side f satisfies $-2\kappa - 1$ orthogonal conditions, and the corresponding homogeneous problem has no nontrivial solutions.*

The proof of this theorem follows from representations like (2.44), (2.48) by means of results on the Riemann-Hilbert problem on the plane domain G.

The problems $S^{(k)}$ are analogous to the Schwartz problem for the inhomogeneous Cauchy-Riemann equation. There are also several such analogues of the Schwartz problems, where instead of the boundary conditions (2.40) or (2.51) on Γ the following conditions are given:

$$\operatorname{Re} u_1 = 0, \quad \operatorname{Re} u_2 = 0; \tag{2.51^a}$$

$$\operatorname{Im} u_1 = 0, \quad \operatorname{Im} u_2 = 0; \tag{2.51^b}$$

$$\operatorname{Re} u_1 = 0, \quad \operatorname{Im} u_2 = 0; \tag{2.51^c}$$

$$\operatorname{Im} u_1 = 0, \quad \operatorname{Re} u_1 = 0. \tag{2.51^d}$$

In the next section we consider more general boundary conditions on Γ, which includes all of these conditions as a particular case.

At the end of this section we derive some formulae for the solutions of the equation

$$\bar{\partial}_x u = 0$$

in the unit ball $|x| < 1$. Acting to this equation by ∂'_x we see that $u(x) = (u_1, u_2)$ is a harmonic vector in the ball $|x| \leq 1$. Hence we may represent $u(x)$ by the Poisson integral formula

$$u(x) = \frac{1}{2\pi} \int_{|\xi|=1} \frac{1-|x|^2}{|\xi - x|^3} u(\xi)\, d\Gamma_\xi .$$

Moreover it follows from this equation that

$$u_1(x) = -\frac{1}{2\pi} \iint_{|\tau|<1} \frac{\partial u_2}{\partial x_1}(0,\tau) \frac{d\sigma_\tau}{\tau - z} - 2\int_0^{x_1} \frac{\partial u_2}{\partial z}(\sigma, z)\, d\sigma + \varphi_0(z),$$

$$u_2(x) = \frac{1}{2\pi} \iint_{|\tau|<1} \frac{\partial u_1}{\partial x_1}(0,\tau) \frac{\partial \sigma_\tau}{\bar{\tau} - \bar{z}} + 2\int_0^{x_1} \frac{\partial u_1}{\partial \bar{z}}(\sigma, z)\, d\sigma + \overline{\psi_0(z)},$$

where $\varphi_0(z)$, $\psi_0(z)$ are arbitrary functions, holomorphic in the disc $|z| \leq 1$, $z = x_2 + ix_3$. Calculating integrals, we have

$$\int_0^{x_1} u_j(\sigma, z)\, d\sigma =$$

$$\int_{|\xi|=1} \left((1-|z|^2) \int_0^{x_1} \frac{d\sigma}{\left((\sigma_1 - \xi_1)^2 + |\zeta - z|^2\right)^{3/2}} - \int_0^{x_1} \frac{\sigma^2\, d\sigma}{\left((\sigma - \xi_1)^2 + |\xi - z|^2\right)^{3/2}} \right) u_j(\xi)\, d\Gamma_\xi$$

$$= \int_{|\xi|=1} K_0(x, \xi)\, u_j(\xi)\, d\Gamma_\xi ,$$

where

$$K_0(x,\xi) = \frac{(1-\xi_1^2-|z|^2)}{|\zeta-z|^2}\left(\frac{\xi_1}{\sqrt{(\xi_1^2+|\zeta-z|^2)}} - \frac{\xi_1-x_1}{|\xi-x|}\right)$$

$$+\frac{\xi_1+x_1}{|\xi-x|} - \frac{\xi_1}{\sqrt{(\xi_1^2+|\zeta-z|^2)}} + \log\left(\sqrt{\xi_1^2+|\zeta-z|^2}-\xi_1\right) - \log\left(|\xi-x|-(\xi_1-x_1)\right),$$

because of

$$\int_0^{x_1} \frac{d\sigma}{\left((\sigma-\xi_1)^2+|\zeta-z|^2\right)^{3/2}} = \frac{1}{|\zeta-z|^2}\left(\frac{\xi_1}{\sqrt{(\xi_1^2+|\zeta-z|^2)}} - \frac{\xi_1-x_1}{|\xi-x|}\right),$$

$$\int_0^{x_1} \frac{\sigma^2\, d\sigma}{\left((\sigma-\xi_1)^2+|\xi-z|^2\right)^{3/2}} = \frac{\xi_1^2}{|\zeta-z|^2}\left(\frac{\xi_1}{\sqrt{(\xi_1^2+|\zeta-z|^2)}} - \frac{\xi_1-x_1}{|\xi-x|}\right)$$

$$+\frac{\xi_1}{\sqrt{(\xi_1^2+|\zeta-z|^2)}} - \frac{\xi_1+x_1}{|\xi-x|}$$

$$+\log(|\xi-x|+x_1-\xi_1) - \log\left(\sqrt{\xi_1^2+|\zeta z|^2)}-\xi_1\right).$$

Also calculating all derivatives, we may write

$$u_1(x) = \frac{1}{4\pi}\int_{|\xi|=1} H_2(x,\xi)\, u_2(\xi)\, d\Gamma_\xi + \varphi_0(z),$$

$$u_2(x) = \frac{1}{4\pi}\int_{|\xi|=1} H_1(x,\xi)\, u_1(\xi)\, d\Gamma_\xi + \overline{\psi_0(z)}, \tag{2.53}$$

where

$$H_2(x,\xi) = -\frac{3\xi_1}{2\pi}\iint_{|\tau|<1} \frac{(1-|\tau|^2)\, d\sigma_\tau}{(1-2\operatorname{Re}\bar{\tau}\zeta+|\tau|^2)^{5/2}(\tau-z)} + H_0(x,\zeta),$$

$$H_1(x,\xi) = \frac{3\xi_1}{2\pi} \iint_{|\tau|<1} \frac{(1-|\tau|^2)\,d\sigma_\tau}{(1-2\operatorname{Re}\bar{\tau}\zeta+|\tau|^2)^{5/2}(\bar{\tau}-\bar{z})} + \overline{H_0(x,\xi)},$$

$$H_0(x,\xi) = \frac{|\zeta|^2-|z|^2}{\xi-z}\Big(\frac{\xi_1}{(1-2\operatorname{Re}\bar{\tau}\zeta+|\tau|^2)^{3/2}} - \frac{\xi_1-x_1}{(1-2\operatorname{Re}\bar{\tau}\zeta+|\tau|^2)^{3/2}}\Big)$$

$$+\frac{2\zeta}{(\zeta-z)^2}\Big(\frac{\xi_1}{\sqrt{1-2\operatorname{Re}\bar{\zeta}z+|z|^2}} - \frac{\xi_1-x_1}{\sqrt{1-2(\xi,x)+|x|^2}}$$

$$+(\bar{\zeta}-\bar{z})\Big(\frac{\xi_1+x_1}{(1-2(\xi,x)+|x|^2)^{3/2}} - \frac{\xi_1}{(1-2\operatorname{Re}\bar{\zeta}z+|z|^2)^{3/2}}$$

$$+\frac{1}{1-2\operatorname{Re}\bar{\zeta}z+|z|^2-\xi_1\sqrt{(1-2\operatorname{Re}\bar{\zeta}z+|z|^2}}$$

$$-\frac{1}{1-2(\xi,x)+|x|^2-(\xi_1-x_1)\sqrt{(1-2(\xi,x)+|x|^2}}\Big).$$

Let γ be the circle $x_2^2+x_3^2=1$. Since $H_0(x,\xi)\,|_{x_1=0}=0$, then we may find holomorphic functions $\varphi_0(z)$ and $\psi_0(z)$ and express them through value of u_1 and u_2 on γ. Indeed from (2.53) we have

$$\operatorname{Re}\varphi_0(z)\,|_{z\in\gamma} = \operatorname{Re}u_1(0,z)\,|_{z\in\gamma} - \operatorname{Re}\frac{1}{4\pi}\int_{|\xi|=1} H_2(0,z;\xi)\,u_2(\xi)\,d\Gamma_\xi\,|_{z\in\gamma},$$

$$\operatorname{Re}\psi_0(z) = \operatorname{Re}u_2(0,z)\,|_{z\in\gamma} - \operatorname{Re}\frac{1}{4\pi}\int_{|\xi|=1} H_1(0,z;\xi)\,u_1(\xi)\,d\Gamma_\xi\,|_{z\in\gamma}.$$

Hence by the Schwartz formula we obtain

$$\varphi_0(z) = \frac{1}{2\pi}\int_\gamma \operatorname{Re}u_1(0,t)\frac{t+z}{t-z}d\theta$$

$$-\int_{|\xi|=1}\frac{3\xi_1 z}{8\pi}\iint_{|\tau|<1}\frac{(1-|\tau|^2\,d\sigma_\tau}{(1-2\operatorname{Re}\bar{\tau}\zeta+|\tau|^2)^{5/2}}\frac{\overline{u_2(\xi)}\,d\Gamma_\xi}{(1-\bar{\tau}z)}+i\,c_0,$$

$$\psi_0(z) = \frac{1}{2\pi}\int_\gamma \operatorname{Re} u_2(0,t)\,\frac{t+z}{t-z}\,d\theta$$

$$+\int_{|\xi|=1}\frac{3\xi_1 z}{8\pi}\iint_{|\tau|<1}\frac{(1-|\tau|^2\,d\sigma_\tau}{(1-2\operatorname{Re}\bar{\tau}\zeta+|\tau|^2)^{5/2}}\,\frac{\overline{u_1(\xi)\,d\Gamma_\xi}}{(1-\bar{\tau}z)}+i\,c_0,$$

where c_0, c^0 are arbitrary real constants.

Thus we obtained the following formulas:

$$u_1(x) = \frac{1}{2\pi}\int_\gamma \operatorname{Re} u_1(0,t)\,\frac{t+z}{t-z}\,d\theta + \frac{1}{4\pi}\int_{|\xi|=1}\left[H_2(x,\xi)\,u_2(\xi)+h(z,\zeta)\,\overline{u_2(\xi)}\right]d\Gamma_\xi + ic_0$$

$$u_2(x) = \frac{1}{2\pi}\int_\gamma \operatorname{Re} u_2(0,t)\,\frac{\bar{t}+\bar{z}}{\bar{t}-\bar{z}}\,d\theta + \frac{1}{4\pi}\int_{|\xi|=1}\left[H_1(x,\xi)\,u_1(\xi)-\overline{h(z,\xi)}\,\overline{u_1(\xi)}\right]d\Gamma_\xi - ic_0, \tag{2.54}$$

where

$$h(z,\xi) = -\frac{3\xi_1 z}{2\pi}\iint_{|\tau|<1}\frac{(1-|\tau|^2)\,d\sigma_\tau}{(1-2\operatorname{Re}\bar{\tau}\zeta+|\tau|^2)^{5/2}(1-\bar{\tau}z)}.$$

As it follows from formulae (2.54) if u_2 is given on the sphere $\Gamma : |x| = 1$:

$$u_2(x) = f_2(x), \quad |x| = 1$$

and u_1 is given on the circle $\gamma : |z| = 1$:

$$u_1(0,z) = g_1(z), \quad |z| = 1,$$

then the holomorphic $u = (u_1, u_2)$ is uniquely determined (apart from an imaginary constant) in ball $|x| < 1$ by equalities

$$u_1(x) = \frac{1}{2\pi i}\int_\Gamma g_1(t)\,\frac{t+z}{t-z}\,\frac{dt}{t} + \frac{1}{4\pi}\int_{|\xi|=1}\left[H_2(x,\xi)\,f_2(\xi)+h_2(z,\xi)\,\overline{f_2(\xi)}\right]d\Gamma_\xi + ic_0,$$

$$u_2(x) = \frac{1}{4\pi}\int_{|\xi|=1}\frac{1-|x|^2}{|\xi-x|^3}\,f_2(\xi)\,d\Gamma_\xi$$

and if u_1 is given on the sphere Γ:

$$u_1(x) = f_1(x), \quad |x| = 1$$

and u_2 is given on the circle γ:

$$u_2(0, z) = g_2(z), \quad |z| = 1,$$

then the holomorphic vector $u = (u_1, u_2)$ uniquely (apart from an imaginary constant) determines in ball $|x| < 1$ by equalities

$$u_1(x) = \frac{1}{4\pi} \int_{|\xi|=1} \frac{1 - |x|^2}{|\xi - x|^3} f_1(\xi)\, d\Gamma_\xi ,$$

$$u_2(x) = -\frac{1}{2\pi} \int_\gamma g_2(t) \frac{\bar{t} + \bar{z}}{\bar{t} - \bar{z}} \frac{d\bar{t}}{\bar{t}} + \int_{|\xi|=1} \left[H_1(x, \xi) f_1(\xi) - \overline{h(z,\xi)}\, \overline{f_1(\xi)} \right] d\Gamma_\xi - ic^0.$$

As a consequence of these formulae we derive also analogous formulae for the solution $\varphi(x) = (\varphi_1, \varphi_2, \varphi_3)$ of the overdetermined Maxwell system

$$\operatorname{div} \varphi = 0, \quad \operatorname{curl} \varphi = 0 \tag{2.55}$$

in the unit ball $|x| < 1$. This system is equivalent to the equation $\bar{\partial}_x u = 0$ for the vector $u = (u_1, u_2)$ with $u_1(x) = \varphi_1(x) + i\varphi_0(x)$, $u_2(x) = \varphi_2(x) + i\varphi_3(x)$ and with complementary boundary condition $\varphi_0 = 0$ on the sphere $|x| = 1$. Hence from the second formula (2.54) we obtain

$$\varphi_2(x) = \frac{1}{4\pi} \int_{|\xi|=1} \operatorname{Re} \left(H_1(x, \xi) - h(z, \xi) \right) \varphi_1(\xi)\, d\Gamma_\xi + \frac{1}{2\pi} \int_\gamma \frac{1 - |z|^2}{|t - z|^2} \varphi_2(0, t)\, d\theta,$$

$$\varphi_3(x) = \frac{1}{4\pi} \int_{|\xi|=1} \operatorname{Im} \left(H_1(x, \xi) - h(z, \xi) \right) \varphi_1(\xi)\, d\Gamma_\xi + \frac{1}{\pi} \int_\gamma \frac{\operatorname{Im} t\bar{z}}{|t - z|^2} \varphi_2(0, t)\, d\theta - c^0.$$

As it follows from these formulae if φ_1 is given on the sphere Γ:

$$\varphi_1(x) = f_1(x), \quad |x| = 1$$

and φ_2 is given on the circle γ:

$$\varphi_2(0, z) = g_2(z), \quad |z| = 1,$$

then the solution $\varphi(x) = (\varphi_1, \varphi_2, \varphi_3)$ of the system (2.55) uniquely (apart from a real constant) determines in ball $|x| < 1$ by the equalities

$$\varphi_1(x) = \frac{1}{4\pi} \int_{|\xi|=1} \frac{1 - |x|^2}{|\xi - x|^3} f_1(\xi)\, d\Gamma_\xi ,$$

$$\varphi_2(x) = \frac{1}{4\pi} \int_{|\xi|=1} \operatorname{Re}\left(H_1(x, \xi) - h(z, \xi)\right) f_1(\xi)\, d\Gamma_\xi + \frac{1}{2\pi} \int_\gamma \frac{1 - |z|^2}{|t - z|^2}\, g_2(t)\, d\theta, \quad (2.56)$$

$$\varphi_3(x) = \frac{1}{4\pi} \int_{|\xi|=1} \operatorname{Im}\left(H_1(x, \xi) - h(z, \xi)\right) f_1(\xi)\, d\Gamma_\xi + \frac{1}{\pi} \int_\gamma \frac{\operatorname{Im} t\bar{z}}{|t - z|^2}\, g_2(t)\, d\theta - c^0.$$

From these formulae as well as from (2.54) we may derive the relations between components of vectors φ and u on the sphere $|x| = 1$.

2.1.4 Riemann - Hilbert-type boundary-value problems

Now we consider for equation (2.23) the boundary-value problem (H) in the bounded domain Ω with a given Riemann-Hilbert type boundary condition

$$\operatorname{Re} Gu = 0, \quad (2.57)$$

on $\Gamma = \partial\Omega$ with given complex-valued (2×2) matrix such that[2]

$$\Delta = \det G = \det \begin{pmatrix} G_{11} & G_{12} \\ G_{21} & G_{22} \end{pmatrix} \neq 0$$

on Γ.

Here we assume for this matrix to have constant elements G_{ij}. Also we assume for the domain Ω to be such that the intersection of Ω with the plane

$$\operatorname{Re}(G_{11}\cdot\bar{G}_{22} - \bar{G}_{12}\cdot G_{21})x_3 - \operatorname{Im}(G_{11}\cdot\bar{G}_{22} - \bar{G}_{12}G_1)x_2 = 0 \quad (2.58)$$

is a simply connected plane domain G with boundary $\gamma = \partial G$, being closed smooth curve without cross points.

Problem *H*. Find in Ω a solution $u = (u_1, u_2)$ of equation (2.23), continuous in $\bar{\Omega} = \Omega + \Gamma$, satisfying the boundary condition (2.57) on Γ and the boundary condition

2 Without loss of generality therefore we may assume that $|G_j1|^2 + |G_j2|^2 \equiv 1$

$$\operatorname{Re} \lambda u = 0 \tag{2.59}$$

on Γ with given Hölder continuous vector-function $\lambda(t) = (\lambda_1, \lambda_2)$.

Introducing new unknown vector function $v = (v_1, v_2)$ by

$$v = G \cdot u \tag{2.60}$$

the condition (2.57) becomes

$$\operatorname{Re} V|_{\Gamma} = 0 \tag{2.61}$$

and the equation (2.33) reduces to

$$G_{22} \frac{\partial v_1}{\partial x_1} - G_{12} \frac{\partial v_2}{\partial x_1} - 2\left(G_{21} \frac{\partial v_1}{\partial z} - G_{11} \frac{\partial v_2}{\partial z} \right) = f_1 \cdot \Delta,$$

$$G_{21} \frac{\partial v_1}{\partial x_1} - G_{11} \frac{\partial v_2}{\partial x_1} - 2\left(G_{22} \frac{\partial v_1}{\partial \bar{z}} - G_{12} \frac{\partial v_2}{\partial \bar{z}} \right) = f_2 \cdot \Delta. \tag{2.62}$$

Moreover the vector function V satisfies in Ω the inhomogeneous Laplace equation

$$\Delta V = G \cdot \partial_x' f, \tag{2.63}$$

because we have relation

$$\Delta u = \partial_x' \bar{\partial}_x u = \partial_x' f,$$

obtained from (2.23) by action ∂_x' from the left.

From (2.63) and (2.61) the real part of the vector is uniquely determined immediately as the solution of Dirichlet problem for the inhomogeneous Laplace equation

$$\Delta \operatorname{Re} v = \operatorname{Re} G \cdot \partial_x' f \tag{2.63'}$$

Thus $\operatorname{Re} v$ is uniquely determined through the right-hand side f by the formula

$$\operatorname{Re} v = \frac{1}{|s_3|} \int_\Omega G(x, \xi) \operatorname{Re} G \partial_\xi' f d\xi, \tag{2.64}$$

where $G(x, \xi)$ is the Green function of the domain Ω.

It remains to find the imaginary part of V. Conjugating we may write (2.62) as

$$-\mathrm{i}G_{21}(\operatorname{Im} v_1)_{x_2} - G_{21}(\operatorname{Im} v_1)_{x_3} + \mathrm{i}G_{11}(\operatorname{Im} v_2)_{x_2} + G_{11}(\operatorname{Im} v_2)_{x_3}$$

$$= -\mathrm{i}G_{22}(\operatorname{Im} v_1)_{x_1} + \mathrm{i}G_{12}(\operatorname{Im} v_2)_{x_1} + F_1,$$

$$\mathrm{i}\bar{G}_{21}(\operatorname{Im} v_1)_{x_2} - \bar{G}_{21}(\operatorname{Im} v_1)_{x_3} - \mathrm{i}\bar{G}_{11}(\operatorname{Im} v_2)_{x_2} + \bar{G}_{11}(\operatorname{Im} v_2)_{x_3}$$

$$= \mathrm{i}\bar{G}_{22}(\operatorname{Im} v_1)_{x_1} - \mathrm{i}\bar{G}_{12}(\operatorname{Im} v_2)_{x_1} + \bar{F}_1,$$

$$-\mathrm{i}G_{22}(\operatorname{Im} v_1)_{x_2} + G_{22}(\operatorname{Im} v_1)_{x_3} + \mathrm{i}G_{12}(\operatorname{Im} v_2)_{x_2} + G_{11}(\operatorname{Im} v_2)_{x_3}$$

$$= \mathrm{i}G_{21}(\operatorname{Im} v_1)_{x_1} + \mathrm{i}G_{11}(\operatorname{Im} v_2)_{x_3} + F_2,$$

$$\mathrm{i}\bar{G}_{22}(\operatorname{Im} v_1)_{x_2} - \bar{G}_{22}(\operatorname{Im} v_1)_{x_3} - \mathrm{i}\bar{G}_{12}(\operatorname{Im} v_2)_{x_2} + \bar{G}_{12}(\operatorname{Im} v_2)_{x_3}$$

$$= \mathrm{i}\bar{G}_{21}(\operatorname{Im} v_1)_{x_1} - \mathrm{i}\bar{G}_{11}(\operatorname{Im} v_2)_{x_1} + \bar{F}_2, \tag{2.65}$$

where

$$F_1 = G_{21}(\operatorname{Re} v_1)_{x_2} - \mathrm{i}G_{21}(\operatorname{Re} v_1)_{x_3} - G_{11}(\operatorname{Re} v_2)_{x_2} + \mathrm{i}G_{22}(\operatorname{Re} v_2)_{x_3} + f_1 \cdot \Delta,$$

$$F_2 = G_{22}(\operatorname{Re} v_1)_{x_2} - \mathrm{i}G_{22}(\operatorname{Re} v_1)_{x_3} - G_{12}(\operatorname{Re} v_2)_{x_2} + \mathrm{i}G_{12}(\operatorname{Re} v_2)_{x_3} + f_2 \cdot \Delta,$$

Considering (2.65) as an inhomogeneous algebraic system of equations with respect to $(\operatorname{Im} v_1)_{x_2}$, $(\operatorname{Im} v_1)_{x_3}$, $(\operatorname{Im} v_2)_{x_2}$, $(\operatorname{Im} v_2)_{x_3}$ we observe that its determinant is equal to

$$\det \begin{pmatrix} -\mathrm{i}G_{21} & -G_{21} & \mathrm{i}G_{11} & G_{11} \\ \mathrm{i}\bar{G}_{21} & -\bar{G}_{21} & -\mathrm{i}\bar{G}_{11} & \bar{G}_{11} \\ -\mathrm{i}G_{22} & G_{22} & \mathrm{i}G_{12} & -G_{12} \\ \mathrm{i}\bar{G}_{22} & \bar{G}_{22} & -\mathrm{i}\bar{G}_{12} & -\bar{G}_{12} \end{pmatrix} = 4\,|G_{11} \cdot \bar{G}_{22} - \bar{G}_{12}G_{21}|^2.$$

Hence assuming that

$$G^0 \equiv G_{11} \cdot \bar{G}_{22} - \bar{G}_{12}G_{21} \neq 0, \tag{2.66}$$

we obtain from (2.65) the following overdetermined system of equations with respect to the real functions $\operatorname{Im} v_1$, $\operatorname{Im} v_2$:

$$(\operatorname{Im} v_1)_{x_2} - a_1 (\operatorname{Im} v_1)_{x_1} - b_1 (\operatorname{Im} v_2)_{x_2} = F_3,$$

$$(\operatorname{Im} v_2)_{x_2} - a_2 (\operatorname{Im} v_1)_{x_1} - b_2 (\operatorname{Im} v_2)_{x_1} = F_4, \tag{2.67}$$

$$(\operatorname{Im} v_1)_{x_3} - c_1 (\operatorname{Im} v_1)_{x_1} - d_1 (\operatorname{Im} v_2)_{x_1} = F_5,$$

$$(\operatorname{Im} v_2)_{x_3} - c_2 (\operatorname{Im} v_1)_{x_1} - d_2 (\operatorname{Im} v_2)_{x_1} = F_6, \tag{2.68}$$

where

$$a_1 = \frac{\operatorname{Re}[G_{11} \cdot G_{12} (\bar{G}_{21}^2 - \bar{G}_{22}^2) - G_{21} \cdot \bar{G}_{22} (|G_{11}|^2 - |G_{12}|^2)]}{|G^0|^2},$$

$$b_1 = \frac{\operatorname{Re}[G^0 (|G_{11}|^2 + |G_{12}|^2)]}{|G^0|^2},$$

$$G_0 = G_{11} \cdot \bar{G}_{21} + \bar{G}_{12} \cdot G_{22},$$

$$a_2 = -\frac{\operatorname{Re} G^0}{|G^0|^2}, \quad c_1 = -\frac{\operatorname{Im}[G_{11} \cdot G_{12} (\bar{G}_{21}^2 + \bar{G}_{22}^2) + G_{21} \cdot \bar{G}_{22} (|G_{11}|^2 - |G_{12}|^2)]}{|G^0|^2},$$

$$d_1 = \frac{\operatorname{Im} G^0}{|G^0|^2}, \quad b_2 = \frac{\operatorname{Re} G^0 \cdot \operatorname{Re} G_0 - \operatorname{Im} G^0 \cdot \operatorname{Im} G_0}{|G^0|^2},$$

$$c_2 = -\frac{\operatorname{Im} G^0}{|G^0|^2},$$

$$d_2 = \frac{\operatorname{Im} G^0 \cdot \operatorname{Re} G_0 + \operatorname{Re} G^0 \cdot \operatorname{Im} G_0}{|G^0|^2}. \tag{2.69}$$

We may also put (2.67) and (2.68) into complex form with $w = \operatorname{Im} v_1 + \mathrm{i} \operatorname{Im} v_2$:

$$w_{x_2} - p_1 w_{x_1} - p_2 \bar{w}_{x_1} = F_3 + \mathrm{i} F_4 = F^{(1)}, \tag{2.67'}$$

$$w_{x_3} - q_1 w_{x_1} - q_2 \bar{w}_{x_1} = F_5 + \mathrm{i} F_6 = F^{(2)}, \tag{2.68'}$$

where

$$p_1 = \frac{1}{2}(a_1 + b_2 + \mathrm{i}\,(a_2 - b_1)),\ p_2 = \frac{1}{2}(a_1 - b_2 + \mathrm{i}\,(a_2 + b_1)),$$

$$q_1 = \frac{1}{2}(c_1 + d_2 + \mathrm{i}\,(c_2 - d_1)),\ q_2 = \frac{1}{2}(c_1 - d_2 + \mathrm{i}\,(c_2 + d_1)).$$

Now we observe that

$$a_2 + b_1 = 0,\quad c_2 + d_1 = 0,$$

$$a_1 - b_2 = -\frac{2}{|G^0|^2}\ \mathrm{Re}\, G^0 \cdot \mathrm{Re}\, G_0,$$

$$c_1 - d_2 = -\frac{2}{|G^0|^2}\ \mathrm{Im}\, G^0 \cdot \mathrm{Im}\, G_0,$$

$$a_1 + b_2 = -\frac{2}{|G^0|^2}\ \mathrm{Im}\, G^0 \cdot \mathrm{Re}\, G_0,$$

$$c_1 + d_2 = \frac{2}{|G^0|^2}\ \mathrm{Re}\, G^0 \cdot \mathrm{Im}\, G_0$$

and so

$$p_1 = -\frac{2}{|G^0|^2}\ (\mathrm{Im}\, G^0 \cdot \mathrm{Im}\, G_0 + \mathrm{i}\ \mathrm{Re}\, G^0),$$

$$p_2 = -\frac{2}{|G^0|^2}\ \mathrm{Re}\, G^0 \cdot \mathrm{Re}\, G_0,$$

$$q_1 = \frac{2}{|G^0|^2}\ (\mathrm{Re}\, G^0\ \mathrm{Im}\, G_0 - \mathrm{i}\ \mathrm{Im}\, G^0),$$

$$q_2 = -\frac{1}{|G^0|^2}(\mathrm{Im}\, G^0 \cdot \mathrm{Re}\, G_0). \tag{2.70}$$

Moreover, since

$$\mathrm{Re}^2 G_0 = \mathrm{Re}^2(G_{11}\cdot\bar{G}_{21}+\bar{G}_{12}G_{22}) \le |G_{11}\cdot\bar{G}_{21}+\bar{G}_{12}G_{22}|^2$$

$$= |G_{11}|^2\cdot|G_{21}|^2+|G_{12}|^2|G_{22}|^2+2\,\mathrm{Re}\,G_{11}\cdot\bar{G}_{21}G_{12}\cdot\bar{G}_{22},$$

$$1 = (|G_{11}|^2+|G_{12}|^2)(|G_{21}|^2+|G_{22}|^2) = |G_{11}|^2|G_{21}|^2+|G_{12}|^2|G_{22}|^2$$

$$+|G_{11}{}^2\cdot|G_{22}|^2+|G_{12}|^2|G_{21}|^2,$$

then

$$\mathrm{Re}^2 G_0 = (|G_{11}|^2+|G_{12}|^2)(|G_{21}|^2+|G_{22}|^2)$$

$$+2\,\mathrm{Re}\,G_{11}\cdot\bar{G}_{21}G_{12}\cdot\bar{G}_{22}-|G_{11}|^2|G_{22}|^2-|G_{12}|^2|G_{21}|^2$$

$$\le (|G_{11}|^2+|G_{12}|^2)(|G_{21}|^2+|G_{22}|^2) = 1,$$

i.e. always

$$1-\mathrm{Re}^2 G_0 \ge 0 \tag{2.71}$$

and we assume that

$$1-\mathrm{Re}^2 G_0 > 0. \tag{2.71'}$$

In this case at least one of the equation (2.67′) or (2.68′) is elliptic, because the one of the quadratic form

$$\xi_2^2-2\,\mathrm{Re}\,p_1\cdot\xi_1\xi_2+(|p_1|^2-|p_2|^2)\xi_1^2,$$

$$\xi_3^2-2\,\mathrm{Re}\,q_1\cdot\xi_1\xi_2+(|q_1|^2-|q_2|^2)\xi_1^2$$

is determined, because of

$$(\mathrm{Im}\,p_1)^2-|p_2|^2 = \frac{\mathrm{Re}^2 G_0}{|G^0|^4}(1-\mathrm{Re}^2 G_0),$$

$$(\mathrm{Im}\,q_1)^2-|q_2|^2 = \frac{\mathrm{Im}^2 G_0}{|G^0|^4}(1-\mathrm{Re}^2 G_0).$$

Introducing the new unknown function

$$\omega = (1 + \sqrt{(1 - \mathrm{Re}^2 G_0)}\, w + \mathrm{i}\,\mathrm{Re}\, G_0 \cdot \bar{w}$$

$$= (1 + \sqrt{(1 - \mathrm{Re}^2 G_0} + \mathrm{i}\,\mathrm{Re}\, G_0)\,\mathrm{Im}\, v_1 + \mathrm{i}(1 + \sqrt{(1 - \mathrm{Re}^2 G_0)} - \mathrm{i}\,\mathrm{Re}\, G_0)\,\mathrm{Im}\, v_2 \quad (2.72)$$

by means (2.67′), (2.68′) we obtain the following equations:

$$\omega_{x_2} - \lambda_2\, \omega_{x_1} = \tilde{F}^{(1)}, \quad (2.73)$$

$$\omega_{x_3} - \lambda_3\, \omega_{x_1} = \tilde{F}^{(2)}, \quad (2.74)$$

with right-hand sides expressed through the right-hand side of the original equation (2.23) and with

$$\lambda_2 = -\frac{1}{|G^0|^2}(\mathrm{Im}\, G^0 \cdot \mathrm{Im}\, G_0 + \mathrm{i}\,\mathrm{Re}\, G^0 \cdot \sqrt{(1 - \mathrm{Re}^2 G_0)}),$$

$$\lambda_3 = -\frac{1}{|G^0|^2}(\mathrm{Re}\, G^0 \cdot \mathrm{Im}\, G_0 - \mathrm{i}\,\mathrm{Im}\, G^0 \cdot \sqrt{(1 - \mathrm{Re}^2 G_0)}).$$

Since

$$\delta^* = (1 + \sqrt{(1 - \mathrm{Re}^2 G_0)})^2 - \mathrm{Re}^2 G_0 = 2(1 - \mathrm{Re}^2 G_0 + \sqrt{(1 - \mathrm{Re}^2 G_0)}) \neq 0,$$

then the linear transformation (2.72) is not degenerate and hence we obtain

$$\mathrm{Im}\, v_1 = \frac{1}{2\,\delta^*}(1 + \sqrt{(1 - \mathrm{Re}^2 G_0)} + \mathrm{i}\,\mathrm{Re}\, G_0)\omega + \frac{1}{2\,\delta^*}(1 + \sqrt{(1 + \mathrm{Re}^2 G_0)} - \mathrm{i}\,\mathrm{Re}\, G_0)\bar{\omega},$$

$$\mathrm{Im}\, v_2 = \frac{1}{2\,\mathrm{i}\,\delta^*}(1 + \sqrt{(1 - \mathrm{Re}^2 G_0)} - \mathrm{i}\,\mathrm{Re}\, G_0)\omega - \frac{1}{2\,\mathrm{i}\,\delta^*}(1 + \sqrt{(1 + \mathrm{Re}^2 G_0)} + \mathrm{i}\,\mathrm{Re}\, G_0)\omega.$$

(2.75)

The condition (2.59) in terms of function ω is the Riemann-Hilbert condition on γ:

$$\mathrm{Re}\, \lambda^* \omega = h, \quad (2.76)$$

where the right-hand side h is expressed through the right-hand side of the equation (2.28) and

$$\lambda^* = \frac{1}{\Delta}[\mathrm{i}\,(\lambda_1(t)G_{22} - \lambda_2(t)G_{21})\,(1 + \sqrt{(1 - \mathrm{Re}^2\, G_0)} + \mathrm{i}\,\mathrm{Re}\, G_0)$$

$$+ (\lambda_2(t)G_{11} - \lambda_1(t)G_{12})\,(1 + \sqrt{(1 - \mathrm{Re}^2\, G_0)} - \mathrm{i}\,\mathrm{Re}\, G_0)]$$

$$- \frac{1}{\bar{\Delta}}[\mathrm{i}\,(\overline{\lambda_1(t)}\,\bar{G}_{22} - \overline{\lambda_2(t)}\,\bar{G}_{21})\,(1 + \sqrt{(1 - \mathrm{Re}^2\, G_0)} + \mathrm{i}\,\mathrm{Re}\, G_0)$$

$$+ (\overline{\lambda_2(t)}\,\bar{G}_{11} - \overline{\lambda_1(t)}\,\bar{G}_{12})\,(1 + \sqrt{(1 - \mathrm{Re}^2\, G_0)} - \mathrm{i}\,\mathrm{Re}\, G_0)].$$

Making the linear transformation

$$\frac{\mathrm{Re}\, G^0}{|G^0|^2}\, x_2 + \frac{\mathrm{Im} G^0}{|G^0|^2}\, x_3 = \eta, \tag{2.77}$$

$$- \frac{\mathrm{Im} G^0}{|G^0|^2}\, x_2 + \frac{\mathrm{Re}\, G^0}{|G^0|^2}\, x_3 = \xi,$$

and taking into account (2.73), (2.74), we have

$$\omega_\xi = - \frac{1}{|G^0|^2}\left\{ \omega_{x_2} \frac{\mathrm{Im} G^0}{|G^0|^2} - \omega_{x_3} \cdot \frac{\mathrm{Re}\, G^0}{|G^0|^2} \right\}$$

$$= - \frac{1}{|G^0|^2}(\lambda_2\, \mathrm{Im}\, G^0 - \lambda_3\, \mathrm{Re}\, G^0)\, \omega_{x_1}$$

$$- \frac{1}{|G^0|^2}\left\{ \tilde{F}^{(1)} \frac{\mathrm{Im} G^0}{|G^0|^2} - \tilde{F}^{(2)} \frac{\mathrm{Re}\, G^0}{|G^0|^2} \right\}, \ \omega_\eta$$

$$= \frac{1}{|G^0|^2}\left\{ \omega_{x_2} \frac{\mathrm{Re}\, G^0}{|G^0|^2} + \omega_{x_3} \frac{\mathrm{Im} G^0}{|G^0|^2} \right\}$$

$$+ \frac{1}{|G^0|^2}\left\{ \tilde{F}^{(1)} \frac{\mathrm{Re}\, G^0}{|G^0|^2} + \tilde{F}^{(2)} \frac{\mathrm{Im} G^0}{|G^0|^2} \right\}.$$

But since

$$\lambda_2 \operatorname{Im} G^0 - \lambda_3 \operatorname{Re} G^0 = - \operatorname{Im} G_0,$$

$$\lambda_2 \operatorname{Re} G^0 + \lambda_3 \operatorname{Im} G^0 = - i \sqrt{(1 - \operatorname{Re}^2 G_0)},$$

then it follows that

$$\omega_\xi - \frac{\operatorname{Im} G_0}{|G^0|^4} \omega_{x_1} = \tilde{G}^{(1)}, \tag{2.78}$$

$$\omega_\eta + \frac{i\sqrt{1 - \operatorname{Re}^2 G_0}}{|G^0|^4} \omega_{x_1} = \tilde{G}^{(2)}.$$

Making a further linear transformation

$$x_1 + \frac{\operatorname{Im} G_0}{|G^0|^4} \xi = \eta', \ \xi = \xi', \tag{2.77'}$$

we obtain

$$\omega_{x_1} = \omega_{\eta'}, \quad \omega_\xi = \frac{\operatorname{Im} G_0}{|G^0|^4} \omega_{\eta'} + \omega_{\xi'}$$

and hence we obtain the following overdetermined system for $\omega(\xi', \eta, \eta')$:

$$\omega_{\xi'} = \omega_\xi - \frac{\operatorname{Im} G_0}{|G^0|^4} \omega_{x_1} = \tilde{G}^{(1)},$$

$$\omega_\eta + \frac{i\sqrt{1 - \operatorname{Re}^2 G_0}}{|G^0|^4} \omega_{\eta'} = \tilde{G}^{(3)}. \tag{2.79}$$

Integrating the first equation, we obtain

$$\omega = \Omega(\eta, \eta') + \int_0^{\xi'} \tilde{G}^{(1)} \, d\xi', \tag{2.80}$$

where $\Omega(\eta, \eta')$ is an arbitrary function, depending only on (η, η'). The second equation in (2.79) gives an inhomogeneous Cauchy-Riemann equation in the variables $\left[\sqrt{(1 - \operatorname{Re}^2 G_0)} / |G^0|^4\right] \eta + i\eta'$:

$$\Omega_\eta + \frac{i\sqrt{1-\mathrm{Re}^2 G_0}}{|G^0|^4}\Omega_{\eta'} = \tilde{G}. \tag{2.81}$$

This means that

$$\Omega = \varphi\,(\eta' - \frac{i\sqrt{1-\mathrm{Re}^2 G_0}}{|G^0|^4}\eta) + \tilde{G}_0, \tag{2.82}$$

where $\tilde{G}_0$ is expressed through the right-hand side f of (2.23) and φ is an arbitrary function, holomorphic in its argument. Now taking into account (2.80), (2.77′), (2.77), we obtain

$$\omega = \varphi(x_1 + \frac{\mathrm{Im}\, G_0}{|G^0|^4}\xi - \frac{i\sqrt{1-\mathrm{Re}^2 G_0}}{|G^0|^4}\eta) + \tilde{G}_0$$

$$= \varphi(x_1 + \lambda_2 x_2 + \lambda_3 x_3) + G_0(x_1, x_2, x_3), \tag{2.83}$$

where G_0 is expressed through the right-hand side f of (2.23) and φ is arbitrary function, holomorphic in the complex variables $x_1 + \lambda_2 x_2 + \lambda_3 x_3$.

Substituting (2.83) into the boundary condition (2.76) we obtain the Riemann-Hilbert boundary-value problem for the holomorphic function φ in the plane domain G with the boundary condition

$$\mathrm{Re}\,\lambda^* \varphi = h^* \tag{2.84}$$

on $\gamma = \partial G$.

Let $\kappa^* = \mathrm{Ind}_\gamma \bar{\lambda}^*$, where

$$\mathrm{Ind}_\gamma \lambda^* = \frac{1}{2\pi i}\int_\gamma d \log \lambda^*$$

Then, using the results on the Riemann-Hilbert problem, we derive the following

Theorem 2.4. *Let* $\det G \neq 0, G^0 \neq 0, \lambda^*(t) \neq 0$ *on* Γ, $G_0 \neq 0$, *then in case* $\kappa^* \geq 0$ *the inhomogeneous problem (2.23), (2.57), (2.59) is solvable for any right-hand side* $f \in L_p(\Omega), p > 3$, *and the corresponding homogeneous problem has exactly* $2\kappa^* + 1$ *nontrivial solutions, but in case* $\kappa^* < 0$ *the inhomogeneous problem is solvable if and only if its right-hand side satisfies* $-2\kappa^* - 1$ *conditions and the corresponding homogeneous problem has no nontrivial solutions.*

2.1.5 Application to the problem with oblique component for the stationary Maxwell equation and with oblique derivative for the inhomogeneous Laplace equation

In this section we give an immediate application of the results of preceding section. We consider the stationary Maxwell equation (2.11) in a simply-connected bounded domain Ω with smooth boundary. Since $\operatorname{div} F = 0$ follows from the second equation (2.11), then the equation (2.11) is equivalent to the inhomogeneous Moisil-Theoderescu equation:

$$\operatorname{div} u = f,$$

$$\operatorname{grad} u_0 + \operatorname{curl} u = F, \quad \operatorname{div} F = 0, \tag{2.85}$$

provided the complementary unknown function $u_0(x)$ satisfies the homogeneous Dirichlet condition

$$u_0 = 0 \tag{2.86}$$

on Γ.

Let $k = (k_1, k_2, k_3)$ be given on Γ real vector such that $k \neq 0$ on Γ.

We seek a solution of (2.11) in Ω, continuous in $\bar{\Omega}$, satisfying the condition

$$(u, k) = 0 \tag{2.87}$$

on Γ. This is an analogue of the Riemann-Hilbert problem for overdetermined stationary Maxwell (2.11).

Let us assume for $k \neq 0$ be a constant vector. Multiplying the second (fourth) equation by i and adding to the first (third) equation in (2.85) we obtain the equations (2.17) with $u = u_1 + \mathrm{i}\, u_0$, $v = u_2 + \mathrm{i}\, u_3$. The boundary conditions (2.86) and (2.87), in case $k_1 = 1$, $k_2 = k_3 = 0$, reduces to

$$u = 0$$

on Γ and so according to the assertion of Section 2.1.3 a solution of (2.11) with given $u\,|_{\Gamma} = 0$ is uniquely determined through its right-hand side, apart from an arbitrary antiholomorphic function in the plane domain $G_1 = \Omega \cap (x_1 = 0)$. Therefore in this case the well-posed problem for the system (2.11) will be the following: find in Ω a solution of (2.11), continuous in $\bar{\Omega}$, satisfying the boundary condition

$$u_1 = 0 \tag{2.88}$$

on Γ and the boundary condition

$$\alpha u_2 + \beta u_2 = 0 \tag{2.89}$$

on $\gamma_1 = \partial G_1$ with given Hölder continuous real functions such that $\alpha^2 + \beta^2 \neq 0$ on γ_1.

Theorem 2.5. *If* $\kappa = \mathrm{Ind}_\gamma (\alpha + \mathrm{i}\,\beta) \geq 0$, *then the problem (2.11), (2.88), (2.89) is solvable for any right-hand side* $(f, F) \in L_p(\Omega)$ *with* $p > 3$, div $F = 0$, *and the corresponding homogeneous problem* $(f = F = 0)$ *has exactly* $2\kappa + 1$ *nontrivial solutions, but if* $\kappa < 0$, *then the inhomogeneous problem (2.11), (2.88), (2.89) is solvable if and only if its right-hand side satisfy* $-2\kappa - 1$ *orthogonal conditions with* div $F = 0$, *and the corresponding homogeneous problem* $(f = F = 0)$ *has nontrivial solutions.*

In the general case $k_2^2 + k_3^2 \neq 0$ for constant vector k, we assume the domain Ω to be such that the intersection of Ω with the plane

$$k_2 x_2 + k_3 x_3 = 0 \tag{2.90}$$

is a simply connected plane domain G with the boundary Γ being a closed smooth curve without cross points. In this case we state the following.

Problem *M*. Find in Ω a solution of (2.11), continuous in $\bar{\Omega}$, satisfying the condition (2.87) on Γ and the condition

$$(u, \ell) = 0 \tag{2.91}$$

on Γ where $\ell = (\ell_1, \ell_2, \ell_3)$ is a Hölder continuous real vector.

We may put this problem into complex form as the problem (H) with

$$G = \begin{pmatrix} \mathrm{i} & 0 \\ k_1 & k_2 - \mathrm{i}\,k_3 \end{pmatrix}, \tag{2.92}$$

and

$$\lambda = (\ell_1, \ell_2 - \mathrm{i}\,\ell_3). \tag{2.93}$$

Since in this case $\det G = k_3 + \mathrm{i}\,k_2 \neq 0$, $G^0 = -\ell_3 + \mathrm{i}\,\ell_2 \neq 0$ and

$$\lambda^*(t) = \frac{2}{k_3 + \mathrm{i}\, k_2}((k_2 - \mathrm{i}\, k_3)\, \ell_1 - k_1\, (\ell_2 - \mathrm{i}\, \ell_3) + \ell_3 + \mathrm{i}\, \ell_2)$$

$$- \frac{2\mathrm{i}}{k_3 - \mathrm{i}\, k_2}\, ((k_2 + \mathrm{i}\, k_3)\, \ell_1 - k_1\, (\ell_2 + \mathrm{i}\, \ell_3) - \ell_2 + \mathrm{i}\, \ell_3)), \tag{2.94}$$

then by means of Theorem 2.4 we obtain, in case $\lambda^*(t) \neq 0$ on Γ, the following:

Theorem 2.6. *If* $\kappa^* = \mathrm{Ind}_\gamma \lambda^* \geq 0$, *then the inhomogeneous problem (M) is solvable for any right-hand side* $(f, F) \in L_p(\Omega)$ *with* $p > 3$, div $F = 0$ *and corresponding homogeneous problem* $(f = F = 0)$ *has exactly* $2\kappa^* + 1$ *nontrivial solutions, but if* $\kappa^* < 0$, *then the inhomogeneous problem (M) is solvable if and only if its right-hand side* (f, F) *satisfy* $-2\kappa^* - 1$ *'orthogonality', conditions for* f, F *with* div $F = 0$, *and the corresponding homogeneous problem* $(f = F = 0)$ *has no nontrivial solutions.*

Corollary 2.1. *If* k *and* ℓ *are constant vectors, then the homogeneous problem (M) has a unique nontrivial solution, equal*

$$u_1 = k_2\, \ell_3 - k_3\, \ell_2,$$

$$u_2 = k_3\, (\ell_1 - \frac{k_1}{k_2^2 + k_3^2}\, (k_2\, \ell_2 + k_3\, \ell_3) - k_1\, k_2 \cdot u_1), \tag{2.95}$$

$$u_3 = -\, k_2\, (\ell_1 - \frac{k_1}{k_2^2 + k_3^2}\, (k_2\, \ell_2 + k_3\, \ell_3) - k_1\, k_3 \cdot u_1)$$

in case $k_2^2 + k_3^2 \neq 0$ and equal to

$$u_1 = 0,\;\; u_2 = \ell_3,\;\; u_3 = -\, \ell_2 \text{ in case } k_2 = k_3 = 0. \tag{2.96}$$

Let $\varphi(x)$ be a solution in Ω of the real inhomogeneous Laplace equation

$$\varphi_{x_1 x_1} + \varphi_{x_2 x_2} + \varphi_{x_3 x_3} = f, \tag{2.97}$$

satisfying the condition

$$\frac{\partial \varphi}{\partial k} = (\mathrm{grad}\, \varphi, k) = 0 \tag{2.98}$$

on $\Gamma = \partial\Omega$, where $k = (k_1, k_2, k_3)$ a constant vector.

Let G denote the simply connected plane domain $\Omega \cap (x_1, = 0)$ in case $k_2 = k_3 = 0$ and denotes the simply connected plane domain $\Omega \cap (k_2 x_2 + k_3 x_3 = 0)$ in case $k_2^2 + k_3^2 \neq 0$ with boundary $\gamma = \partial G$ being a smooth closed curve without crossing points. We state the following:

Problem *N*. Find in Ω a solution of (2.97), continuous in $\bar{\Omega}$ with its first derivatives, satisfying the condition (2.98) on Γ and the Dirichlet condition

$$\varphi = 0 \tag{2.99}$$

on Γ.

Theorem 2.7. *The problem N is uniquely solvable for any $f \in L_p(\Omega)$ with $p > 3$.*

Proof. Introducing the functions

$$u_1 = \varphi_{x_1}, \quad u_2 = \varphi_{x_2}, \quad u_3 = \varphi_{x_3}$$

we obtain in Ω

$$\operatorname{div} u = f, \quad \operatorname{curl} u = 0 \tag{2.100}$$

and on Γ

$$(u, k) = 0. \tag{2.101}$$

It is enough to prove the uniqueness. If for instance $k_2^2 + k_3^2 \neq 0$, then a solution (2.100), (2.101) (in case $f \equiv 0$) is represented as

$$\varphi_{x_1} = \operatorname{Re} \varphi, \quad \varphi_{x_2} = -\frac{1}{k_2^2 + k_3^2}(k_1 k_2 \operatorname{Re} \varphi - k_3 \operatorname{Im} \varphi),$$

$$\varphi_{x_3} = -\frac{1}{k_2^2 + k_3^2}(k_1 k_3 \operatorname{Re} \varphi - k_3 \operatorname{Im} \varphi),$$

where φ is an arbitrary function, holomorphic in the complex variable

$$x_1 + \lambda_2 x_2 + \lambda_3 x_3 = x_1 - \frac{\kappa_1}{k_2^2 - k_3^2}\xi + \frac{i}{k_2^2 + k_3^2}\eta,$$

where

$$\xi = -k_2 x_2 - k_3 x_3, \quad \eta = -k_3 x_2 + k_2 x_3.$$

Since

$$\varphi_{x_2} = -\varphi_\xi \cdot k_2 - \varphi_\eta \cdot k_3, \quad \varphi_{x_3} = -\varphi_\xi \cdot k_3 + \varphi_\eta \cdot k_2,$$

then

$$\varphi_\xi = -\frac{k_2 \varphi_{x_2} + k_3 \varphi_{x_3}}{k_2^2 + k_3^2}, \quad \varphi_\eta = -\frac{k_3 \varphi_{x_2} - k_2 \varphi_{x_3}}{k_2^2 + k_3^2},$$

i.e.

$$\varphi_\xi = \frac{k_1}{k_2^2 + k_3^2} \operatorname{Re} \varphi, \quad \varphi_\eta = \frac{1}{k_2^2 + k_3^2} \operatorname{Im} \varphi.$$

Making the transformation

$$x_1 + \frac{1}{k_2^2 + k_3^2} \xi = \eta', \quad \xi = \xi'$$

we obtain

$$\varphi_{x_1} = \varphi_{\eta'}, \quad \varphi_\xi = \varphi_{\eta'} \frac{1}{k_2^2 + k_3^2} + \varphi_{\xi'}$$

and hence

$$\varphi_\xi + i\, k_1 \varphi_\eta = \frac{k_1}{k_2^2 + k_3^2} \Phi(\eta' - \frac{i}{k_2^2 + k_3^2}\eta),$$

i.e.

$$k_1\,(\varphi_{\eta'}\frac{1}{k^{2}+k_3^2}+\mathrm{i}\,\varphi_\eta)+\varphi_{\xi'}=\frac{k_1}{k_2^2+k_3^2}\;\Phi(\eta'-\frac{\mathrm{i}}{k_2^2+k_3^2}\eta)$$

or if we put $\tau=(k_2^2+k_3^2)\eta'$, then

$$\frac{1}{2}(\varphi_\tau+\mathrm{i}\,\varphi_\eta)=\frac{1}{2\,(k_2^2+k_3^2)}\;\Phi\left(\frac{\tau-\mathrm{i}\eta}{k_2^2+k_3^2}\right)-\frac{1}{2\,k_1}\;\varphi_{\xi'}\,.$$

But

$$\varphi_{\xi'}=\varphi_\xi-\frac{k_1}{k_2^2+k_3^2}\;\varphi_{x_1}$$

$$=\frac{k_1}{k_2^2+k_3^2}\;\mathrm{Re}\,\Phi\left(\frac{\tau-\mathrm{i}\eta}{k_2^2+k_3^2}\right)-\frac{k_1}{k_2^2+k_3^2}\;\mathrm{Re}\,\Phi\left(\frac{\tau-\mathrm{i}\eta}{k_2^2+k_3^2}\right)=0$$

and we obtain the inhomogeneous Cauchy–Riemann equation:

$$\frac{1}{2}(\varphi_\tau+\mathrm{i}\,\varphi_\eta)=\frac{1}{2\,(k_2^2+k_3^2)}\;\Phi\left(\frac{\tau-\mathrm{i}\eta}{k_2^2+k_3^2}\right).$$

Integrating this equation, we obtain:

$$\varphi=\frac{1}{k_2^2+k_3^2}\;\mathrm{Re}\,\Phi\left(\frac{\tau-\mathrm{i}\eta}{k_2^2+k_3^2}\right).$$

Thus φ is the harmonic function at the domain G expressed in two variables (τ,η) (i.e. G is the intersection of the domain Ω by plane $\xi'=0$). Since $\varphi\,|_{\partial G}=0$, then $\varphi\equiv 0$ in case $f\equiv 0$. This proves the uniqueness.

2.1.6 On the boundary-value problem for the equation curl $u+\lambda\,u=h$.

In this section we consider the same boundary-value problem (2.87), (2.89) with boundary conditions

$$(u, k)_{\Gamma} = 0 \tag{2.87}$$

on Γ and

$$(u, \ell)_{\gamma} = 0 \tag{2.91}$$

on γ, where Ω, G, Γ and γ the same domains and boundaries as in previous section, but for the system $(u = (u_1, u_2, u_3))$:

$$\operatorname{curl} u + \lambda u = h \tag{A.1}$$

with given real constant $\lambda \neq 0$ and with (given real vector function $h = (h_1, h_2, h_3)$). The equation is not elliptic, but it is equivalent to the overdetermined system (elliptic)

$$\operatorname{div} u = \lambda^{-1} \operatorname{div} h,$$

$$\operatorname{curl} u = -\lambda u + h.$$

Hence the problem (A.1), (2.87), (2.91) is equivalent to the Riemann-Hilbert type problem $\operatorname{Re} GU|_{\Gamma} = 0$, $U = (u, v)$, for the system

$$u_{x_1} + 2v_z = -\mathrm{i}\lambda \operatorname{Re} u + \lambda^{-1} \operatorname{div} h + \mathrm{i}\, h_1,$$

$$-2u_{\bar{z}} + v_{x_1} = \mathrm{i}\lambda v + h_3 + \mathrm{i}\, h_2 \tag{A.2}$$

in Ω with $u = u_1 + \mathrm{i}\, u_0$, $v = u_2 + \mathrm{i}\, u_3$ and with matrix G, having view

$$G = \begin{pmatrix} \mathrm{i} & 0 \\ k_1 & k_2 - \mathrm{i} k_3 \end{pmatrix}. \tag{2.92}$$

We investigate this problem for the case $k_2^2 + k_3^2 \neq 0$. For the case $k_2^2 + k_3^2 = 0$ it investigates analogously. From (A.1) it follows that $\Delta u + \lambda^2 u = \lambda h + \operatorname{grad} \operatorname{div} h + \operatorname{curl} h$ in Ω. Hence the function (u, k) satisfies the equation $\Delta(u, k) + \lambda^2(u, k) = h^*$ with $h^* = \lambda(h, k) + (\operatorname{grad} \operatorname{div} h, k) + (\operatorname{curl} h, k)$ in Ω and since $(u, k) = 0$ on Γ, it follows that this function is the solution of Fredholm's integral equation:

$$(u, k)(x) - \frac{\lambda^2}{4\pi} \int_{\Omega} G(x, \xi)\, (u, k)(\xi)\, d\xi = -\frac{1}{4\pi} \int_{\Omega} G(x, \xi)\, h^*(\xi)\, d\xi, \tag{A.3}$$

where $G(x, \xi)$ is a harmonic Green's function on the domain Ω. It means that the real part of the vector function $V(x) = (v_1, v_2) = GU$, where the matrix G determined by (2.92) is determined by right-hand side h, provided h satisfies a finite number of conditions, because $\operatorname{Re} v_1 = -u_0$, $\operatorname{Re} v_2 = k_1u_1 + k_2u_2 + k_3u_3 = (u, k)$.

By the same arguments as in previous sections we obtain the following overdetermined system of equations for the complex-valued function[3] $w = \operatorname{Im} v_1 + i \operatorname{Im} v_2$:

$$w_{x_2} + \frac{k_1k_2 - ik_3}{1-k_1^2} w_{x_1} + \frac{\lambda(k_1k_3 + ik_2)}{1-k_1^2} w = f^{(1)},$$

$$w_{x_3} + \frac{k_1k_3 + ik_2}{1-k_1^2} w_{x_1} - \frac{\lambda(k_1k_2 + ik_3)}{1-k_1^2} \bar{w} = f^{(2)} \tag{A.4}$$

with known right-hand sides $f^{(j)}$ such that this system is compatible.

Making the linear transformation

$$\frac{1}{1-k_1^2}(-k_3x_2 + k_2x_3) = \eta, \quad -\frac{1}{1-k_1^2}(k_2x_2 + k_3x_3) = \xi$$

and also

$$x_1 + \frac{k_1}{1-k_1^2}\xi = \eta', \quad \xi = \xi'$$

we obtain the following system in variables ξ', η, η':

$$w_{\xi'} - \frac{\lambda}{(1-k_1^2)^3}\left(k_2(k_1k_3 + ik_2)w - (k_3(k_1k_2 + ik_3)\bar{w}\right) = \tilde{f}^{(1)},$$

$$w_\eta + \frac{i}{(1-k_1^2)^2} w_{\eta'} - \frac{\lambda}{(1-k_1^2)^3}\left(k_3(k_1k_3 + ik_2)w + k_2(k_1k_2 + ik_3)\bar{w}\right) = \tilde{f}^{(2)}. \tag{A.5}$$

---

3 Without loss of generality we assume that $|k| = 1$.

Let us assume first that $(\operatorname{Im}\mu_1)^2 > |\mu_2|^2$, where

$$\mu_1 = \frac{\lambda k_2(k_1k_3 + ik_2)}{(1-k_1^2)^3}, \quad \mu_2 = -\frac{\lambda k_2(k_1k_2 + ik_3)}{(1-k_1^2)^3}.$$

In this case we introduce the function ψ by the equality

$$\psi = -i(\operatorname{Im}\mu_1 + \sqrt{((\operatorname{Im}\mu_1)^2 - |\mu_2|^2)}\, w - \mu_2\bar{w},$$

and obtain instead of the first equation in (A.5) the following:

$$\psi_{\xi'} - \lambda_0\psi = \tilde{g}, \text{ where } \lambda_0 = \operatorname{Re}\mu_1 + i\sqrt{[(\operatorname{Im}\mu_1)^2 - |\mu_2|^2]}.$$

Hence $\psi = e^{\lambda_0\xi'}\omega_0(\eta, \eta') + g_0$ with arbitrary function ω_0 of two variables η, η' and so

$$w(\xi', \eta, \eta') = \frac{i(\operatorname{Im}\mu_1 + \sqrt{[(\operatorname{Im}\mu_1)^2 - |\mu_2|^2]})\psi + \mu_2\bar{\psi}}{J}$$

$$= \frac{i(\operatorname{Im}\mu_1 + \sqrt{[(\operatorname{Im}\mu_1)^2 - |\mu_2|^2)]}}{J} e^{\lambda_0\xi'}\omega_0(\eta, \eta') \tag{A.6}$$

$$+ \frac{i\mu_2}{J} e^{\bar{\lambda}_0\xi'}\overline{\omega_0(\eta, \eta')} + g_0(\xi', \eta, \eta'),$$

where

$$J = 2\left[(\operatorname{Im}\mu_1)^2 - |\mu_2|^2 - \operatorname{Im}\mu_1\sqrt{[(\operatorname{Im}\mu_1)^2 - |\mu_2|^2]}\right].$$

From the equality, we obtain

$$\omega_0(\eta, \eta') = -i\operatorname{Im}\mu_1 + \sqrt{[(\operatorname{Im}\mu_1)^2 - |\mu_2|^2]})\,\omega(0, \eta, \eta') - i\mu_2\overline{\omega(0, \eta, \eta')}. \tag{A.7}$$

But as $\xi' = 0$ the second equation in (A.5) gives

$$\omega(0, \eta, \eta') = \varphi(\eta, \eta') + \tilde{f}^*(\eta, \eta'),$$

where $\varphi(\eta, \eta')$ is a generalized analytic function in the image of the plane domain $G = \Omega \cap (k_2x_2 + k_3x_3 = 0)$, i.e. it is a solution of the homogeneous equation

$$\varphi_\eta + \frac{i}{(1-k_1^2)^2}\varphi_{\eta'} - \nu_1\varphi - \nu_2\bar{\varphi} = 0, \tag{A.8}$$

where

$$\nu_1 = \frac{\lambda k_3(k_1k_3 + ik_2)}{(1-k_1^2)^3}, \quad \nu_2 = \frac{\lambda k_2(k_1k_2 + ik_3)}{(1-k_1^2)^3}.$$

Hence (A.6), (A.7) gives

$$\omega(\xi', \eta, \eta') = \Gamma_1(\xi')\,\varphi(\eta, \eta') + \Gamma_2(\xi')\,\overline{\varphi(\eta, \eta')} + \tilde{h}^*(\xi', \eta, \eta'), \tag{A.9}$$

where $\tilde{h}^*(\xi', \eta, \eta')$ is a known function, expressed by the right-hand side h and

$$\Gamma_1(\xi') = \frac{(\operatorname{Im}\mu_1 + \sqrt{[(\operatorname{Im}\mu_1)^2 - |\mu_2|^2]})\,e^{\lambda_0\xi'} - |\mu_2|^2\,e^{\bar{\lambda}_0\xi'}}{J},$$

$$\Gamma_2(\xi') = \frac{\mu_2(\operatorname{Im}\mu_1 + \sqrt{[(\operatorname{Im}\mu_1)^2 - |\mu_2|^2]})\,e^{\lambda_0\xi'} - \mu_2(\operatorname{Im}\mu_1 + \sqrt{[(\operatorname{Im}\mu_1)^2 - |\mu_2|^2})\,e^{\bar{\lambda}_0\xi'}}{J}.$$

From these relations we obtain as an evident consequence that

$$\Gamma_1|_{\xi'=0} = 1, \quad \Gamma_2|_{\xi'=0} = 0.$$

Hence we have proved the following:

Lemma 2.1. *Any solution of the overdetermined system (A.5) is expressed in the form (A.9) with* $\varphi(\eta, \eta')$ *being an arbitrary solution of the equation (A.7) and with known* $\Gamma_1(\xi'), \Gamma_2(\xi'), \tilde{h}^*(\xi', \eta, \eta')$ *such that* $\Gamma_1|_{\xi'=0} = 1,\ \Gamma_2|_{\xi'=0} = 0.$

We have proved this lemma in case $(\operatorname{Im}\mu_1)^2 > |\mu_2|^2$, but it holds also for the case $(\operatorname{Im}\mu_1)^2 < |\mu_2|^2$. For instance if $k_2 = 0$, $k_3 = 1$, then the first equation in (A.5) becomes

$$\omega_{\xi'} + \frac{i}{(1-k_1^2)^3}\bar{w} = \tilde{f}^{(1)}$$

and its general solution is expressed as

$$w(\xi', \eta, \eta') = \sin h \frac{\xi'}{(1-k_1^2)^3} \cdot \omega_0(\eta, \eta') + \cos h \frac{\xi'}{(1-k_1^2)^3} \overline{\omega_0(\eta, \eta')} + \tilde{f}_*^{(1)} \tag{A.10}$$

where $\Gamma_1(\xi')$ is the hyperbolic sine and $\Gamma_2(\xi')$ the hyperbolic cosine, i.e. $\Gamma_1|_{\xi'=0} = 1$, $\Gamma_2|_{\xi'=0} = 0$. From the second equation in (A.5) we derive that we may take instead $\omega_0(\eta, \eta')$ in (A.10) an arbitrary solution of the equation (A.8). The additional boundary condition, which is given on the curve $\gamma^* = \partial G^*$ (where G^* is the plane domain obtained from Ω in variables ξ', η, η' by intersecting with the plane $\xi' = 0$) $(u, \ell)_{\gamma^*} = 0$ leads to the condition $\mathrm{Re}\,(\lambda^* w)_{\gamma^*} = \tilde{h}^*$ and so by (A.9) it leads to the Riemann-Hilbert boundary-value problem:

$$\mathrm{Re}\,(\lambda^* \varphi)_{\gamma^*} = \tilde{h}^* \tag{A.11}$$

with the function λ^* determined as

$$\lambda^* = \ell_1 (k_2^2 + k_3^2) - k_1 (\ell_2 k_2 + \ell_3 k_3) - \mathrm{i}\,(\ell_2 k_3 - \ell_3 k_2)$$

for solutions of equation (A.8), i.e. for generalized analytic functions, completely investigated in [58]. Hence denoting

$$\kappa^* = \mathrm{Ind}_{\gamma^*} \overline{\lambda^*} = \frac{1}{2\pi \mathrm{i}} \int_\gamma \mathrm{d} \log \overline{\lambda^*}$$

we come to the following:

Theorem 2.8. *If* $\lambda^* \neq 0$ *then the problem (2.87), (2.91) for the (A.1) has Noetherian property with index equal to* $2\kappa^* + 1$.

Indeed, if we denote by N the number of solvability conditions of the Fredholm equation (A.3), then the real part of the vector function $V(x)$ will consist of N real constants. Consequently the right-hand side $\hat{h}^*$ of the boundary condition (A.11) consists also of N real constants. The necessary and sufficient solvability conditions for the problem (A.11) gives ℓ' inhomogeneous algebraic equations for N real constants:

$$\sum_{j=1}^{N} a_{ij} c_j = b_i \,.$$

If $r = \mathrm{rank}\,(a_{ij})$, then we have from here $\ell' - r$ conditions. Hence the number $\overset{\circ}{k}$ of the solvability conditions of the original problem is equal to $\overset{\circ}{k}' = \ell' + N + r$ and the number $\overset{\circ}{k}$ of the homogeneous of original problem is equal $\overset{\circ}{k} = \ell + N - r$, where ℓ is the number of linearly independent solutions of the problem (A.11) with $\tilde{h}^* \equiv 0$. Since it is known that $\ell - \ell' = 2\kappa^* + 1$, then it follows that $\overset{\circ}{k} - \overset{\circ}{k}' = 2\kappa^* + 1$.

In conclusion we note that in case $k_2^2 + k_3^2 = 0$ the function λ^* should be taken as $\lambda^* = \ell_2 + i\,\ell_3$.

2.2 Second-order systems

2.2.1. Conjugation problems for second-order systems

Let $\Omega^{3+}, \Omega^{3-}, \Gamma^{(3)}$ denote the same as in Section 2.1.2. We state the following problems:

Problem ($\mathbf{C}_x$). Find in $\Omega^{(3)+} \cup \Omega^{(3)-}$ a piecewise continuous complex-valued solution $u = (u_1, u_2)$ of the inhomogeneous Laplace equation:

$$\Delta u = f(x_1, z) \tag{2.102}$$

with piecewise continuous first derivatives with Laplacian in $L_p^{\text{loc}}(\Omega^{(3)+} \cup \Omega^{(3)-})$, vanishing at infinity with the first-order derivatives and satisfying on $\Gamma^{(3)}$ the boundary conditions

$$(\partial_x' u)^+ - G(\partial_x' u)^- = 0, \quad u^+ - Hu^- = 0. \tag{2.103}$$

Problem $\bar{\mathbf{C}}_x$. Find in $\Omega^{(3)+} \cap \Omega^{(3)-}$ a solution of (2.102) with the same properties as in Problem (C_x) except that they satisfy on $\Gamma^{(3)}$ instead of (2.103) the boundary conditions

$$(\bar{\partial}_x u)^+ - G(\bar{\partial}_x u)^- = 0, \quad u^+ - Hu^- = 0, \tag{2.104}$$

where G, H are given on $\Gamma^{(3)}$ nondegenerating (2×2) complex-valued matrices.

Also we state the same kind of problems in $\Omega^{4+} \cup \Omega^{4-}$:

Problem $\mathbf{C}_z$. Find in $\Omega^{4+} \cup \Omega^{4-}$ a piecewise continuous complex-valued solution $u(z) = (u_1, u_2)$ of the inhomogeneous Laplace equation

$$\Delta u = f(z) \tag{2.105}$$

with piecewise continuous first derivatives, with Laplacian in $L_p^{\text{loc}}(\Omega^{4+} \cup \Omega^{4-}), p > 4$ vanishing at infinity with first-order derivatives and satisfying on $\Gamma^{(4)}$ the boundary conditions

$$(\partial_z' u)^+ - G(\partial_z' u)^- = 0, \quad u^+ - Hu^- = 0. \tag{2.106}$$

Problem $\bar{\text{C}}_z$. Find in $\Omega^{4+} \cup \Omega^{4-}$ a solution of (2.105) with the same properties as in problem (C_z), except that they satisfy on $\Gamma^{(4)}$ instead of (2.106) the boundary conditions

$$(\partial_{\bar{z}} u)^+ - G(\partial_{\bar{z}} u)^- = 0, \; u^+ - Hu^- = 0 \tag{2.107}$$

with nondegenerate complex-valued (2×2) matrices given on $\Gamma^{(4)}$.

Theorem 2.9. *The problems* (C_x), $(\bar{\text{C}}_x)$, (C_z), $(\bar{\text{C}}_z)$ *in case constant matrices* G, H *are uniquely solvable for any right-hand side from* $L_p^{\text{loc}}(\Omega^{k+} \cup \Omega^{k-})$, $p > k$, *vanishing at infinity of the order* $1 + \varepsilon$, $\varepsilon > 0$.

Proof. We prove this theorem only for problem (C_x). For others the proof is analogous. Since

$$\Delta u = \bar{\partial}_x \partial_x' u,$$

then for the vector function $w(x_1, z)$, defined by

$$w(x_1, z) = \partial_x' u, \tag{2.108}$$

we obtain the problem (R_x)

$$\bar{\partial}_x w = f(x) \text{ in } \mathbb{R}^3 \backslash \Gamma^{(3)}, \quad w^+ - G w^- = 0 \text{ on } \Gamma^{(3)}. \tag{2.109}$$

Since w vanishes at infinity and $\det G \neq 0$, then by Theorem 2.1 the problem (2.108), (2.109) has unique solution w, which is expressed by formula (2.35) through the right-hand side f of the equation (2.105).

Now it remains to find the vector $u(x_1, z)$ from (2.108) and from the second condition (2.106). Since $\det H \neq 0$, then vector function

$$v = \begin{cases} u & \text{in } \Omega^{3+} \\ H\cdot u & \text{in } \Omega^{3-} \end{cases} \tag{2.110}$$

will be in $\mathbb{R}^3 \backslash \Gamma^{(3)}$ the solution of inhomogeneous Laplace equation

$$\Delta v = \mathcal{F} . \tag{2.111}$$

continuous on $\mathbb{R}^3$, where

$$\mathcal{F} = \begin{cases} \overline{\partial}_x w & \text{in } \Omega^{3+} \\ H\cdot \overline{\partial}_x w & \text{in } \Omega^{3-} \end{cases} \tag{2.112}$$

and hence

$$v(x_1, z) = -\frac{1}{|s_3|} \frac{\mathcal{F}(\xi_1, \zeta)\, d\xi}{\sqrt{[(\xi_1 - x_1)^2 + |\zeta - z|^2]}} \tag{2.113}$$

Taking into account the definitions (2.110), (2.112) and integrating by parts we obtain

$$u(x_1, z) = \begin{cases} \dfrac{1}{|s_3|}\displaystyle\int_{\Omega^{3+}} \overline{E^{(3)}(\xi - x)\, w(\xi)}\, d\xi + \dfrac{1}{|s_3|}\displaystyle\int_{\Omega^{3-}} \overline{E^{(3)}(\xi - x)\, w(\xi)}\, d\xi & \text{in } \Omega^{3+} \\ \dfrac{H^{-1}}{|s_3|}\displaystyle\int_{\Omega^{3+}} \overline{E'^{(3)}(\xi - x)\, w(\xi)}\, d\xi + \dfrac{1}{|s_3|}\displaystyle\int_{\Omega^{3-}} \overline{E'^{(3)}(\xi - x)\, w(\xi)}\, d\xi & \text{in } \Omega^{3-} \end{cases} \tag{2.114}$$

i.e. $u(x_1, z)$ is uniquely determined by the right-hand side of (2.105).

Considering the equation

$$\Delta u + Au_{x_1} + Bu_{x_2} + cu_{x_3} + au_{x_1} + b\bar{u} = f(x) \tag{2.115}$$

perturbed by (2×2)-matrices A, B, C from $L_p^{\mathrm{loc}}(\Omega^{3+} \cup \Omega^{3-})$ with $p > 3$, vanishing at infinity of order $1 + \varepsilon$, $\varepsilon > 0$ by means of (2.114) integrating by parts we reduce the problem (C_x) for (2.115) into Fredholm's integral equation. Thus we reach the following.

Theorem 2.10. *The problems* (C_x), $(\bar{C}_x)$, (C_z), $(\bar{C}_z)$ *for the general second-order systems of equations with Laplacian in the principal part possess Fredholm's property.*

2.2.2. Boundary-value problems in bounded domains for second-order systems

(a) Let Ω be a bounded simply connected domain in $\mathbb{R}^3$ with smooth boundary Γ. We consider in Ω the inhomogeneous Laplace equation for the complex-valued vector function $u(x_1, z) = (u_1, u_2)$

$$\Delta u = f(x_1, z) \tag{2.114}$$

with given right-hand side $f \in L_p(\Omega), p > 3$. We assume the domain Ω to be such that $\Omega \cap (x_1 = 0)$ is a simply connected plane domain G with boundary γ being a closed smooth curve without crossing points.

Problem P_1. Find in Ω a solution $u(x_1, z)$ of (114) continuous in $\bar{\Omega} = \Omega + \Gamma$, with its first derivatives satisfying the boundary conditions

$$2\frac{\partial u_1}{\partial \bar{z}} + \frac{\partial u_2}{\partial x_1} = 0, \quad u_2 = 0 \tag{2.115}$$

on Γ and the boundary conditions

$$\operatorname{Re}\left[\overline{\lambda(t)}\left(\frac{\partial u_1}{\partial x_1} - 2\frac{\partial u_2}{\partial \bar{z}}\right)\right] = 0, \quad \operatorname{Re}\left[t'(s)\lambda(t)\, u_1\right] = 0 \tag{2.116}$$

with given Hölder continuous function $\lambda(t) \neq 0$ on γ.

Problem P_2. Find in Ω a solution $u(x_1, z)$ of (2.114) continuous in $\bar{\Omega}$ with its first derivatives satisfying the boundary conditions

$$2\frac{\partial u_1}{\partial \bar{z}} + \frac{\partial u_2}{\partial x_1} = 0, \quad u_1 = 0 \tag{2.117}$$

on Γ and the boundary conditions

$$\operatorname{Re}\left[\overline{\lambda(t)}\left(\frac{\partial u_1}{\partial x_1} - 2\frac{\partial u_2}{\partial z}\right)\right] = 0, \quad \operatorname{Re}\left[\overline{t'(s)\lambda(t)}\, u_2\right] = 0. \tag{2.118}$$

Problem Q_1. Find in Ω a solution $u(x_1, z)$ of (2.114) continuous in $\bar{\Omega}$, with its first derivatives satisfying the boundary conditions

$$\frac{\partial u_1}{\partial x_1}+2\frac{\partial u_2}{\partial z}=0,\ u_2=0 \tag{2.119}$$

on Γ and the boundary conditions

$$\mathrm{Re}\left[\overline{\lambda(t)}\left(\frac{\partial u_2}{\partial x_1}-2\frac{\partial u_1}{\partial \bar{z}}\right)\right]=0,\ \mathrm{Re}\left[\lambda(t)\,t'(s)\,u_1\right]=0 \tag{2.120}$$

on γ.

Problem Q_2. Find in Ω a solution $u(x_1, z)$ of (2.114) continuous in $\bar{\Omega}$, with its first derivatives satisfying the boundary conditions

$$\frac{\partial u_1}{\partial x_1}+2\frac{\partial u_2}{\partial z}=0,\ u_1=0 \tag{2.121}$$

on Γ and the boundary conditions

$$\mathrm{Re}\left[\overline{\lambda(t)}\left(\frac{\partial u_2}{\partial x_1}-2\frac{\partial u_1}{\partial \bar{z}}\right)\right]=0,\ \mathrm{Re}\left[\overline{\lambda(t)t'(s)}\,u_2\right]=0. \tag{2.122}$$

Let us consider the problem P_1. Since $\Delta u=\bar{\partial}_x\partial_x' u$, then for the vector function $w(x_1, z)=(w_1, w_2)$ defined by

$$\partial_x' u=w(x_1, z) \tag{2.123}$$

we obtain the problem $S^{(1)}$. Hence by (2.42)

$$w_2(x_1, z)=\frac{2}{|s_3|}\int_\Omega G_\xi(x_1, z;\xi)f_1(\xi)\,d\xi$$

$$+\frac{1}{|s_3|}\int_\Omega G_{\xi_1}(x_1, z;\xi)f_2(\xi)\,d\xi$$

and by (2.48)

$$w_1(x_1, z) = \varphi(z) + K^{(1)} f, \tag{2.124}$$

where $\varphi(z)$ is an arbitrary function, holomorphic in G. If $\kappa = \mathrm{Ind}_\gamma \lambda \geq 0$, then by means of the first boundary condition (2.116) the holomorphic function determines and we obtain

$$w_1(x_1, z) = \sum_{k=1}^{2\kappa+1} a_k \varphi_k(z) + K^{(2)} f, \tag{2.125}$$

where $\{\varphi_k(z)\}$ is a complete system of nontrivial solutions of the homogeneous Riemann-Hilbert problem

$$\mathrm{Re}\, \bar{\lambda}\, \varphi = 0 \tag{2.126}$$

and $a_1,...,a_{2\kappa+1}$ are arbitrary real constants.

If $\kappa < 0$, then the Riemann-Hilbert problem

$$\mathrm{Re}\, \bar{\lambda}\, \varphi = h,$$

with right-hand side h expressed through the right-hand side f of (2.114), is solvable if and only if $-2\kappa - 1$ conditions

$$\mathrm{Re} \int_\Omega f(x)\, \overline{v_j(x)}\, dx = 0, \quad j = 1,..., -2\kappa - 1$$

are fulfilled. In this case the function φ determines uniquely through f.

From (2.123) and the second condition (2.115) we uniquely determine $u_2(x_1, z)$ by a formula like (2.42) through $w(x_1, z)$:

$$u_2(x_1, z) = \frac{2}{|s_3|} \int_\Omega G_{\bar{\zeta}}(x_1, z; \xi)\, w_1(\xi)\, d\xi + \frac{1}{|s_3|} \int_\Omega G_{\xi_1}(x_1, z; \xi)\, w_2(\xi)\, d\xi \tag{2.127}$$

and by a formula like (2.48):

$$u_1(x_1, z) = \varphi(z) + K^{(1)} w \tag{2.128}$$

where φ is a function holomorphic in G. Using the second condition (2.116) we obtain the Riemann-Hilbert problem:

$$\mathrm{Re}\, t' \lambda \varphi = \tilde{h}. \tag{2.129}$$

In case $\kappa \geq 0$ the right-hand side in (2.129) depends on constants $a_1,\dots,a_{2\kappa+1}$:

$$\tilde{h} = h^* + \sum_{j=1}^{2\kappa+1} a_j h_j^* . \tag{2.130}$$

Since the index of the problem (2.129) is equal to $\tilde{\kappa} = \mathrm{Ind}_\gamma \overline{t'(\lambda)} = -\kappa - 1 < 0$, then this problem is solvable if and only if $-2\tilde{\kappa} - 1 = 2\kappa + 1$ conditions

$$\int_\gamma \tilde{h}(t)\,\overline{\lambda(t)}\,\psi_j(t)\,dt = 0, \quad j = 1,\dots,2\kappa+1 \tag{2.131}$$

are fulfilled.

Substituting (2.130) into (2.131) we obtain for $a_1,\dots,a_{2\kappa+1}$ an inhomogeneous algebraic system of equations

$$\sum_{j=1}^{2\kappa+1} A_{ij}\, a_j = b_i$$

with $\det(A_{ij}) \neq 0$. This means that in case $\kappa \geq 0$, the problem P_1 is solvable for any right-hand side $f \in L_p(\Omega), p > 3$. If $\kappa < 0$, then the problem (2.129) is solvable for any right-hand side $\tilde{h}$ and in this case $u_1(x_1, z)$ by (2.128) depends on $-2\kappa - 1$ arbitrary constants $a_1,\dots,a_{-2\kappa-1}$. This means that in case $\kappa < 0$ the homogeneous problem P_1 is solvable if and only if $-2\kappa - 1$ conditions (2.126) are fulfilled, i.e. the index of the problem P_1 is equal to zero. The same arguments we use for other problems P_2, Q_1 and Q_2 to prove the following:

Theorem 2.11. *If $\kappa \geq 0$, then the problems P_1, P_2, Q_1, Q_2 are uniquely solvable for any right-hand side $f \in L_p(\Omega)$, $p > 3$, but if $\kappa < 0$, then these problems possess Fredholm's property.*

(b) Now we state the following boundary-value problems in the bounded domain Ω for the second-order elliptic, but not strongly elliptic, system[4]

4 The homogeneous system (2.132) has a family of solutions $u_1 = x_1 w_{x_1}, u_2 = 2x_1 w_z$ in the half-space $x_1 > 0$, satisfying the homogeneous Dirichlet condition $u|_{x_1=0} = 0$ and the inhomogeneous Dirichlet problem is solvable, provided its right-hand side $f = (f_1, f_2)$ satisfies an infinite number of conditions, so the Dirichlet problem for (2.132) is ill posed.

$$\frac{\partial^2 u_1}{\partial x_1^2} - 4\frac{\partial^2 u_1}{\partial z \partial \bar{z}} + 4\frac{\partial^2 u_2}{\partial x_1 \partial z} = f_1,$$

$$\frac{\partial^2 u_2}{\partial x_1^2} - 4\frac{\partial^2 u_2}{\partial \bar{z} \partial z} - 4\frac{\partial^2 u_1}{\partial x_1 \partial \bar{z}} = f_2 \tag{2.132}$$

Problem P^1. Find in Ω a solution of (2.132), continuous in $\bar{\Omega}$, with its first derivatives satisfying the boundary conditions

$$\frac{\partial u_2}{\partial x_1} - 2\frac{\partial u_1}{\partial \bar{z}} = 0, \quad u_2 = 0 \tag{2.133}$$

on Γ and the boundary conditions

$$\mathrm{Re}\left[\overline{\lambda(t)}\left(\frac{\partial u_1}{\partial x_1} + 2\frac{\partial u_2}{\partial z}\right)\right] = 0, \ \mathrm{Re}\left[\lambda(t)\, t'(s)\, u_1\right] = 0 \tag{2.134}$$

on γ with Hölder continuous function $\lambda(t) \neq 0$ on γ.

Problem P^2. This is the same, but instead of (2.133) the conditions

$$\frac{\partial u_2}{\partial x_1} - 2\frac{\partial u_1}{\partial \bar{z}} = 0, \quad u_1 = 0 \tag{2.135}$$

are given on Γ and instead of (2.134)

$$\mathrm{Re}\left[\overline{\lambda(t)}\left(\frac{\partial u_1}{\partial x_1} + 2\frac{\partial u_2}{\partial z}\right)\right] = 0, \ \mathrm{Re}\left[\overline{\lambda(t)}\,\overline{t'(s)}\, u_2\right] = 0 \tag{2.136}$$

are given on γ.

Problem Q^1. This is the same, but the conditions

$$\frac{\partial u_1}{\partial x_1} + 2\frac{\partial u_2}{\partial z} = 0, \quad u_2 = 0 \tag{2.137}$$

are given on Γ

$$\operatorname{Re}\Big[\overline{\lambda(t)}\Big(\frac{\partial u_2}{\partial x_1}-2\frac{\partial u_1}{\partial \bar z}\Big)\Big]=0,\ \operatorname{Re}\big[\lambda(t)\,t'(s)\,u_1\big]=0 \tag{2.138}$$

are given on Γ.

Problem Q^2. This is the same, but the conditions

$$\frac{\partial u_1}{\partial x_1}+2\frac{\partial u_2}{\partial z}=0,\ u_1=0 \tag{2.139}$$

are given on Γ and

$$\operatorname{Re}\Big[\overline{\lambda(t)}\Big(\frac{\partial u_2}{\partial x_1}-2\frac{\partial u_1}{\partial \bar z}\Big)\Big]=0,\ \operatorname{Re}\big[\overline{\lambda(t)\,t'}\,u_2\big]=0 \tag{2.140}$$

are given on γ.

Let us consider Problem P^1. The system (2.132) is

$$\bar\partial_x^2\, u=f. \tag{2.141}$$

Hence if $w(x_1,z)$ is the vector function defined by

$$\bar\partial_x\, u=w, \tag{2.142}$$

then by the first condition in (2.133) and the first condition in (2.134) we have Problem $P_1 : \bar\partial_x w = f$ in Ω and

$$w_2=0 \tag{2.143}$$

on Γ and (2.144)

$$\operatorname{Re}\bar\lambda\, w_1=0 \tag{2.145}$$

on γ.

Hence in case $\kappa=\operatorname{Ind}_\gamma\lambda\ge 0$, w is determined by the right-hand side f apart from $2\kappa+1$ nontrivial solutions of the homogeneous problem and in case $\kappa<0$, the problem (2.143)-(2.145) is solvable if and only if its right-hand side satisfies $-2\kappa-1$ conditions. The second conditions in (2.133) and (2.134) together with equation (2.142) again amounts to Problem P_1, but with negative index in case $\kappa > 0$ and with nonnegative index in case $\kappa \leq 0$. In the first case we obtain $2\kappa + 1$ complementary conditions, which gives inhomogeneous algebraic systems for $2\kappa + 1$ constants, arising from problem (2.143)-(2.145) in case $\kappa \geq 0$ and appearing in the right-hand side of (2.142). From this system we uniquely determine all constants through the right-hand side f as functionals, and thus in this case Problem P^1 will be uniquely solvable for any right-hand side $f \in L_p(\Omega)$, $p > 3$. If $\kappa < 0$, then the problem (2.142), with $u_2 \mid_\Gamma = 0$, $\operatorname{Re}[\lambda\, t' u_1]_\gamma = 0$, will be solvable for any right-hand side w in (2.142), and the solution u of this problem will be expressed through w apart from $-2\kappa - 1$ solutions of the homogeneous problem. Since in this case the problem (2.143)-(2.145) is solvable if and only if $-2\kappa - 1$ condition are fulfilled, then it follows that the index of the problem P^1 is equal to zero. We use the same arguments to prove the following:

Theorem 2.12. *If* $\kappa \geq 0$, *then the problems* P^1, P^2, Q^1, Q^2 *are uniquely solvable for any right-hand side* $f \in L_p(\Omega), p > 3$, but if $\kappa > 0$, *then these problems possess Fredholm's property.*

Now it is not too difficult to extend these results for systems with perturbed lower parts.

Some well-posed problems for second-order real systems including the inhomogeneous Laplace equation and non-strongly elliptic equations such as

$$\operatorname{grad}\operatorname{div} u + \operatorname{curl}\operatorname{curl} u = f$$

we stated and investigated in [25].

3 Degenerate elliptic problems

3.1 Introduction

(a) As we saw in Chapter I any first-order elliptic system of two real equations may be put into complex form

$$aw_{\bar{z}} + cw_z + b\bar{w}_z + d\bar{w}_{\bar{z}} + a_0 w + b_0 \bar{w} = f \tag{3.1}$$

with ellipticity condition:

$$4|a\bar{c} - \bar{b}d|^2 - (|a|^2 + |c|^2 - |b|^2 - |d|^2)^2 > 0. \tag{3.2}$$

We consider (3.1) in the whole complex plane $\mathbb{C}$ without contour $\Gamma = \Gamma_0 + \Gamma_1 + \cdots + \Gamma_m$, which divides this plane into bounded multiple connected domain G^+ and its exterior G^-. If the coefficients in G^+ and G^- satisfy the condition

$$||a|^2 - |b|^2| > |a\bar{c} - \bar{b}d| + |a\bar{d} - \bar{b}c|,$$

then we may eliminate $\bar{w}_z$ and write (3.1) as

$$w_{\bar{z}} - q_1 w_z - q_2 \overline{w_z} + a_1 w + b_1 \bar{w} = f_1 \tag{3.3}$$

with

$$|q_1| + |q_2| < 1$$

on $G^+ \cup G^-$. In this case, if q_1, q_2 are measurable bounded functions and a_1, a_2 are $L_{p,2}(\mathbb{C})$-functions with $p > 2$, then equation (3.3) has a unique solution continuous on the whole plane for any $f \in L_{p,2}(\mathbb{C})$, $p > 2$.

But if the coefficients of (3.1) satisfy the condition

$$||a|^2 - |b|^2| > |a\bar{c} - \bar{b}d| + |a\bar{d} - \bar{b}c| \text{ in } G^+$$

$$||d|^2 - |c|^2| > |a\bar{c} - \bar{b}d| + |a\bar{d} - \bar{b}c| \text{ in } G^-,$$

then we will have (3.3) in G^+ and

$$w_z - p_1 w_{\bar z} - p_2 \overline{w_{\bar z}} + a_2 w + b_2 \bar w = f_2 \tag{3.4}$$

with

$$|p_1| + |p_2| < 1$$

in G^- and so equation (3.1) in general is not solvable in the class of functions continuous on the whole plane $\mathbb{C}$. Indeed, the equation

$$Lw \equiv (1 + \operatorname{sign}(1 - |z|))\, w_{\bar z} + (1 - \operatorname{sign}(1 - |z|))\, w_z = f \tag{3.5}$$

is elliptic inside and outside the disc $|z| < 1$ and has no solution continuous on the plane $\mathbb{C}$ for arbitrary right-hand side $f \in L_{p,2}(\mathbb{C})$, $p > 2$: the necessary and sufficient conditions for (3.5) to have solution, continuous on $\mathbb{C}$ are (see [18]).

$$\iint\limits_{|z|<1} z^k f(z)\, dx\, dy + \iint\limits_{|z|>1} \frac{f(z)}{\bar z^{k+2}}\, dx\, dy = 0, \quad = 0,1,\ldots, \tag{3.6}$$

i.e. the dimension of the cokernel of the operator (3.5) is infinite. Also we see that the functions $w_k(z)$, $k = 0,1,\ldots,$

$$w_k(z) = \begin{cases} z^k, & |z| < 1, \\ \bar z^{-k}, & |z| > 1 \end{cases} \tag{3.7}$$

are continuous on the whole plane $\mathbb{C}$ and satisfy the homogeneous equation (3.5) ($f \equiv 0$), i.e. the kernel of the operator (3.5) is also equal infinite. Of course equation (3.5) has coefficients, discontinuous on the circle $|z| = 1$. But for the equation with continuous coefficients on the finite part of $\mathbb{C}$

$$w_x + i\, x^m w_y = f \tag{3.8}$$

the dimension of the kernel and cokernel is also equal infinite for odd m (for even m (3.8) reduces to the Cauchy-Riemann equation and so has unique solution continuous on $\mathbb{C}$).

Another equation with coefficient continuous on the finite part of $\mathbb{C}$ is the following Cauchy-Riemann equation perturbed by the lower part:

$$w_{\bar z} - \lambda z w = f. \tag{3.9}$$

If $\operatorname{Re}\lambda > 0$, then the necessary and sufficient conditions for (3.9) to be solvable are

$$\iint_{\mathbb{C}} f(z)\, e^{-\bar{\lambda}|z|^2} z^k \, dx\, dy = 0, \quad k = 0, 1, \ldots,$$

i.e. codim = ∞.

These and other examples which arise further produce ill-posed problems with dim, codim = ∞. All such problems are called degenerate problems. Our main aim is to describe those subspaces defined by given information or to introduce additional terms to the problem to make them well posed.

Let us consider the conditions (3.6) more closely. Note that the functions

$$v_k(z) = \begin{cases} \bar{z}^k, & |z| < 1, \\ z^{-k}, & |z| > 1 \end{cases} \tag{3.10}$$

are solutions of the homogeneous equation

$$L^* v \equiv -(1 - \operatorname{sign}(1 - |z|))\, v_z - (1 + \operatorname{sign}(1 - |z|))\, v_{\bar{z}} = 0$$

adjoint to (3.5) (in the sense of Hilbert scalar product $(f, g) = \iint_{\mathbb{C}} f \bar{g}\, dx\, dy$), satisfying the boundary conditions

$$v^+ - t^2 v^- = 0 \text{ on } |t| = 1.$$

Take an arbitrary function $f_0(z)$ from the Hilbert space $L_2(\mathbb{C})$ and consider the function

$$f(z) = f_0(z) - \sum_{\ell=0}^{\infty} a_\ell\, v_\ell(z) \tag{3.11}$$

with $v_\ell(z)$ as in (3.10). The function (3.11) will satisfy the conditions (3.6) if we choose constants a_ℓ such that

$$\iint_{|z|<1} f_0(z) z^k \, dx\, dy + \iint_{|z|>1} \frac{f_0(z)\, dx\, dy}{\bar{z}^{k+2}} - \sum_{\ell=0}^{\infty} a_\ell \iint_{|z|<1} z^k \bar{z}^\ell \, dx\, dy$$

$$- \sum_{\ell=0}^{\infty} a_\ell \iint_{|z|>1} \frac{dx\, dy}{\bar{z}^{k+2} z^{\ell+2}} = 0,$$

i.e.

$$a_k = \frac{k+1}{2\pi} \iint_{\mathbb{C}} f_0(z)\, \overline{v_k(z)}\, dx\, dy, \tag{3.12}$$

because

$$\iint_{|z|<1} z^k \bar{z}^{\ell}\, dx\, dy = \begin{cases} \dfrac{\pi}{k+1}, & \ell = k, \\ 0, & \ell \neq k \end{cases}$$

$$\iint_{|z|>1} \frac{dx\, dy}{\bar{z}^{k+2} z^{\ell+2}} = \begin{cases} \dfrac{\pi}{k+1}, & \ell = k, \\ 0, & \ell \neq k \end{cases}.$$

Hence the functions $f(z)$ from $L_2(\mathbb{C})$ for which the equation (3.5) is solvable are expressed as

$$f(z) = f_0(z) - \iint_{\mathbb{C}} K_0(\zeta, z) f_0(\zeta)\, d\xi\, d\eta, \quad \forall f_0 \in L_2(\mathbb{C}), \tag{3.13}$$

where

$$K_0(\zeta, z) = \frac{1}{2\pi} \sum_{\ell=0}^{\infty} (\ell+1)\, \overline{v_\ell(\zeta)}\, v_\ell(z). \tag{3.14}$$

Note that

$$\iint_{\mathbb{C}} K_0(\zeta, z) f_0(\zeta)\, d\xi\, d\eta = \frac{1}{2\pi} \iint_{|\zeta|<1} \sum_{\ell=1}^{\infty} \ell \zeta^{\ell-1} f_0(\zeta)\, d\xi\, d\eta \cdot v_{\ell-1}(z)$$

$$+ \frac{1}{2\pi} \iint_{|\zeta|>1} \sum_{\ell=1}^{\infty} \frac{\ell f_0(\zeta)}{\bar{\zeta}^{\ell+1}}\, d\xi\, d\eta\, v_{\ell-1}(z),$$

i.e. for $|z| < 1$:

$$\iint_{\mathbb{C}} K_0(\zeta, z) f_0(\zeta)\, d\xi\, d\eta = \frac{1}{2\pi} \iint_{|\zeta|<1} \sum_{\ell=1}^{\infty} \ell(\zeta \bar{z})^{\ell-1} f_0(\zeta)\, d\xi\, d\eta$$

$$+\frac{1}{2\pi}\iint_{|\zeta|>1}\sum_{\ell=1}^{\infty}\frac{\ell\bar{z}^{\ell-1}}{\bar{\zeta}^{\ell+1}}f_0(\zeta)\,d\xi\,d\eta=\frac{1}{2\pi}\iint_{|\zeta|<1}\frac{f_0(\zeta)\,d\xi\,d\eta}{(1-\zeta\bar{z})^2}$$

$$+\frac{1}{2\pi}\iint_{|\zeta|>1}\frac{f_0(\zeta)\,d\xi\,d\eta}{(\bar{\zeta}-\bar{z})^2} \tag{3.15}$$

and for $|z|>1$:

$$\iint_{\mathbb{C}}K_0(\zeta,z)f_0(\zeta)\,d\xi\,d\eta=\frac{1}{2\pi}\iint_{|\zeta|<1}\sum_{\ell=1}^{\infty}\frac{\ell\zeta^{\ell-1}}{z^{\ell+1}}f_0(\zeta)\,d\xi\,d\eta$$

$$+\frac{1}{2\pi}\iint_{|\zeta|>1}\sum_{\ell=1}^{\infty}\frac{\ell f_0(\zeta)\,d\xi\,d\eta}{(\bar{\zeta}z)^{\ell+1}}$$

$$=\frac{1}{2\pi}\iint_{|\zeta|<1}\frac{f_0(\zeta)\,d\xi\,d\eta}{(\zeta-z)^2}+\frac{1}{2\pi}\iint_{|\zeta|>1}\frac{f_0(\zeta)\,d\xi\,d\eta}{(1-\bar{\zeta}z)^2}. \tag{3.16}$$

Let us denote by H_{L^*} the subspace of $L_2(\mathbb{C})$ consisting all piecewise continuous solutions of the homogeneous adjoint equation $L^*v=0$ on $\mathbb{C}\setminus\Gamma$ satisfying the jump conditions

$$v^+-t^2v^-=0$$

on the circle $|z|=1$.

Then (3.15), (3.16) implies that the operator

$$K_0f\equiv\iint_{\mathbb{C}}K_0(\zeta,z)f_0(\zeta)\,d\xi\,d\eta \tag{3.17}$$

is the Hilbert space $L_2(\mathbb{C})$ orthogonal projection onto H_{L^*}, because the right-hand sides in (3.15) and (3.16) are antiholomorphic inside and holomorphic outside of the disc, and moreover, for $|t|=1$, $(K_0f)^+=t^2(K_0f)^-$ and the operator K_0 reproduces H_{L^*}:

$$K_0f\equiv f,\ \forall f\in H_{L^*}.$$

To prove the last assertion we note that any function $f(z)$ from H_{L^*} takes the form

$$f(z)=\begin{cases}\sum_{k=0}^{\infty} c_k \bar{z}^{-k}, & |z|<1,\\ \sum_{k=0}^{\infty} \dfrac{c_k}{z^{k+2}}, & |z|>1\end{cases}$$

then for $|z|<1$:

$$\frac{1}{2\pi}\iint_{|\zeta|<1}\frac{f(\zeta)\,d\xi\,d\eta}{(1-\zeta\bar{z})^2}=\frac{1}{2}f(z),$$

$$\frac{1}{2\pi}\iint_{|\zeta|>1}\frac{f(\zeta)\,d\xi\,d\eta}{(\bar{\zeta}-\bar{z})^2}=-\frac{1}{2\pi}\iint_{|\zeta|>1}\frac{\partial}{\partial\bar{\zeta}}\left(\frac{f(\zeta)}{\bar{\zeta}-\bar{z}}\right)d\xi\,d\eta$$

$$=-\frac{1}{4\pi i}\int_{|t|=1}\frac{f(t)\,dt}{\bar{t}-\bar{z}}=\frac{1}{4\pi i}\int_{|t|=1}\frac{f(t)\,d\bar{t}}{\bar{t}^2(\bar{t}-\bar{z})}$$

$$=\sum_{k=0}^{\infty}\frac{c_k}{4\pi i}\int_{|t|=1}\frac{\bar{t}^k\,d\bar{t}}{\bar{t}-\bar{z}}=\frac{1}{2}\sum_{k=0}^{\infty}c_k\bar{z}^k=\frac{1}{2}f(z).$$

Hence $K_0 f\equiv f$ for $|z|<1$.

If $|z|>1$, then

$$\frac{1}{2\pi}\iint_{|\zeta|<1}\frac{f(\zeta)\,d\xi\,d\eta}{(\zeta-z)^2}=-\frac{1}{2\pi}\iint_{|\zeta|>1}\frac{\partial}{\partial\zeta}\left(\frac{f(\zeta)}{\zeta-z}\right)d\xi\,d\eta$$

$$=\frac{1}{4\pi i}\int_{|t|=1}\frac{f(t)\,d\bar{t}}{t-z}=\frac{1}{2}\sum_{k=0}^{\infty}\frac{c_k}{2\pi i}\int_{|t|=1}\frac{dt}{t^{k+2}(t-z)}$$

$$=\frac{1}{2}\sum_{k=0}^{\infty}\frac{c_k}{z^{k+2}}=\frac{1}{2}f(z),$$

$$\frac{1}{2\pi}\iint_{|\zeta|>1}\frac{f(\zeta)\,d\xi\,d\eta}{(1-\bar{\zeta}z)^2}=\frac{1}{2\pi}\iint_{|\zeta|>1}\frac{\partial}{\partial\bar{\zeta}}\left(\frac{\bar{\zeta}f(\zeta)}{1-\bar{\zeta}z}\right)d\xi\,d\eta$$

$$=\frac{1}{4\pi i}\int_{|t|=1}\frac{\bar{t}f(t)\,dt}{1-\bar{t}z}=\frac{1}{4\pi i}\int_{|t|=1}\frac{f(t)\,dt}{t-z}=\frac{1}{2}f(z),$$

i.e. $K_0 f \equiv f$.

By H. Weyl's orthogonal decomposition lemma $L_2(\mathbb{C}) = H_{L^*} \oplus H_{L^*}^{\perp}$ where $H_{L^*}^{\perp}$ is the orthogonal complement of the subspace H_{L^*}. Thus we have proved the following.

Lemma 3.1. *The equation (3.5) has solution continuous on* $\mathbb{C}$ *if and only if its right-hand side* f *belongs to subspace* $H_{L^*}^{\perp}$.

(b) Let us fix that the general equation (3.1) satisfies the condition

$$||a|^2 - |b|^2| > |a\bar{c} - \bar{b}d| + |a\bar{d} - \bar{b}c| \tag{i}$$

in G^+ and the condition

$$||d|^2 - |c|^2| > |a\bar{c} - \bar{b}d| + |a\bar{d} - \bar{b}c| \tag{ii}$$

in G^-. Then we introduce in G^- a new unknown function $v(z)$ by

$$v(z) = (a\bar{d} - \bar{b}c)w + (|d|^2 - |c|^2)\bar{w}. \tag{3.18}$$

If we assume for functions a, b, c, d to be $H_p^1(\bar{G}^-)$ functions with $p > 2$, then we obtain the following equation in G^-:

$$c_0 v_{\bar{z}} + \alpha_0 v_z + \beta_0 \overline{v_z} + \alpha_3 v + \beta_3 \bar{v} = f_3, \tag{3.19}$$

where

$$c_0(z) = (|d|^2 - |c|^2|^2 - |a\bar{d} - \bar{b}c|^2)^2 - |a\bar{c} - \bar{b}d|^2 |a\bar{d} - \bar{b}c|^2,$$

$$\beta_0(z) = (|d|^2 - |c|^2)|a\bar{c} - \bar{b}d|^2 (a\bar{d} - \bar{b}c).$$

Since $|c_0(z)| > |\alpha_0(z)| + |\beta_0(z)|$, $z \in G^-$, it follows that equation (3.19) is like equation (3.2), but since we are seeking a solution $w(z)$ of the equation (3.1) continuous on the whole plane and in particular $w^+ = w^-$ on Γ, vanishing at infinity, then by (3.18) we obtain on Γ the jump condition

$$\varphi^+ - (a\bar{d} - \bar{b}c)\varphi^- + (|d|^2 - |c|^2)\bar{\varphi}^- = 0 \tag{3.20}$$

for a piecewise continuous function φ defined by

$$\varphi(z) = \begin{cases} w(z), & z \in G^+, \\ v(z), & z \in G^-. \end{cases} \tag{3.21}$$

Since $|a\bar{d} - \bar{b}c| < ||d|^2 - |c|^2|$ in G^- it is clear that this problem has the Noetherian property if and only if

$$a(t)\,\overline{d(t)} - \overline{b(t)}\,c(t) \neq 0 \tag{3.22}$$

on Γ. In this case the index of the problem (3.20) for piecewise continuous solutions of the equation

$$\varphi_{\bar{z}} - q_1\varphi_z - q_2\overline{\varphi_z} + A\varphi + B\bar{\varphi} = F \tag{3.23}$$

with bounded measurable functions such that

$$|q_1| + |q_2| < 1 \tag{3.24}$$

and A, B, F from $L_{p,2}(\mathbb{C})$, $p > 2$ is equal to

$$2\,\mathrm{Ind}_\Gamma(a\bar{d} - \bar{b}c) = \frac{1}{\pi i}\int_\Gamma \mathrm{d}\log(a\bar{d} - \bar{b}c). \tag{3.25}$$

Thus we have proved the following:

Theorem 3.1. *If in G^+ equation (3.1) satisfies condition (i) and in G^- this equation satisfies condition (ii), then in case condition (3.22) on Γ is fulfilled the problem of finding a solution of (3.1) continuous on the whole plane has the Noetherian property with the index equal to* $2\,\mathrm{Ind}_\Gamma(a\bar{d} - \bar{b}c)$.

Thus involving the additional condition (3.22) for coefficients we obtained a well-posed problem to find continuous solutions of the equation (3.1) belonging to different homotopy classes (i) and (ii).

3.2 The Cauchy problem for the inhomogeneous Cauchy - Riemann equation and its higher-order generalization

In this section we consider the Cauchy problems for elliptic systems which are as it turns out also degenerate elliptic problems.

As above let G^+ be the multiple connected bounded domain in G^+. First we consider in G^+ the inhomogeneous Cauchy–Riemann equation

$$w_{\bar{z}} = f(z) \tag{3.26}$$

with $f(z) \in L_2(G^+)$.

The Cauchy problem is to find a solution of (3.26), continuous in $\bar{G}^+$, satisfying the condition

$$w = 0 \tag{3.27}$$

on Γ.

It is clear that this problem has no solution for any right-hand side $f \in L_2(G^+)$, where L_2 is a Hilbert space with the scalar product

$$(f, g) = \iint_{G^+} f\bar{g}\, \mathrm{d}x\, \mathrm{d}y. \tag{3.28}$$

We also denote by $H(G^+)$ the subspace consisting of all functions from $L_2(G^+)$ that are holomorphic in G^+, and by $\bar{H}(G^+)$ the subspace of functions antiholomorphic in G^+.

Lemma 3.2. *The Cauchy problem (3.26), (3.27) is solvable if and only if its right-hand side f is orthogonal to the subspace $\bar{H}(G^+)$.*

Proof. If $f \in L_2(G^+)$, then from (3.26) follows that

$$w(z) = \varphi + T_{G^+}f = \varphi(z) - \frac{1}{\pi} \iint_{G^+} \frac{f(\zeta)\,\mathrm{d}\xi\,\mathrm{d}\eta}{\zeta - z}, \tag{3.29}$$

where $\varphi \in H(G^+)$. Since for $t \in \Gamma$, $\varphi^+(t)$ is the boundary value of a function holomorphic in G^+ and continuous in $\bar{G}^+$, it follows from Cauchy's theorem that

$$\int_\Gamma \frac{\varphi^+(t)\,\mathrm{d}t}{t - z} = 0 \tag{3.30}$$

for $z \in G^-$. Now by condition (3.28) and Green's formula it follows from (3.29), (3.30) that

$$\iint_{G^+} \frac{f(\zeta)\, d\xi\, d\eta}{\zeta - z} = 0 \tag{3.31}$$

for any $z \in G^-$. But if $\zeta \in G^+$, $z \notin \bar{G}^+$ the function $(\zeta - z)^{-1}$ is an element of the subspace $H(G)$, so that it can be represented by the Fourier series

$$(\zeta - z)^{-1} = \sum_{k=1}^{\infty} a_k \omega_k(z), \tag{3.32}$$

where $\{\omega_k(z)\rangle$ is a complete orthonormal system of holomorphic functions in G^+ and a_k are Fourier coefficients of $(\zeta - z)^{-1}$:

$$a_k = ((\zeta - z)^{-1}, \omega_k) = \iint_{G^+} \frac{\overline{\omega_k(\zeta)}\, d\xi\, d\eta}{\zeta - z}.$$

Hence by (3.32), (3.31) holds if and only if

$$\iint_{G^+} f(\zeta)\, \omega_k(\zeta)\, d\xi\, d\eta = 0, \quad k = 1, 2, \ldots . \tag{3.33}$$

Corollary 3.1. *The Cauchy problem (3.27), (3.28) is solvable if and only if its right-hand side f has the form*

$$f(z) = f_0(z) - \iint_{G^+} K(\bar{z}, \zeta)\, f_0(z)\, d\xi\, d\eta$$

for any $f_0 \in L_2(G^+)$, where $K(z, \bar{\zeta})$ is Bergman's kernel function.

Proof. Take an arbitrary element $f_0 \in L_2(G^+)$ and put

$$f(z) = f_0(z) - \sum_{k=1}^{\infty} c_k \overline{\omega_k(z)}.$$

Then (3.33) is automatically satisfied, provided constants c_k are chosen:

$$c_k = \iint_{G^+} f_0(\zeta)\, \omega_k(\zeta)\, d\xi\, d\eta.$$

Hence the Cauchy problem (3.37), (3.38) is solvable iff

$$f(z) = f_0(z) - \iint_{G^+} K(\bar{z}, \zeta) f_0(\zeta)\, d\xi\, d\eta, \tag{3.34}$$

where $K(z, \bar{\zeta})$ is the well-known Bergman's kernel function of the domain G^+:

$$K(z, \bar{\zeta}) = \sum_{k=1}^{\infty} \omega_k(z)\, \overline{\omega_k(z)}. \tag{3.35}$$

Now it is evident that f defined by (3.34) belongs to $\bar{H}^{\perp}(G^+)$, because the second term on the right-hand side is an orthogonal projection of $L_2(G^+)$ onto $\bar{H}(G^+)$.

Among the properties of $K(z, \bar{\zeta})$ we note that it is a holomorphic in $z \in G^+$ for fixed $\zeta \in G^+$ and antiholomorphic in ζ for fixed $z \in G^+$. If $g(z, \zeta)$ is the harmonic Green's function of the domain G^+, then

$$K(z, \bar{\zeta}) = -\frac{2}{\pi} \frac{\partial^2 g(z, \zeta)}{\partial z\, \partial \bar{\zeta}}. \tag{3.36}$$

In [24] the generalization $K_n(z, \bar{\zeta})$ of this function has been given such that $K_n(z, \bar{\zeta})$ is polyanalytic in z, i.e.

$$\frac{\partial^n K_n(z, \bar{\zeta})}{\partial \bar{z}^n} = 0$$

for fixed $\zeta \in G^+$ and antipolyanalytic in ζ, i.e.

$$\frac{\partial^n K_n(z, \bar{\zeta})}{\partial \zeta^n} = 0$$

for fixed $z \in G^+$. This function $K_n(z, \bar{\zeta})$ is expressed, through the polyharmonic Green's function $g_n(z, \zeta)$, by

$$K_n(z, \zeta) = \frac{(-1)^n 2^{2n-1}}{\pi} \frac{\partial^{2n} g_n(z, \zeta)}{\partial z^n\, \partial \bar{\zeta}^n}, \tag{3.37}$$

where

$$g_n(z,\zeta) = \frac{|\zeta - z|^{2(n-1)}}{(2^{n-1}(n-1)!)^2}\log\frac{1}{|\zeta - z|} + h_n(z,\zeta), \tag{3.38}$$

$h_n(z,\zeta)$ is a regular polyharmonic function with boundary values $(\zeta \in \Gamma, z \in G^+)$:

$$\frac{\partial^j h_n(z,\zeta)}{\partial \nu_\zeta^j} = -\frac{\partial^j \omega_n(z,\zeta)}{\partial \nu_\zeta^j},\ j = 0,\dots,n-1,$$

$$\omega_n(z,\zeta) = \frac{|\zeta - z|^{2(n-1)}}{(2^{n-1}(n-1)!)^2}\log\frac{1}{|\zeta - z|},$$

ν_ζ denotes the unit outward normal.

Using $K_n(z,\bar{\zeta})$ we may give an immediate generalization of Lemma 3.2 and Corollary 3.1 for any positive integer n:

Lemma 3.3. *The Cauchy problem*

$$\frac{\partial^n w}{\partial \bar{z}^n} = f(z) \tag{3.39}$$

in G^+,

$$\frac{\partial^j w}{\partial \nu^j} = 0,\ j = 0,\dots,n-1 \tag{3.40}$$

on Γ, *is solvable if and only if its right-hand side* f *has the form*

$$f(z) = f_0(z) - \iint_{G^+} K_n(\bar{z},\zeta) f_0(\zeta)\, d\xi\, d\eta,\quad \forall f_0 \in L_2(G^+). \tag{3.41}$$

Corollary 3.2. *The Cauchy problem (3.39), (3.40) is solvable if and only if its right-hand side* f *is orthogonal to a subspace of the Hilbert space* $L_2(G^+)$, *consisting of antipolyanalytic functions in* G^+.

Proof. Suppose first that $f(z) \in C^n(G^+)$. Then differentiating (3.30) n times with respect to z we obtain the inhomogeneous polyharmonic equation

$$\frac{\partial^{2n} w}{\partial z^n \partial \bar{z}^n} = \frac{\partial^n f}{\partial z^n}. \tag{3.42}$$

Hence by means of conditions (3.40) we may represent $w(z)$ as the solution of the Dirichlet's problem by the formula

$$w(z) = -\frac{2^{2n-1}}{\pi} \iint_{G^+} g_n(z,\zeta) \frac{\partial^n f}{\partial \zeta^n} \, d\xi \, d\eta. \tag{3.43}$$

Since

$$\frac{\partial^j g_n(z,\zeta)}{\partial \nu^j} = 0, \quad j = 0,\dots, n-1 \tag{3.44}$$

for $\zeta \in \Gamma$, $z \in G^+$, then integrating by parts n times in the right-hand side of (3.43), we obtain

$$w(z) = \frac{(-1)^{n-1} 2^{2n-1}}{\pi} \iint_{G^+} \frac{\partial^n g_n(z,\zeta)}{\partial \zeta^n} f(\zeta) \, d\xi \, d\eta, \tag{3.45}$$

because from boundary conditions (3.44) and also from the identities

$$\frac{\partial^j g_n(z,\zeta)}{\partial s^j} = 0, \quad j = 0,\dots,n-1$$

for $\zeta \in \Gamma$, $z \in G^+$ following from (3.44) by differentiation with respect to the arc parameter s, we can derive that

$$\frac{\partial^j g_n(z,\zeta)}{\partial \zeta^j} = 0, \; j = 0,\dots, n-1$$

for $\zeta \in \Gamma$, $z \in G^+$. Taking into account (3.38) we rewrite (3.45) as

$$w(z) = \frac{(-i)^n}{\pi(n-1)!} \iint_{G^+} \frac{(\bar{\zeta}-\bar{z})^{n-1}}{\zeta - z} f(\zeta)\, d\xi\, d\eta$$

$$+ \frac{(-i)^{n+1} 2^{2\nu-1}}{\pi} \iint_{G^+} \frac{\partial^n h_n(z,\zeta)}{\partial \zeta^n} f(\zeta)\, d\xi\, d\eta. \tag{3.46}$$

Now we consider this expression for any $f \in L_2(G^+)$. Since (3.46) obtained form (3.43) satisfies the conditions (3.40), then the necessary and sufficient condition for (3.46) to satisfy the equation $\partial^n w/\partial \bar{z}^n = \tilde{f}$ is for $\tilde{f}$ to be represented as

$$\tilde{f}(z) = f(z) - \frac{(-i)^n 2^{2n-1}}{\pi} \iint_{G^+} \frac{\partial^{2n} h_n(z,\zeta)}{\partial \bar{z}^n \partial \zeta^n} f(\zeta)\, d\xi\, d\eta.$$

But since

$$\frac{(-i)^n 2^{2n-1}}{\pi} \frac{\partial^{2n} h_n(z,\zeta)}{\partial \bar{z}^n \partial \zeta^n} = \frac{(-i)^n 2^{2n-1}}{\pi} \frac{\partial^{2n} g_n(z,\zeta)}{\partial \bar{z}^n \partial \zeta^n} = K_n(\bar{z},\zeta),$$

then

$$\tilde{f}(z) = f(z) - \iint_{G^+} K_n(\bar{z},\zeta) f(\zeta)\, d\xi\, d\eta,$$

so we have proved Lemma 3.3.

Let $\varphi \in \bar{H}_n(G^+)$. Then

$$(\tilde{f},\varphi) = \iint_{G^+} \tilde{f}(z)\overline{\varphi(z)}\, dx\, dy = \iint_{G^+} f(z)\overline{\varphi(z)}\, dx\, dy - \iint_{G^+}\Big(\iint_{G^+} K_n(\bar{z},\zeta) f(\zeta)\, dG_\zeta\Big)\overline{\varphi(z)}\, dG_z.$$

Changing the order of integration we obtain

$$\iint_{G^+}\Big(\iint_{G^+} K_n(\bar{z},\zeta) f(\zeta)\, dG_\zeta\Big)\overline{\varphi(z)}\, dG_z = \iint_{G^+}\Big(\iint_{G^+} K_n(\bar{z},\zeta)\, \overline{\varphi(z)}\, dG_z\Big) f(\zeta)\, dG_\zeta.$$

Using Hermitian symmetry $\overline{K_n(\bar{z},\zeta)} = K_n(\bar{\zeta}, z)$ and reproducing property (see [24]):

$$\iint_{G^+} K_n(\bar{\zeta}, z)\,\overline{\varphi(z)}\,dG_z = \overline{\varphi(\zeta)},\ \forall\,\varphi \in \bar{H}_n(G^+),$$

we obtain $(\tilde{f}, \varphi) = 0$. Hence by orthogonal decomposition of the Hilbert space $L_2(G^+) = \bar{H}_n(G^+) \oplus \bar{H}_n^{\perp}(G^+)$ we have also proved Corollary 3.2.

3.3 Kernel matrices in $\mathbb{R}^3$ and $\mathbb{R}^4$

(a) Let Ω be a bounded domain (or a collection of finite disjoint bounded domains) in $\mathbb{R}^3$ and $\Gamma = \partial\Omega$ the boundary of Ω, which we assume to be sufficiently smooth. For two complex-valued vectors $u = (u_1, u_2)$, $v = (v_1, v_2)$ from $c^1(\Omega)$ we have an identity

$$v\bar{\partial}_x u = -u\bar{\partial}_x' v + (v_1 u_1)_{x_1} + 2(v_1 u_2)_z - 2(v_2 u_1)_{\bar{z}}$$

$$+ (v_2 u_2)_{x_1},$$

integration of which over Ω by means of Stoke's lemma gives

$$\int_\Omega v\bar{\partial}_x u\,dx = -\int_\Omega u\bar{\partial}_x' v\,dx + \int_\Gamma v(\xi)\,\nu(\xi)\,u(\xi)\,d\xi, \tag{3.47}$$

where

$$\nu(\xi) = \begin{pmatrix} \nu_1 & \nu_2 - i\,\nu_3 \\ -(\nu_2 + i\,\nu_3) & \nu_1 \end{pmatrix}. \tag{3.48}$$

Let $\Omega_\varepsilon = \Omega \setminus (|\xi - x| < \varepsilon)$. Since vectors

$$-\frac{1}{|\xi - x|^3}(\xi_1 - x_1, -(\bar{\zeta} - \bar{z})),\quad -\frac{1}{|\xi - x|^3}(\zeta - z, \xi_1 - x_1)$$

satisfy in ξ the equation $\bar{\partial}_\xi' v = 0$, then applying (3.47) to domain Ω_ε we obtain

$$\int_{\Omega_\varepsilon} E(\xi - x)\,\bar{\partial}_\xi u\,d\xi = \int_\Gamma E(\xi - x)\,\nu(\xi)\,u(\xi)\,d\Gamma - \int_{\Gamma_\varepsilon} E(\xi - x)\,\nu(\xi)\,u(\xi)\,d\Gamma_\varepsilon, \tag{3.49}$$

where Γ_ε is a sphere $|\xi - x| = \varepsilon$ and

$$E(x) = \frac{1}{|x|^3}\begin{pmatrix} x_1 & -\bar{z} \\ z & x_1 \end{pmatrix}.$$

Since on $\Gamma_\varepsilon = d\Gamma_\varepsilon = \varepsilon^2\,d\theta$,

$$\nu_1 = -\frac{\xi_1 - x_1}{\varepsilon}, \quad \nu_2 + i\,\nu_3 = -\frac{\zeta - z}{\varepsilon},$$

then

$$\int_{\Omega_\varepsilon} E(\xi - x)\,\nu(\xi)\,u(\xi)\,d\Gamma_\varepsilon$$

$$= \frac{1}{\varepsilon^2}\int_{|\theta|=1}\begin{pmatrix} \xi_1 - x_1 & -(\bar{\zeta} - \bar{z}) \\ \zeta - z & \xi_1 - x_1 \end{pmatrix}\begin{pmatrix} \xi_1 - x_1 & \bar{\zeta} - \bar{z} \\ \zeta - z & \xi_1 - x_1 \end{pmatrix}_u (x + \varepsilon\theta)\,d\theta$$

$$= \int_{|\theta|=1} u(x + \varepsilon\theta)\,d\theta \underset{\varepsilon \to 0}{\to} |s_3| \cdot u(x).$$

Hence as $\varepsilon \to 0$ from (3.49) we obtain the following formula for the vector function $u(x)$ from $C^1(\Omega)$:

$$u(x) = \frac{1}{|s_3|}\int_\Gamma E(\xi - x)\,\nu(\xi)\,u(\xi)\,d\Gamma_\xi - \frac{1}{|s_1|}\int_\Omega E(\xi - x)\,\bar{\partial}_\xi u\,d\xi. \tag{3.50}$$

Let $G(x, \xi)$ be Green's function of the domain Ω, i.e.

$$G(x, \xi) = \frac{1}{|x - \xi|} + h(x, \xi), \tag{3.51}$$

where $h(x, \xi)$ is a regular harmonic function in Ω, having $-\dfrac{1}{|x-\xi|}$ as boundary value:

$$h(x, \xi) = -\frac{1}{|x-\xi|} \tag{3.52}$$

for $\xi \in \Gamma$, $x \in \Omega$.

We introduce the following (2×2) matrix:

$$K(x, \xi) = \frac{1}{|s_3|} \partial_x' \bar{\partial}_\xi G(x, \xi), \tag{3.53}$$

$$\bar{\partial}_\xi = \begin{pmatrix} \dfrac{\partial}{\partial \xi_1} & 2\dfrac{\partial}{\partial \zeta} \\ -2\dfrac{\partial}{\partial \bar{\zeta}} & \dfrac{\partial}{\partial \xi_1} \end{pmatrix}.$$

Since

$$\bar{\partial}_\xi \frac{1}{|x-\xi|} = -\frac{1}{|x-\xi|} \begin{pmatrix} \xi_1 - x_1 & \bar{\zeta} - \bar{z} \\ -(\zeta - z) & \xi_1 - x_1 \end{pmatrix} = \overline{E'(\xi - x)}$$

and

$$\partial_x' \overline{E'(\xi - x)} = 0$$

for $\xi - x \neq 0$, then it follows that

$$K(x, \xi) = \frac{1}{|s_3|} \partial_x' \bar{\partial}_\xi h(x, \xi), \tag{3.54}$$

where $h(x, \xi)$ is the regular part of Green's function, i.e.

$$K(x, \xi) = \frac{1}{|s_3|} (K_{ij}(x, \xi)),$$

$$K_{11}(x, \xi) = h_{x_1 \xi_1} + 4h_{z\bar{\zeta}}, \quad K_{12}(x, \xi) = 2h_{x_1 \zeta} - 2h_{z\xi_1},$$

$$K_{21}(x,\xi) = 2h_{\bar{z}\xi_1} - 2h_{x_1\bar{\zeta}} = -\overline{K_{12}(x,\xi)},$$

$$K_{22}(x,\xi) = 4h_{\bar{z}\zeta} + h_{x_1\xi_1} = \overline{K_{11}(x,\xi)}.$$

Evidently $K(x,\xi)$ has no singularity for x, ξ inner points of Ω. Moreover

$$\frac{\partial K_{11}}{\partial x_1} + 2\frac{\partial K_{21}}{\partial z} = (h_{x_1x_1} + 4h_{z\bar{z}})_{x_1} = 0,$$

$$-2\frac{\partial K_{11}}{\partial \bar{z}} + \frac{\partial K_{21}}{\partial x_1} = -2(4h_{\bar{z}z} + h_{x_1x_1})_{\bar{\zeta}} = 0,$$

$$\frac{\partial K_{12}}{\partial x_1} + 2\frac{\partial K_{22}}{\partial z} = \overline{2\frac{\partial K_{11}}{\partial \bar{z}} - \frac{\partial K_{21}}{\partial x_1}} = 0,$$

$$-2\frac{\partial K_{12}}{\partial \bar{z}} + \frac{\partial K_{22}}{\partial x_1} = \overline{\frac{\partial K_{11}}{\partial x_1} + 2\frac{\partial K_{21}}{\partial z}} = 0,$$

i.e. $K(x,\xi) \in H(\Omega)$ for x, ξ being inner points of Ω, where $H(\Omega)$ is a closed subspace of the Hilbert space[1] $L_2(\Omega)$, consisting a solution of the equation

$$\bar{\partial}_x u = 0. \tag{3.55}$$

Hence the integral operator

$$Kf \equiv \int_\Omega K(x,\xi) f(\xi)\, d\xi \tag{3.56}$$

projects the Hilbert space $L_2(\Omega)$ onto the subspace $H(\Omega)$. Let $f = (f_1, f_2) \in H(\Omega)$, i.e.

$$\frac{\partial f_1}{\partial x_1} + 2\frac{\partial f_2}{\partial z} = 0, \quad -2\frac{\partial f_1}{\partial \bar{z}} + \frac{\partial f_2}{\partial x_1} = 0 \tag{3.57}$$

in Ω. Then we have

[1] Here $L_2(\Omega)$ is the space of complex-valued vectors $u = (u_1, u_2)$, $v = (v_1, v_2)$ in Ω with inner product $(u,v) = \int_\Omega (u_1 \bar{v}_1 + u_2 \bar{v}_2)\, d\Omega$.

$$K_{11}(x,\xi)f_1(\xi)+K_{12}(x,\xi)f_2(\xi)=(G_{x_1\xi_1}+4G_{z\bar{\zeta}})f_1(\xi)+(2G_{x_1\xi}-2G_{z\zeta_1})$$

$$f_2(\xi)=\frac{\partial}{\partial\xi_1}(G_{x_1}f_1)+2\frac{\partial}{\partial\zeta}(G_{x_1}f_2)-G_{x_1}\Big(\frac{\partial f_1}{\partial\xi_1}+2\frac{\partial f_2}{\partial\zeta}\Big)$$

$$+4\frac{\partial}{\partial\bar{\zeta}}(G_zf_1)-2\frac{\partial}{\partial\zeta_1}(G_zf_2)-2G_z\Big(-2\frac{\partial f_1}{\partial\bar{\zeta}}+\frac{\partial f_2}{\partial\xi_1}\Big)$$

$$=\frac{\partial}{\partial\xi_1}(G_{x_1}f_1-2G_zf_2)+2\frac{\partial}{\partial\zeta}G_{x_1}f_2+4\frac{\partial}{\partial\bar{\zeta}}(G_zf_1).$$

Analogously

$$K_{21}(x,\xi)f_1(\xi)+K_{22}(x,\xi)f_2(\xi)=2\frac{\partial}{\partial\xi_1}(G_{\bar{z}}f_1)+4\frac{\partial}{\partial\bar{\zeta}}G_{\bar{z}}f_2$$

$$-2G_{\bar{z}}\Big(\frac{\partial f_1}{\partial\xi_1}+2\frac{\partial f_2}{\partial\zeta}\Big)-2\frac{\partial}{\partial\bar{\zeta}}(G_{x_1}f_1)+\frac{\partial}{\partial\xi_1}(G_{x_1}f_2)+G_{x_1}\Big(2\frac{\partial f_1}{\partial\bar{\zeta}}-\frac{\partial f_2}{\partial\xi_1}\Big)$$

$$=\frac{\partial}{\partial\xi_1}(2G_{\bar{z}}f_1+G_{x_2}f_2)+4\frac{\partial}{\partial\zeta}(G_{\bar{z}}f_2)-2\frac{\partial}{\partial\bar{\zeta}}(G_{x_1}f_1).$$

Now by Stoke's lemma we obtain

$$\int_{\Omega_\varepsilon}\big[K_{11}(x,\xi)f_1(\xi)+K_{12}(x,\xi)f_2(\xi)\big]d\xi=\int_\Gamma\Big\{\big[G_{x_1}\nu_1+2G_z(\nu_2+i\nu_3)\big]f_1(\xi)$$

$$+\big[G_{x_1}(\nu_2-i\nu_3)-2G_z\nu_1\big]f_2(\xi)\Big\}d\Gamma-\int_{\Gamma_\varepsilon}\Big\{\big[G_{x_1}\nu_1+2G_z(\nu_2+i\nu_3)\big]f_1(\xi)$$

$$+\big[G_{x_1}(\nu_2-i\nu_3)-2G_z\nu_1\big]f_2(\xi)\Big\}d\Gamma_\varepsilon=-\int_{\Gamma_\varepsilon}\Big\{\quad\Big\}d\Gamma_\varepsilon, \tag{3.58}$$

because of $G(x,\xi)=0$ for $\xi\in\Gamma$, $x\in\Omega$. Since, on the sphere Ω_ε

$$G_{x_1}\nu_1+2G_z(\nu_2+i\nu_3)=-\frac{(\xi_1-x_1)^2}{\varepsilon^4}-\frac{|\zeta-z|^2}{\varepsilon^4}=-\frac{1}{\varepsilon^2},$$

$$G_{x_1}(\nu_2 - i\nu_3) - 2G_z \cdot \nu_1 = -\frac{(\xi_1 - x_1)(\bar{\zeta} - \bar{z})}{\varepsilon^4} + \frac{(\bar{\zeta} - \bar{z})(\xi_1 - x_1)}{\varepsilon^4} = 0,$$

then as $\varepsilon \to 0$ from (3.58) we obtain

$$\int_\Omega \left[K_{11}(x,\xi) f_1(\xi) + K_{12}(x,\xi) f_2(\xi) \right] d\xi = f_1(x). \tag{3.59}$$

analogously from

$$\int_{\Omega_\varepsilon} \left[K_{21}(x,\xi) f_1(\xi) + K_{22}(x,\xi) f_2(\xi) \right] d\xi =$$

$$-\int_{\Gamma_\varepsilon} \left\{ \left[2G_{\bar{z}} \nu_1 - G_{x_1}(\nu_2 + i\nu_3) \right] f_1(\xi) + \left[2\,G_{\bar{z}}\,(\nu_2 - i\nu_3) + G_{x_1}\,\nu_1 \right] f_2(\xi) \right\} d\Gamma_\varepsilon$$

taking into account that on Γ_ε:

$$2G_{\bar{z}}\,\nu_1 - G_{x_1}(\nu_2 + i\nu_3) = -\frac{(\zeta - z)(\xi_1 - x_1)}{\varepsilon^4} + \frac{(\xi_1 - x_1)(\zeta - z)}{\varepsilon^4} = 0,$$

$$2G_{\bar{z}}(\nu_2 - i\nu_3) + G_{x_1}\,\nu_1 = -\frac{|\zeta - z|^2}{\varepsilon^4} - \frac{(\xi_1 - x_1)^2}{\varepsilon^4} = -\frac{1}{\varepsilon^2},$$

we obtain as $\varepsilon \to 0$:

$$\int_\Omega \left[K_{21}(x,\xi) f_1(\xi) + K_{22}(x,\xi) f_2(\xi) \right] d\xi = f_2(x). \tag{3.60}$$

Thus we have proved

Lemma 3.4. *If $f \in H(\Omega)$, then $Kf \equiv f$, i.e. the operator (3.56) is an orthogonal Hilbert space projection in $L_2(\Omega)$.*

Corollary 3.3. *The matrix $K(x,\xi)$ has strong singularity at the boundary Γ of domain Ω.*

Indeed, Lemma 3.4 asserts that the integral equation $f - Kf = 0$ cannot be Fredholm's equation because it has infinitely many nontrivial solutions, so its kernel has a strong singularity. Since $K(x, \xi)$ has no singularity inside Ω, this means that the singularity of $K(x, \xi)$ lies on the boundary Γ.

Since

$$K_{11}(\xi, x) = h_{\xi_1 x_1} + 4h_{\zeta\bar{z}} = \overline{K_{11}(x, \xi)}\,,$$

$$K_{12}(\xi, x) = 2h_{\xi_1\bar{z}} - 2h_{\zeta x_1} = -K_{12}(x, \xi)$$

then it follows that

$$K(\xi, x) = \frac{1}{|s_3|}\begin{pmatrix} K_{11}(\xi, x) & K_{12}(\xi, x) \\ -K_{12}(\xi, x) & K_{11}(\xi, x) \end{pmatrix} = \overline{\frac{1}{|s_3|}\begin{pmatrix} K_{11}(x, \zeta) & K_{12}(x, \xi) \\ \overline{-K_{12}(x, \xi)} & \overline{K_{11}(x, \xi)} \end{pmatrix}'} = \overline{K'(x, \xi)},$$

i.e. $K(x, \xi)$ has Hermitian symmetry. Therefore if we take $f(\xi) = K(\xi, x)$ for fixed $x \in \Omega$, then by Lemma 3.4,

$$\int_\Omega K(x, \xi)\, K(\xi, x)\, d\xi = K(x, x),$$

because $K(\xi, x) \in H(\Omega)$ in ξ for fixed $x \in \Omega$. But since $K(\xi, x) = \overline{K'(x, \xi)}$. then

$$K(x, \xi)\, K(\xi, x) = K(x, \xi)\, \overline{K'(x, \xi)}$$

$$= \begin{pmatrix} K_{11}(x, \xi) & K_{12}(x, \xi) \\ \overline{-K_{12}(x, \xi)} & \overline{K_{11}(x, \xi)} \end{pmatrix} \begin{pmatrix} \overline{K_{11}(x, \xi)} & -K_{12}(x, \xi) \\ \overline{K_{12}(x, \xi)} & K_{11}(x, \xi) \end{pmatrix}$$

$$= \begin{pmatrix} |K_{11}(x, \xi)|^2 + |K_{12}(x, \xi)|^2 & 0 \\ 0 & |K_{12}(x, \xi)|^2 + |K_{11}(x, \xi)|^2 \end{pmatrix}.$$

Hence

$$K(x, x) = \int_\Omega \begin{pmatrix} |K_{11}(x, \xi)|^2 + |K_{12}(x, \xi)|^2 & 0 \\ 0 & |K_{12}(x, \xi)|^2 + |K_{11}(x, \xi)|^2 \end{pmatrix} d\xi$$

i.e.

$$K_{12}(x,x)=0,\quad K_{11}(x,x)=K_{22}(x,x)=\int_\Omega (|K_{11}(x,\xi)|^2+|K_{12}(x,\xi)|^2\,d\xi \quad (3.61)$$

Now from

$$\varphi(x)=\int_\Omega K(x,\xi)\,\varphi(\xi)\,d\xi,\quad \forall\varphi\in H(\Omega)$$

by means of (3.61) and Schwartz's inequality we obtain

$$|\varphi_k(x)|^2\le K_{11}(x,x)\;\|\varphi\|^2_{L_2(\Omega)},\quad k=1,2.$$

Hence we have proved the following:

Corollary 3.4. *Among all vectors from* $H(\Omega)\subset L_2(\Omega)$, *which are equal to (1.0) at the fixed point* $x\in\Omega$, *the minimal possible norm* $(K_{11}(x,x)^{-1/2}$ *has the vector* $(K_{11}(x,x))^{-1}\,(K_{11}(\xi,x),-\overline{K_{12}(\xi,x)})$ *and among all vectors from* $H(\Omega)$, *which are equal to (0.1) at the fixed point* $x\in\Omega$, *the minimal possible norm* $(K_{11}(x,x))^{-1}$ *has the vector* $(K_{11}(x,x))^{-1}\,(K_{12}(\xi,x),\overline{K_{11}(\xi,x)}$.

As we see the matrix $K(x,\xi)$ possesses all the properties of Bergman's kernel. There is also another kernel matrix,

$$\overset{\circ}{K}(x,\xi)=\frac{1}{|s_3|}\,\bar\partial_x\partial'_\xi\,G(x,\xi)=(\overset{\circ}{K}_{ij}),$$

$$\overset{\circ}{K}_{11}(x,\xi)=h_{x_1\xi_1}+4h_{z\bar\zeta},\quad \overset{\circ}{K}_{12}(x,\xi)=-2h_{x_1\zeta}+2h_{z\xi_1},$$

$$\overset{\circ}{K}_{21}(x,\xi)=-2h_{\bar z\xi_1}+2h_{\bar\zeta x_1}=-\overset{\circ}{K}_{12}(x,\xi),$$

$$\overset{\circ}{K}_{22}(x,\xi)=4h_{\bar z\zeta}+h_{x_1\xi_1}=\overline{\overset{\circ}{K}_{11}(x,\xi)}.$$

The integral operator

$$\overset{\circ}{K}f=\int_\Omega \overset{\circ}{K}(x,\xi)\,f(\xi)\,d\xi$$

projects the Hilbert space $L_2(\Omega)$ onto the subspace $\bar H'(\Omega)$ of all elements from $L_2(\Omega)$, which satisfy the equation

$$\partial'_x u=0. \quad (3.62)$$

It is easy to verify that $\overset{\circ}{K}(x,\xi)$ possesses Hermitian symmetry $\overset{\circ}{K}'(\xi,x)=\overline{\overset{\circ}{K}(x,\xi)}$

and that $\overset{\circ}{K}'(x, \xi)$ is the solution of equation $\bar{\partial}'_\xi \overset{\circ}{K}' = 0$ in ξ when x is fixed in Ω, and $\overset{\circ}{K}(x, \xi)$ is the solution of equation $\partial'_x \overset{\circ}{K} = 0$ in x when ξ is fixed in Ω.

(b) We introduce also the matrix

$$L(x, \xi) = \frac{1}{|s_3|} \partial'_x \partial'_\xi G(x, \xi) = \frac{1}{|s_3|} (L_{ij}(x, \xi)). \tag{3.63}$$

Evidently

$$L(x, \xi) = \frac{6J(\xi - x)}{|s_3||\xi - x|^5} - \ell(x, \xi), \tag{3.63'}$$

where

$$J(\xi - x) = \begin{pmatrix} 0 & (\bar{\zeta} - \bar{z})(\xi_1 - x_1) \\ -(\zeta - z)(\xi_1 - x_1) & 0 \end{pmatrix} \tag{3.64}$$

$$\ell(x, \xi) = -\frac{1}{|s_3|} \begin{pmatrix} h_{x_1\xi_1} - 4h_{z\bar{\zeta}} & -2h_{x_1\zeta} - 2h_{z\xi_1} \\ 2\bar{h}_{z\xi_1} + 2\bar{h}_{\zeta x_1} & -4\bar{h}_{z\zeta} + h_{x_1\xi_1} \end{pmatrix}. \tag{3.65}$$

The matrix $\ell(x, \xi)$ is the solution of the equation $\bar{\partial}_x u = 0$ in x when ξ is fixed in Ω, as follows from the harmonicity of $h(x, \xi)$.

Let

$$\overset{\circ}{L}(x, \xi) = \frac{1}{|s_3|} \partial_x \partial_\xi G(x, \xi)$$

$$= \frac{6J'(\xi - x)}{|s_3||\xi - x|^5} - \ell'(\xi, x) \tag{3.66}$$

and consider the following integral for $f \in L_2(\Omega)$:

$$\int_\Omega \overset{\circ}{L}(x, \xi) f(\xi)\, d\xi.$$

If $f \in H(\Omega)$, i.e.

$$\frac{\partial f_1}{\partial \xi_1} + 2\frac{\partial f_2}{\partial \zeta} = 0, \ -2\frac{\partial f_1}{\partial \bar{\zeta}} + \frac{\partial f_1}{\partial \xi_1} = 0,$$

then

$$(G_{x_1\xi_1} - 4G_{z\bar{\zeta}})f_1(\xi) + 2(G_{x_1\zeta} + G_{z\xi_1})f_2(\xi) = (G_{x_1}f_1 + 2G_z f_2)_{\xi_1}$$

$$+ 2(G_{x_1}f_2)_\zeta - 4(G_z f_1)_{\bar{\zeta}} - G_{x_1}\left(\frac{\partial f_1}{\partial \xi_1} + 2\frac{\partial f_2}{\partial \zeta}\right)$$

$$- 2G_z\left(-2\frac{\partial f_1}{\partial \bar{\zeta}} + \frac{\partial f_2}{\partial \xi_1}\right),$$

$$-2(G_{\bar{z}\xi_1} + G_{x_1\bar{\zeta}})f_1(\xi) + (-4G_{\bar{z}\zeta} + G_{x_1\xi_1})f_2(\xi)$$

$$= (-2G_{\bar{z}}f_1 + G_{x_1}f_2)_{\xi_1}$$

$$-4(G_{\bar{z}}f_2)_\zeta - 2(G_{x_1}f_1)_{\bar{\zeta}} + 2G_{\bar{z}}\left(\frac{\partial f_1}{\partial \xi_1} + 2\frac{\partial f_2}{\partial \zeta}\right)$$

$$+ G_{x_1}\left(2\frac{\partial f_1}{\partial \bar{\zeta}} - \frac{\partial f_2}{\partial \xi_1}\right).$$

Hence, if we take into account that $G(x, \xi) = 0$ for $\xi \in \Gamma$, $x \in \Omega$, then in the usual way we obtain

$$\int_{\Omega_\varepsilon} \mathring{L}(x, \xi) f(\xi)\, d\xi = -\int_{\Gamma_\varepsilon} \Big\{\big[G_{x_1}\nu_1 + G_z \nu_1\big]f_1(\xi) + \big[G_{x_1}(\nu_2 - i\nu_3)$$

$$- G_z(\nu_2 + i\nu_3)\big]f_2(\xi)\Big\}\, d\Gamma_\varepsilon$$

$$+ \Big\{\big[-2G_{\bar{z}}\nu_1 - G_{x_1}(\nu_2 + i\nu_3)\big]f_1(\xi) + \big[-2G_{\bar{z}}(\nu_2 + i\nu_3) + G_{x_1}\nu_1\big]f_2(\xi)\Big]\, d\Gamma_\varepsilon$$

and so as $\varepsilon \to 0$ we obtain

$$\int_\Omega \overline{\mathring{L}(x, \xi)} f(\xi)\, d\xi = 0, \ \forall f \in H(\Omega).$$

It means that

$$\frac{6}{|s_3|}\int_\Omega \frac{J'(\xi-x)}{|\xi-x|^5} f(\xi)\,d\xi = \int_\Omega \overline{\ell'(\xi,x)} f(\xi)\,d\xi, \quad \forall f\in H(\Omega).$$

Taking here $f(\xi) = K(\xi,\eta) = \overline{K'(\eta,\xi)}$, we have

$$\frac{6}{|s_3|}\int_\Omega \frac{J'(\xi-x)\overline{K'(\eta,\xi)}}{|\xi-x|^5}\,d\xi = \int_\Omega \overline{\ell'(\xi,x)}\;\overline{K'(\eta,\xi)}\,d\xi$$

or, conjugating and transposing,

$$\frac{6}{|s_3|}\int_\Omega \frac{K(\eta,\xi)\overline{J(\xi-x)}}{|\xi-x|^5}\,d\xi = \int_\Omega K(\eta,\xi)\,\ell(\xi,x)\,d\xi.$$

Since $\ell(\xi,x)$ is the solution of the equation $\bar{\partial}_\xi\,\ell(\xi,x) = 0$ when x fixed in Ω, then by the reproducing property,

$$\int_\Omega K(\eta,\xi)\,\ell(\xi,x)\,d\xi = \ell'(\eta,x)$$

i.e.

$$\frac{6}{|s_3|}\int_\Omega \frac{K(\eta,\xi)J(\xi-x)}{|\xi-x|^5}\,d\xi = \ell(\eta,x),$$

or, transposing, we have

$$\ell'(\eta,x) = \frac{6}{|s_3|}\int_\Omega \frac{\overline{J'(\xi-x)}\,K'(\eta,\xi)}{|\xi-x|^5}\,d\xi.$$

Taking into account (3.66) we obtain the following representation of the matrix $\overset{\circ}{L}(x,\xi)$ through the kernel matrix $K(x,\xi)$:

$$\overset{\circ}{L}(x,\xi) = \frac{6J'(\xi-x)}{|s_3||\xi-x|^5} - \frac{6}{|s_3|}\int_\Omega \frac{\overline{J'(\eta-x)}\,K'(\xi,\eta)}{|\eta-x|^5}\,d\eta. \tag{3.67}$$

Now we try to derive the representation for the kernel matrix. For this we note that since $G(x, \xi) = 0$ for $\xi \in \Gamma$, $x \in \Omega$, its 'tangential derivative' also vanishes for $\xi \in \Gamma$, $x \in \Omega$: $[\text{grad}_\xi\, G(x, \xi) \times \nu] = 0$, i.e.

$$G_{\xi_2}\nu_3 - G_{\xi_3}\nu_2 = 0,\quad G_{\xi_3}\nu_1 - G_{\xi_1}\nu_3 = 0,\quad G_{\xi_1}\nu_2 - G_{\xi_2}\nu_1 = 0,$$

or

$$G_{\xi_1}(\nu_2 + \mathrm{i}\,\nu_3) - 2\,G_{\bar\zeta}\nu_1 = 0,\quad G_{\bar\zeta}(\nu_2 - \mathrm{i}\,\nu_3) - G_{\zeta}(\nu_2 + \mathrm{i}\,\nu_3) = 0. \tag{3.68}$$

Since $K(\eta, x)$ is an element of $H(\Omega)$ in η for fixed $x \in \Omega$, then from (3.50) we obtain

$$K'(x, \eta) = \overline{K(\eta, x)} = \frac{1}{|s_3|}\,\overline{\int_\Gamma E(\xi - \eta)\,\nu(\xi)\,K(\xi, x)\,d\Gamma_\xi}\,. \tag{3.69}$$

Now

$$\nu(\xi)\,K(\xi, x) = \begin{pmatrix} k_{11}(\xi, x) & k_{12}(\xi, x) \\ k_{21}(\xi, x) & k_{22}(\xi, x) \end{pmatrix},$$

where

$$k_{11}(\xi, x) = (G_{x_1\xi_1} + 4G_{\bar z\zeta})\nu_1 + 2(G_{x_1\bar\zeta} - G_{\bar z\xi_1})\,(\nu_2 - \mathrm{i}\,\nu_3),$$

$$k_{12}(\xi, x) = 2(G_{z\xi_1} - G_{x_1\zeta})\nu_1 + (4G_{z\bar\zeta} + G_{x_1\xi_1})\,(\nu_2 + \mathrm{i}\,\nu_3),$$

$$k_{21}(\xi, x) = 2(G_{x_1\bar\zeta} - G_{\bar z\xi_1})\nu_1 - (G_{x_1\xi_1} + 4G_{\bar z\zeta})\,(\nu_2 + \mathrm{i}\,\nu_3),$$

$$k_{22}(\bar\xi, x) = (4G_{z\bar\zeta} + G_{x_1\xi_1})\nu_1 - 2(G_{z\xi_1} - G_{x_1\zeta})\,(\nu_2 + \mathrm{i}\,\nu_3).$$

But according to (3.68)

$$k_{11}(\xi, x) = \nu_1\,G_{x_1\xi_1} + 2(\nu_2 - \mathrm{i}\,\nu_3)G_{x_1\bar\zeta} - 2\left[(\nu_2 - \mathrm{i}\,\nu_3)G_{\bar z\xi_1} - 2\nu_1\,G_{\bar z\zeta}\right]$$

$$= \nu_1\,G_{x_1\xi_1} + 2(\nu_2 - \mathrm{i}\,\nu_3)\,G_{x_1\bar\zeta}\,,$$

$$k_{12}(\xi, x) = 2\nu_1\,G_{z\xi_1} + 4(\nu_2 - \mathrm{i}\,\nu_3)G_{z\bar\zeta} + \left[G_{x_1}(\nu_2 - \mathrm{i}\,\nu_3) - 2\nu_1\,G_{x_1\zeta}\right]$$

$$= 2(\nu_1 G_{z\xi_1} + 2(\nu_2 - i\,\nu_3) G_{z\bar{\zeta}}),$$

$$k_{21}(\xi, x) = -2\nu_1 G_{\bar{z}\xi_1} - 4(\nu_2 + i\,\nu_3) G_{\bar{z}\zeta} - \left[G_{x_1\xi_1} - (\nu_2 - i\,\nu_3) - 2G_{x_1\bar{\zeta}}\nu_1 \right]$$

$$= -2(\nu_1 G_{\bar{z}\xi_1} + 2(\nu_2 + i\,\nu_3) G_{\bar{z}\zeta}),$$

$$k_{22}(\xi, x) = \nu_1 G_{x_1\xi_1} + 2(\nu_2 + i\,\nu_3) G_{x_1\zeta} - 2\left[G_{z\xi_1}(\nu_2 + i\,\nu_3) - 2G_{z\bar{\zeta}}\nu_1 \right]$$

$$= \nu_1 G_{x_1\xi_1} + 2(\nu_2 + i\,\nu_3) G_{x_1\zeta}.$$

Hence

$$(\nu(\xi) K(\xi, x))' = \begin{pmatrix} k_{11}(\xi,x) & k_{21}(\xi,x) \\ k_{12}(\zeta,x) & k_{22}(\xi,x) \end{pmatrix}.$$

Moreover

$$\overline{\nu'(\xi) L'(x,\xi)} = \begin{pmatrix} \lambda_{11}(x,\xi) & \lambda_{12}(x,\xi) \\ \lambda_{21}(x,\xi) & \lambda_{22}(x,\xi) \end{pmatrix},$$

where

$$\lambda_{11}(x, \xi) = 2\,(G_{\bar{z}\xi_1}(\nu_2 - i\,\nu_3) - 2G_{z\bar{\zeta}}\nu_1) + \nu_1 G_{x_1\xi_1}$$

$$+ 2(\nu_2 - i\,\nu_3) G_{x_1\bar{\zeta}} = k_{11}(\xi, x),$$

$$\lambda_{12}(x, \xi) = 2(\nu_1 G_{z\xi_1} + 2(\nu_2 - i\,\nu_3) G_{z\bar{\zeta}})$$

$$- (G_{x_1\xi_1}(\nu_2 - i\,\nu_3) - 2\,G_{x_1\zeta}\nu_1) = K_{12}(\xi, x),$$

$$\lambda_{21}(x, \xi) = k_{21}(\xi, x), \quad \lambda_{22}(x, \xi) = k_{22}(\xi, x),$$

i.e.

$$\overline{\nu'(\xi) L'(x,\xi)} = (\nu(\xi) K(\xi, x))'$$

and by (3.69)

$$K(x,\eta) = \frac{1}{|s_3|}\int_\Gamma \overline{(\nu(\xi)K(\xi,x))'}\ \overline{E'(\xi-\eta)}\, d\Gamma_\xi$$

$$= \frac{1}{|s_3|}\int_\Gamma \nu'(\xi)\, L'(x,\xi)\, \overline{E'(\xi-\eta)}\, d\Gamma_\xi. \tag{3.70}$$

(c) If Ω is a bounded domain in $\mathbb{R}^4$ with smooth boundary $\Gamma = \partial\Omega$, then for a complex-valued vector function $u(z) = (u_1, u_2)$ from $C^1(\Omega)$ we may obtain an analogue of formula (3.50):

$$u(z) = \frac{1}{2\pi^2}\int_\Gamma E(\zeta - z)\nu(\zeta)\, u(\zeta)\, d\Gamma_\zeta - \frac{1}{\pi^2}\int_\Omega E(\zeta - z)\,\partial_{\bar\zeta} u\, d\Omega_\zeta, \tag{3.71}$$

where

$$E(z) = \frac{1}{|z|^4}\begin{pmatrix} \bar z_1 & -z_2 \\ \bar z_2 & z_1 \end{pmatrix}.$$

Also if $G(z,\zeta)$ is the Green function of the domain Ω, i.e.

$$G(z,\zeta) = \frac{1}{|z-\zeta|^2} + h(z,\zeta), \tag{3.72}$$

where $h(z,\zeta)$ is a function that is regular harmonic in Ω, having boundary value

$$h(z,\zeta) = -\frac{1}{|z-\zeta|^2}$$

for $\zeta \in \Gamma$, $z \in \Omega$, then we introduce the kernel matrix

$$K(z,\zeta) = -\frac{2}{\pi^2}\,\partial_z'\,\partial_{\bar\zeta}\, G(z,\zeta) \tag{3.73}$$

which is an element of the subspace $H(\Omega) \subset L_2(\Omega)$ of solutions of the equation

$$\partial_{\bar z} u = 0 \tag{3.74}$$

in z for fixed $\zeta \in \Omega$. $K(z, \zeta)$ is Hermitian symmetric.

The integral operator

$$Kf \equiv \int_\Omega K(z, \zeta) f(\zeta)\, d\Omega_\zeta \tag{3.75}$$

projects the space $L_2(\Omega)$ onto $H(\Omega)$ and

$$Kf \equiv f, \ \forall f \in H(\Omega). \tag{3.76}$$

So K is an orthogonal projection in Hilbert space, and is such that $K(z, \zeta)$ has a strong singularity at the boundary Γ.

(d) At the end of this section we calculate the kernel matrices of unit balls in $\mathbb{R}^3$ and in $\mathbb{R}^4$.

Since the Green's function of the unit ball $|x| < 1$ in $\mathbb{R}^3$ is

$$G(x, \zeta) = |x - \xi|^{-1} + h(x, \xi),$$

where

$$h(x, \zeta) = |\xi|\, |\xi - |\xi|^2 x|^{-1},$$

then

$$K(x, \xi) = \frac{1}{4\pi |\xi - |\xi|^2 x|} \begin{pmatrix} \omega_{11}(x, \zeta) & \omega_{12}(x, \zeta) \\ -\overline{\omega_{12}(x, \zeta)} & \overline{\omega_{11}(x, \zeta)} \end{pmatrix},$$

$$\omega_{11}(x, \xi) = |\xi|^{-4} (2|\xi|^2 - 3|\xi|^2 + |\xi|^{-2}(x_1 \xi_1 + 2\zeta \bar{z})$$

$$+ 3\Big[2\Big(\frac{\xi}{|\xi|}\Big)^2 (\bar{\zeta} - \bar{z}) - \xi_1^2 \Big(\frac{\xi}{|\xi|^2} - z\Big)\Big]\Big(\frac{\zeta}{|\xi|^2} - \bar{z}\Big)$$

$$- \Big(\frac{\xi_1}{|\xi|^2} - x_2\Big)\Big(|\xi|^2 - 2\xi_1^2\Big)\Big) \Big|\frac{\xi}{|\xi|^2} - x\Big|^{-2},$$

$$\omega_{12}(x, \xi) = |\xi|^{-4} (\bar{\zeta} |\xi|^2 \Big(\frac{\xi_1}{|\xi|^2} + x_1\Big) - 2\xi_1 |\xi|^2 \Big(\frac{\zeta}{|\xi|^2} + \bar{z}\Big)$$

$$+3\left[(2|\xi|^2-5\xi_1^2)\left(\frac{\xi_1}{|\xi|^2}-x_1\right)\left(\frac{\zeta}{|\xi|^2}-x_1\right)(\bar{\zeta}/|\xi|^2-\bar{z})\right.$$

$$\left.+\bar{\zeta}^2(\xi_1/|\xi|^2-x_1)(\zeta/|\xi|^2-z)\right]|\xi/|\xi|^2-x|^{-2},$$

if $\xi\in\Gamma$,

i.e.

$$|\xi|=1,$$

then

$$\omega_{11}(x,\xi)=2-3|\zeta|^2+\xi_1x_1+2\zeta\bar{z}+3\left[2\left[\zeta^2(\bar{\zeta}-\bar{z})^2\right]-\xi_1^2|\zeta-z|^2\right.$$

$$\left.-(\xi_1-x_1)^2\right)|\xi-x|^{-2},$$

$$\omega_{12}(x,\zeta)=\bar{\zeta}(\xi_1+x_1)-2\xi_1(\bar{\zeta}+\bar{z})+3\left((\xi_1-x_1)(\bar{\zeta}-\bar{z})(2-5\xi_1^2)\right.$$

$$+\bar{\zeta}^2(\xi_1-x_1)(\zeta-z)|\xi-x|^{-2}$$

and

$$K(x,\xi)=\frac{\omega(x,\xi)}{\pi^2|\xi-x|^3}.$$

The Green's function of the unit ball $|z|<1$ in $\mathbb{R}^4\sim\mathbb{C}^2$ is

$$G(z,\zeta)=|\zeta-z|^{-2}+h(z,\zeta)$$

where

$$h(z,\zeta)=|\zeta|^2|\zeta-|\zeta|^2z|^{-2}.$$

Hence

$$K(z,\zeta)=\frac{\nu(z,\zeta)}{|\zeta-|\zeta|^2 z|^4},\ \nu(z,\zeta)=(\nu_{ij}(z,z)),$$

$$\nu_{11}(z,\zeta)=\frac{|\zeta|^2}{\pi^2}\Big\{2(1-|\zeta|^2)-3|\zeta|^2(\zeta\cdot\bar{z})$$

$$+2|\zeta|^2\Big[\frac{\zeta_1 z_1(\bar{\zeta}_1-|\zeta|^2\bar{z}_1)^2}{|\zeta-|\zeta|^2 z|^2}+\frac{\zeta_2 z_2(\bar{\zeta}_2-|\zeta|^2\bar{z}_2)^2}{|\zeta-|\zeta|^2 z|^2}$$

$$+\frac{\zeta_1\bar{z}_1|\zeta_1-|\zeta|^2 z_1|^2}{|\zeta-|\zeta|^2 z|^2}+\frac{\zeta_2\bar{z}_2|\zeta_2-|\zeta|^2 z_2|^2}{|\zeta-|\zeta|^2 z|^2}$$

$$+\frac{(\zeta_1 z_2+\zeta_2 z_1)(\bar{\zeta}_1-|\zeta|^2\bar{z}_1)(\bar{\zeta}_2-|\zeta|^2\bar{z}_2)}{|\zeta-|\zeta|^2 z|^2}+\frac{\zeta_1\bar{z}_2(\bar{\zeta}_1-|\zeta|^2\bar{z}_1)(\zeta_2-|\zeta|^2 z_2)}{|\zeta-|\zeta|^2 z|^2}$$

$$+\frac{\zeta_2\bar{z}_1(\zeta_1-|\zeta|^2 z_1)(\bar{\zeta}_2-|\zeta|^2\bar{z}_2)}{|\zeta-|\zeta|^2 z|^2}\Big]\Big\},$$

$$\nu_{12}(z,\zeta)=\frac{|\zeta|^2}{\pi^2}$$

$$\Big\{\bar{\zeta}_1(2(\bar{\zeta}_2-|\zeta|^2\bar{z}_2)-\bar{z}_2|\zeta|^2)-\bar{\zeta}_2(2(\bar{\zeta}_1-|\zeta|^2\bar{z}_1)-\bar{z}_1|\zeta|^2)$$

$$+2|\zeta|^2\Big[\bar{\zeta}_1 z_2(\bar{\zeta}_2-|\zeta|^2\bar{z}_2)^2-\bar{\zeta}_2 z_1(\bar{\zeta}_1-|\zeta|^2\bar{z}_1)^2$$

$$+\bar{\zeta}_1\bar{z}_2|\zeta_2-|\zeta|^2 z_2|^2-\bar{\zeta}_2\bar{z}_1|\zeta_1-|\zeta|^2 z_1|^2$$

$$-\frac{(\bar{\zeta}_1 z_1-\zeta_2\bar{z}_2)(\bar{\zeta}_1-|\zeta|^2\bar{z}_1)(\bar{\zeta}_2-|\zeta|^2\bar{z}_2)}{|\zeta-|\zeta|^2 z|^2}$$

$$+\frac{\bar{\zeta}_1 z_1(\zeta_1-|\zeta|^2 z_1)(\bar{\zeta}_2-|\zeta|_2\bar{z}_2)}{|\zeta-|\zeta|^2 z|^2}-\frac{\bar{\zeta}_2 z_2(\bar{\zeta}_1-|\zeta|^2\bar{z}_1)(\zeta_2-|\zeta|^2 z_2)}{|\zeta-|\zeta|^2 z|^2}\Big]\Big\}.$$

3.4 Cauchy problems for the inhomogeneous equations $\bar{\partial}_x u = f$, and $\partial_{\bar{z}} u = f$

(a) We consider the Cauchy problem: Find in $\Omega \subset \mathbb{R}^3$ a solution $u = (u_1, u_2)$ of the inhomogeneous equation

$$\bar{\partial}_x u = f \tag{3.77}$$

in Ω, continuous in the closure $\bar{\Omega} = \Omega + \Gamma$, satisfying the condition

$$u = 0 \tag{3.78}$$

on $\Gamma = \partial\Omega$ with given $f \in L_2(\Omega)$.

It is clear that this problem is not solvable for any $f \in L_2(\Omega)$.

Lemma 3.5. *The Cauchy problem (3.77), (3.78) is solvable if and only if its right-hand side f has view*

$$f(x) = f^0(x) - \int_\Omega K(x, \xi)\, f^0(\xi)\, d\zeta, \quad \forall f^0 \in L_2(\Omega), \tag{3.79}$$

where $K(x, \zeta)$ is the kernel matrix, introduced above by (3.53).

Proof. Since $\bar{\partial}_x \partial'_x \equiv I\,\Delta$, where Δ is the Laplacian in $\mathbb{R}^3$, then acting on (3.77) from the left by $\bar{\partial}_x$ we obtain[2]

$$\Delta u_1 = \frac{\partial f_1}{\partial x_3} + 2\frac{\partial f_2}{\partial z}, \quad \Delta u_2 = -2\frac{\partial f_1}{\partial \bar{z}} + \frac{\partial f_2}{\partial x_3}. \tag{3.80}$$

Hence

$$u_1(x) = \frac{1}{4\pi}\int_\Omega G(x, \xi)\left(\frac{\partial f_1}{\partial \xi_1} + 2\frac{\partial f_2}{\partial \zeta}\right) d\xi,$$

$$u_2(x) = \frac{1}{4\pi}\int_\Omega G(x, \xi)\left(-2\frac{\partial f_1}{\partial \bar{\zeta}} + \frac{\partial f_2}{\partial \xi_1}\right) d\xi,$$

where $G(x, \xi)$ is the Green function of the domain Ω. Integrating by parts and

[2] We assume for a moment that $f \in C^1(\Omega)$.

taking into account that $G(x, \xi) = 0$ for $\xi \in \Gamma$, $x \in \Omega$, we obtain

$$u_1(x) = -\frac{1}{4\pi}\int_\Omega \left[G_{\xi_1}(x, \zeta) f_1(\zeta) + 2G_\zeta(x, \xi) f_2(\xi) \right] d\xi,$$

$$u_2(x) = -\frac{1}{4\pi}\int_\Omega \left[-2\, G_{\bar{\zeta}}(x, \xi) f_1(\xi) + G_{\xi_1}(x, \xi) f_2(\xi) \right] d\xi,$$

i.e.

$$u(x) = -\frac{1}{4\pi}\int_\Omega \partial'_\xi G(x, \xi) f(\xi)\, d\xi. \tag{3.81}$$

Taking into account (3.51) we have

$$\partial'_\xi G(x, \xi) = \overline{E'(\xi - x)} + \partial'_\xi h(x, \xi).$$

Hence acting to (3.81) from left by ∂'_x and taking into account that $\bar{\partial}_x \overline{E'(\xi - x)} = 0$ and so

$$\partial'_x\left(-\frac{1}{4\pi}\int_\Omega \overline{E'(\xi - x)} f(\xi)\, d\xi \right) = f(x).$$

for $x \in \Omega$, we obtain

$$\partial'_x u = f(x) - \frac{1}{4\pi}\int_\Omega \partial'_x \bar{\partial}_\xi h(x, \xi) f(\xi)\, d\xi$$

$$= f(x) - \int_\Omega K(x, \xi) f(\xi)\, d\xi$$

for any $f \in L_2(\Omega)$. It is evident that the Cauchy problem (3.77), (3.78) with right-hand side (3.79) is solvable.

Corollary 3.5. *The Cauchy problem (3.77), (3.78) is solvable if and only if its right-hand side f is orthogonal to subspace $H'(\Omega) \subset L_2(\Omega)$ of all vectors v from $L_2(\Omega)$, which are solutions of the homogeneous equation*

$$\partial'_x v = 0. \tag{3.82}$$

Proof. Consider all vectors represented as (3.79) for any $f^0 \in L_2(\Omega)$ and let v be a solution of (3.82). Then we obtain

$$(f, v) = \int_\Omega f(x)\, \overline{v(x)}\, dx = \int_\Omega f^0(x)\, \overline{v(x)}\, dx$$

$$-\int_\Omega \left(\int_\Omega K(x, \xi) f^0(\xi)\, d\xi \right) \overline{v(x)}\, dx.$$

Changing the order of integration and using the reproducing property we obtain

$$\int_\Omega \left(\int_\Omega K(x, \xi) f(\xi)\, d\xi \right) \overline{v(x)}\, dx = \int_\Omega \left(\int_\Omega f(\xi)\, K'(x, \xi)\, d\xi \right) \overline{v(x)}\, dx$$

$$= \int_\Omega f(\xi) \left(\int_\Omega K'(x, \xi)\, \overline{v(x)}\, dx \right) d\xi = \int_\Omega f(\xi)\, \overline{v(\xi)}\, d\xi$$

and hence $(f, v) = 0$.

(2) Now we consider the Cauchy problem: find in $\Omega \subset \mathbb{R}^4$ a solution $u = (u_1, u_2)$ of the inhomogeneous equation

$$\partial_z' u = f \tag{3.83}$$

in Ω, continuous in the closure $\bar{\Omega} = \Omega + \Gamma$, satisfying the condition

$$u = 0 \tag{3.84}$$

on Γ with given $f \in L_2(\Omega)$.

Lemma 3.6. *The Cauchy problem (3.83), (3.84) is solvable if and only if its right-hand side f has the form*

$$f(z) = f^0(z) - \int_\Omega K(z, \zeta) f^0(\zeta)\, d\Omega_\zeta, \ \forall f^0 \in L_2(\Omega)$$

where $K(z, \zeta)$ is the kernel matrix, determined by (3.73).

Corollary 3.6. The Cauchy problem (3.83), (3.84) is solvable if and only if its right-hand side f is orthogonal to subspace $H(\Omega)$ of all vectors u from $L_2(\Omega)$, which are solutions of the homogeneous equation

$$\partial_{\bar{z}} u = 0.$$

The proof of this lemma and corollary are the same as we used for Lemma 3.5 and Corollary 3.5.

We introduce also the matrix

$$\overset{\circ}{K}(z,\zeta) = \frac{1}{4\pi}\,\partial_{\bar z}\partial'_{\zeta}G(z,\xi).$$

Note that as above we may show that this matrix is a solution of the equation

$$\partial'_z \overset{\circ}{K}(z,\zeta) = 0$$

in z, when ζ is fixed in Ω, but the transposed matrix $\overset{\circ}{K}'(z,\zeta)$ is a solution of the equation

$$\partial'_{\bar\zeta}\, \overset{\circ}{K}'(z,\zeta) = 0$$

in ζ, when z is fixed in Ω. The matrix $\overset{\circ}{K}(z,\zeta)$ has Hermitian symmetry:

$$\overline{\overset{\circ}{K}'(\zeta,z)} = \overset{\circ}{K}(z,\zeta).$$

If $\overline{H'(\Omega)}$ is the subspace of the Hilbert space $L_2(\Omega)$ consisting of a solution of the equation $\partial'_z u = 0$, then an integral operator

$$\overset{\circ}{K}f \equiv \int_\Omega \overset{\circ}{K}(z,\zeta)\,f(\zeta)\,d\Omega_\zeta$$

is an orthogonal projection of the Hilbert space $L_2(\Omega)$ onto $\overline{H'(\Omega)}$. From this it follows also that $\overset{\circ}{K}(z,\zeta)$ is continuous inside domain Ω and has strong singularity at the boundary, when $z = \zeta \in \Gamma$.

Lemma 3.7. *The Cauchy problem (3.84) for the equation*

$$\partial_{\bar z} u = f(z), \quad z \in \Omega \tag{3.83'}$$

is solvable if and only if its right-hand side takes the form

$$f(z) = f^0(z) - \int_\Omega \overset{\circ}{K}(z,\zeta)\,\overset{\circ}{f}(\zeta)\,d\Omega_\zeta$$

for any $f^0 \in L_2(\Omega)$.

Corollary 3.7. The Cauchy problem (3.84) for the equation (3.83′) is solvable if and only if its right-hand side f is orthogonal to the subspace $\overline{H'(\Omega)}$.

Now we apply these results to the Cauchy problem for the inhomogeneous Cauchy-Riemann equation in $\mathbb{C}^2$. It is known (see [41], [62]) that if Ω is a bounded domain in $\mathbb{C}^n$, $n \geq 2$, with equation $\rho(z) < 0$ with connected C^3 boundary $\Gamma = \partial\Omega$ with equation $\rho(z) = 0$, grad $\rho \neq 0$, then the Cauchy problem

$$u|_{\partial\Omega} = f$$

with a C^1-function f given on $\partial\Omega$ for the Cauchy-Riemann equation

$$\bar{\partial} u = 0,$$

where

$$\bar{\partial} u \equiv \sum_{k=1}^{n} \frac{\partial u}{\partial \bar{z}_k} d\bar{z}_k$$

is solvable if and only if its right-hand side f satisfies the tangent Cauchy-Riemann equation

$$\bar{\partial} \rho \wedge \bar{\partial} f = 0.$$

We consider the homogeneous Cauchy problem

$$u|_{\partial\Omega} -= 0 \tag{3.0}$$

for the inhomogeneous Cauchy-Riemann equation

$$\overline{\partial u} = f$$

in $\mathbb{C}^2$ with (0,1) form $f \in C^1(\Omega) \cap L_2(\Omega)$, i.e. we are considering two overdetermined Cauchy-Riemann equations with respect to the function $u(z)$, $z = (z_1, z_2)$:

$$u_{\bar{z}_1} = f_1, \quad u_{\bar{z}_2} = f_2 \tag{3.0'}$$

with $C^1(\Omega) \cap L_2(\Omega)$ functions $f_j(z)$.

Theorem 3.2. *The Cauchy problem (3.0), (3.0') is solvable if and only if its right-hand side is expressed*

$$f_1(z) = f(z) - \int_\Omega \left[\overset{\circ}{K}_{11}(z,\zeta) f(\zeta) + \overset{\circ}{K}_{12}(z,\zeta)\, g(\zeta) \right] d\Omega_\zeta ,$$

$$f_2(z) = g(z) - \int_\Omega \left[\overset{\circ}{K}_{21}(z,\zeta) f(\zeta) + \overset{\circ}{K}_{22}(z,\zeta)\, g(\zeta) \right] d\Omega_\zeta$$

for any vector $(f, g) \in C^1(\Omega) \cap L_2(\Omega)$, satisfying the condition

$$f_{\bar{z}_2} = g_{\bar{z}_1} ,$$

where

$$\overset{\circ}{K}_{11}(z,\zeta) = \frac{1}{4\pi}\left(G_{\bar{z}_1\zeta_1} + G_{z_2\bar{\zeta}_2} \right),\ \overset{\circ}{K}_{12}(z,\zeta) = \frac{1}{4\pi}\left(G_{\bar{z}_1\zeta_2} - G_{z_2\bar{\zeta}_1} \right),$$

$$\overset{\circ}{K}_{21}(z,\zeta) = -\overline{K_{12}(z,\zeta)},\ \overset{\circ}{K}_{22}(z,\zeta) = \overline{K_{11}(z,\zeta)} .$$

Corollary 3.8. *The Cauchy problem (3.0), (3.0') is solvable if and only if its right-hand side* (f_1, f_2) *is orthogonal to the subspace* $H(\Omega)$ *and satisfies the condition* $\partial f_1/\partial\bar{z}_2 = \partial f_2/\partial\bar{z}_1$.

The proof of the theorem and the corollary easily follows from the above results on the Cauchy problem (3.84) for the equation (3.83') if we take into account the following assertion:

Lemma 3.8. *Let vector* $u = (u, v)$ *be a solution of the Cauchy problem*

$$\partial_{\bar{z}} u = F \text{ in } \Omega, \quad u|_{\partial\Omega} = 0$$

with right-hand side $F = (f, g) \in C^1(\Omega) \cap L_2(\Omega)$. *Then in case the condition* $f_{\bar{z}_2} = g_{\bar{z}_1}$ *is satisfied the function* u *will be a solution of the Cauchy problem (3.0), (3.0')..*

Proof. Indeed, differentiating the second equation (3.0′) with respect to $\bar{z}_1$ and the first equation with respect to $\bar{z}_2$ we obtain

$$u_{\bar{z}_1\bar{z}_2} + v_{z_1\bar{z}_1} = g_{\bar{z}_1},\quad u_{\bar{z}_1\bar{z}_2} - v_{z_2\bar{z}_2} = f_{\bar{z}_2} .$$

Subtracting this second equation from the first and taking into account that $g_{\bar{z}_1} - f_{\bar{z}_2} = 0$ we see that v is harmonic in Ω and so $v \equiv 0$, because $v|_{\partial\Omega} = 0$.

Now the equation $\partial_{\bar{z}} v = F$ reduces to the equation (3.0′) and the condition $u \,|\, \partial\Omega = 0$ reduces to (3.0).

3.5 First-order elliptic systems degenerate at the boundary

(a) Let G^+ denote, as above a multiple connected bounded domain in plane $\mathbb{C}$. As we saw at the beginning of Chapter 1 any first-order linear elliptic systems of two real equations in G^+ may be put into the following complex form:

$$Lw \equiv w_{\bar{z}} + q_1(z)w_z + q_2(z)\,\overline{w_z} + a(z)w + b(z)\bar{w} = f(z) \tag{3.85}$$

with a, b, f from $L_p(G^+)$, $p > 2$ and $q_1(z), q_2(z) \in H_p^1(G^+)$ such that

$$|q_1(z)| + |q_2(z)| \leq q_0 < 1 \tag{3.86}$$

for $z \in \bar{G}^+ = G^+ + \Gamma$.

In this section we allow for equation (3.85) to degenerate at the boundary, i.e. we assume that ellipticity condition (3.86) holds at the inner points of G^+ and degenerates at the boundary

$$|q_1(t)| + |q_2(t)| \equiv 1, \quad t \in \Gamma. \tag{3.87}$$

We seek a solution of (3.85) continuous in the closure $\bar{G}^+$. We know that in the pure elliptic case a set of solutions in G^+ of equation (3.85) comparable with the set of analytic functions in G^+, so to determine a solution of this equation we need a boundary condition. What happens to a set of solutions of (3.85) if (3.86) holds in G^+ and (3.87) on Γ? Let us look at examples. The equation

$$w_{\bar{z}} + z^2 w_z = f \tag{3.88}$$

is elliptic inside the disc $|z| < 1$ and degenerate at the circle $|t| = 1$. By the transformation

$$\zeta = \frac{2z}{1 + |z|^2}, \tag{3.89}$$

which maps the disc $|z| \leq 1$ into the disc $|\zeta| \leq 1$, we obtain

$$w_{\bar{\zeta}} = \frac{f}{2\sqrt{(1 - |\zeta|^2}} \tag{3.90}$$

in the disc $|\zeta|<1$. This means that a set of solutions of the equation is the same as in the pure elliptic case and so to determine a solution of this equation we need boundary conditions. Another example in the disc $|z|<1$,

$$w_{\bar{z}} - z^2 w_z = f, \tag{3.91}$$

has a quite different property although it is also elliptic inside the disc $|z|<1$ and degenerate at the circle. Making the transformation

$$\zeta = \frac{z}{1-|z|^2}, \tag{3.92}$$

which maps the disc $|z| \leq 1$ into the whole complex plane $\mathbb{C}$ of the variable ζ so that circle $|z|=1$ transforms to infinity, we obtain

$$w_{\bar{\zeta}} = \frac{f}{\sqrt{(1+2|\zeta|^2)}} \tag{3.93}$$

in the plane $\mathbb{C}$.

Consequently a solution of the equation (3.91) is uniquely determined, without any boundary condition, by its right-hand side apart from a constant for any right-hand side such that

$$(1+2|\zeta|^2)^{-1/2} f \in L_{p,2}(\mathbb{C}), \quad p>2.$$

Note that in contrast to (3.88) the circle $|z|=1$ is the characteristic set for equation (3.91). On the other hand this circle is not characteristic for the equation

$$w_{\bar{z}} - z^2 \overline{w_z} = f, \tag{3.94}$$

which is also elliptic inside the disc and degenerate at the circle. For this equation a regular solution is determined without any boundary condition for arbitrary right-hand side $f \in L_p(|z|<1)$, $p>2$, so the general solution is

$$w(z) = a + \frac{i c_0 z}{1+|z|^2}$$

$$-\frac{1}{\pi(1-|z|^4)} \iint_{|\zeta|<1} [\left(\frac{1}{\zeta - z} + \frac{\bar{z}|z|^2}{1-\zeta\bar{z}}\right) f(\zeta)$$

$$+ z^2\left(\frac{z}{1-\bar{\zeta} z} + \frac{1}{\bar{\zeta} - \bar{z}}\right)\overline{f(\zeta)}\,\Big]\,d\xi\, d\eta, \tag{3.95}$$

where c_0 and a are arbitrary real and complex constants. These examples show that different situations arise for degenerate elliptic systems.

Now we turn to general equation (3.85). Let $t = t(s)$ be the parametric equation of the curve Γ. We introduce the function

$$\kappa(t) = t'(s) - (\theta(t)/\overline{\theta(t)} \cdot \overline{t'(s)}, \tag{3.96}$$

where $t'(s) = dt/ds$ and

$$\theta(t) = 2\,\mathrm{Im}\, q_1(t) + i\,(|1 - q_1(t)|^2 - |q_2(t)|^2). \tag{3.97}$$

we note that $\kappa(t) = 0$ along the characteristic of equation (3.85).

Theorem 3.3. *If $q_2(t) \neq 0$ on Γ and Γ is not the characteristic for (3.85), then the equation possesses Noetherian property: inhomogeneous equation (3.85) is solvable if and only if its right-hand side satisfies the finite number of orthogonality conditions*

$$\mathrm{Re} \iint_{G^+} f(z)\,\overline{v_j(z)}\, dx\, dy = 0, \quad j = 1, \ldots, k', \tag{3.98}$$

where $\{v_j(z)\}$ is a complete system of solutions, linearly independent in G^+, of the homogeneous adjoint equation

$$-v_z - (\bar{q}_1 v)_{\bar{z}} - (q_2 \bar{v})_z + \bar{a}\, v + b\, \bar{v} = 0 \tag{3.99}$$

continuous in $\bar{G}^+$, satisfying the boundary condition

$$\overline{(1 - q_1(t)\, t'^2(s))}\, \overline{v^+(t)} - \overline{q_2(t)\, t'^2(s)}\, v^+(t) = 0 \tag{3.100}$$

on Γ. The corresponding homogeneous equation (3.85) ($f \equiv 0$) may have only a finite number k of linearly independent nontrivial solutions $k = k' + \kappa_2 - m + 1$, where

$$\kappa_2 = \frac{1}{2\pi i}\int_\Gamma d \log q_2(t). \tag{3.101}$$

Proof. First we consider the case $q_1(z) \equiv 0$. In this case evidently Γ is not characteristic. Introducing a new unknown function $u(z)$ by

$$u(z) = w(z) + q_2(z)\,\overline{w(z)} \tag{3.102}$$

we obtain the following equation for u:

$$u_{\bar{z}} + \frac{a(z)}{1-|q_2|^2}(u - q_2\,\bar{u}) + \frac{b_1(z)}{1-|q_2|^2}(\bar{u} - \bar{q}_2\,u) = f, \tag{3.103}$$

$$b_1(z) = b(z) - \partial q_2/\partial\bar{z}.$$

Moreover since we are seeking a solution of (3.85) continuous in closure $\bar{G}^+$, then it follows from (3.102) that the function satisfies the boundary condition

$$u^+(t) - q_2(t)\,\overline{u^+(t)} = 0 \tag{3.104}$$

at the boundary Γ.

Let us denote by $C_\Gamma(G^+)$ the subspace of the space of complex-valued functions, continuous in $\bar{G}^+$, satisfying the condition (3.104) on Γ.

Equation (3.103) is equivalent to the integral equation

$$u + \hat{T}\,u = \varphi(z) + T_{G^+}f, \tag{3.105}$$

where $\varphi(z)$ is arbitrary function, holomorphic in G^+,

$$\hat{T}\,u \equiv T_{G^+}\left(\frac{a}{1-|q_2|^2}(u - q_2\,\bar{u}) + \frac{b_1}{1-|q_2|^2}(\bar{u} - \bar{q}_2\,u)\right) \equiv T_{G^+}\,\hat{u},$$

$$T_{G^+}f = -\frac{1}{\pi}\iint_{G^+}\frac{f(\zeta)\,dG^+_\zeta}{\zeta - z}.$$

We seek a solution of the equation (3.105) in $C_\Gamma(G^+)$. In this space the operator $\hat{T}$ is compact as it can be shown following [58], because

$$|\hat{T}\,u| \le A_p\,\hat{u} \le A_p\,L_p\,\check{u}\cdot|u|,$$

$$\check{u} = \frac{a}{1-|q_2|^2}(1 - q_2\frac{\bar{u}}{u}) + \frac{b_1}{1-|q_2|^2}(\frac{\bar{u}}{u} - \bar{q}_2),$$

$$\left|\frac{(1-q_2\frac{\bar{u}}{u})}{1-|q_2|^2}\right|, \left|\frac{\frac{\bar{u}}{u} - \bar{q}_2}{1-|q_2|^2}\right|$$

are bounded according (3.104), and for $g(z) = T_{G^+}\hat{u}$, $z_1, z_2 \in G^+$,

$$|g(z_1) - g(z_2)| \le A_p L_p \hat{u} |z_1 - z_2|^{1-2/p}.$$

Thus the equation is Fredholm's equation on $C_\Gamma(G^+)$.

Let $u^0(z)$ be a solution of the corresponding homogeneous equation

$$u^0 - \hat{T}^0 = 0, \tag{3.106}$$

from $C_\Gamma(G^+)$. From (3.106) it follows that u^0 is holomorphic in G^- and vanishes at infinity, but in G^+ u^0 is a solution of the homogeneous differential equation

$$u^0_{\bar{z}} + \frac{a}{1-|q_2|^2}(u^0 - q_2\bar{u}^0) + \frac{b_1}{1-|q_2|^2}(\bar{u}^0 - \bar{q}_2 u^0) = 0 \tag{3.103'}$$

obtained from (3.106) by differentiation with respect to $\bar{z}$. If we denote by $\hat{\omega}$ the function

$$\hat{\omega}(z) = \begin{cases} -\dfrac{a(1-q_2\frac{\overline{u^0}}{u^0})}{1-|q_2|^2} - \dfrac{b_1(\frac{\overline{u^0}}{u^0} - \bar{q}_2)}{1-|q_2|^2} & \text{if } u^0 \neq 0, \\ -a - b_1 & \text{if } u^0 = 0, \end{cases}$$

then it follows that $|\hat{\omega}(z)| \le |a(z)| + |b(z)|$, because $1 - q_2\frac{\overline{u^0}}{u^0}$, $\frac{\overline{u^0}}{u^0} - \bar{q}_2$ are bounded in $\bar{G}^+$. Hence from (3.103') by the similarity principle (see [58]) we have

$$u^0(z) = \varphi^0(z)e^{T_{G^+}\hat{\omega}}, \tag{3.107}$$

where $\varphi^0(z)$ is an arbitrary function, holomorphic in G^+. Since $u^0(z)$ is holomorphic in G^- together with $T_{G^+}\hat{\omega}$ and vanishes at infinity, then from (3.107) it follows that $\varphi 0(z)$ holomorphically extends into G^- and vanishes at infinity and hence by Liouville's theorem $\varphi^0 \equiv 0$, i.e. $u^0 \equiv 0$. It means that the inhomogeneous equation (3.105) is uniquely solvable in $C_\Gamma(G^+)$ for any right-hand side, so we may represent its solution by

$$u(z) = \varphi(z) + R\varphi + Qf, \tag{3.108}$$

where R is some integral operator (the resolvent operator) and $Qf \equiv T_{G^+}f + RT_{G^+}f$. Now from (3.104) it follows that the holomorphic function $\varphi(z)$ in (3.108) satisfies the condition

$$\varphi^+(t) - q_2(t)\,\overline{\varphi^+(t)} + \underset{\sim}{K}\,\varphi = H(t) \tag{3.109}$$

on Γ, where $H(t) = -(Qf - q_2(t)\,\overline{Qf}\,)$ and $\underset{\sim}{K}$ is a compact integral operator

$$\underset{\sim}{K}\,\varphi \equiv R\varphi - q_2(t)\,\overline{R\,\varphi}\ .$$

If $q_2(t) \equiv 1$ on Γ, then (3.109) is the Schwartz problem, perturbed by the compact term

$$\operatorname{Re} i\,(\varphi + R\varphi) = \operatorname{Im} Qf\,, \tag{3.110}$$

but if $q_2(t) \neq 1$ on Γ, then it is the Riemann-Hilbert problem, perturbed by compact term

$$\operatorname{Re}\left[\,(1 - \overline{q_2(t)})\,(\varphi + R\varphi)\,\right] = -\operatorname{Re}\,(1 - \overline{q_2(t)})\,Qf\,, \tag{3.111}$$

because since $|q_2(t)| \equiv 1,\ t \in \Gamma$ and so

$$-q_2(t) \equiv \frac{1 - q_2(t)}{1 - \overline{q_2(t)}},\quad t \in \Gamma.$$

Consider now the homogeneous adjoint problem:

$$\overset{*}{L_0}\,v \equiv -v_z - (q_2\,\bar{v})_{\bar{z}} + \bar{a}\,v + b\bar{v} = 0 \tag{3.112}$$

in G^+ and

$$v^+ - \overline{q_2(t)}\,\overline{t'^2(s)}\,\overline{v^+} = 0 \tag{3.113}$$

on Γ.

Note that if equation (3.85) (with $q_1 \equiv 0$) is solvable, then Green's g formula leads to the equality

$$\operatorname{Re} \iint_{G^+} f(z)\,\overline{v(z)}\,dx\,dy = \operatorname{Re} \frac{1}{2i} \int_\Gamma u(t)\,(\overline{v^+(t)} - \overline{q_2(t)\,t'^2(s)\,v^+(t)})\,dt = 0$$

for any solution of the adjoint problem (3.112), (3.113). From (3.12) it follows that

$$(\overline{a(z)} - q_2(z)\,\overline{b_1(z)})\,v(z) + (\overline{b_1(z)} - q_2(z)\,a(z))\,\overline{v(z)} = (1 - |q_2(z)|^2)\,v_z,$$

i.e. besides (3.113) the function $v(z)$ satisfies also the condition

$$[\overline{a(t)} - q_2(t)\,(\overline{b(t)} - \overline{q_{2\bar z}(t)})]\,v^+(t) + [\overline{b(t)} - \overline{q_{2\bar z}(t)} - q_2(t)\,a(t)]\,\overline{v^+(t)} = 0,\ t \in \Gamma. \tag{3.114}$$

Note that (3.112) is

$$v_z - \frac{(\bar a - q_2(\bar b - \overline{q_{2\bar z}}))}{1 - |q_2|^2}\,v - \frac{(\bar b - \overline{q_{2\bar z}} - q_2 a)}{1 - |q_2|^2}\,\bar v = 0 \tag{3.112}$$

Hence the equation (3.112) is equivalent to the integral equation

$$v - \bar T_{G^+} \hat v = \overline{\psi(z)}, \tag{3.115}$$

where $\psi(z)$ is an arbitrary function, holomorphic in G^+,

$$\hat v = \frac{a - q_2(\bar b - \overline{q_{2\bar z}})}{1 - |q_2|^2}\,v + \frac{(\bar b - \overline{q_{2\bar z}} - q_2 a)}{1 - |q_2|^2}\,\bar v,$$

$$\bar{T}\hat{v} = -\frac{1}{\pi}\iint_{G^+} \frac{\hat{v}(\zeta)\,dG_\zeta^+}{\bar{\zeta}-\bar{z}}. \tag{3.116}$$

Let $C_{\Gamma'}(G^+)$ be the subspace of complex-valued functions, continuous in the closure $\bar{G}^+$, satisfying the condition (3.114). Then as above we prove that the operator $\check{T}v = \bar{T}_{G^+}\hat{v}$ is compact in $C_{\Gamma'}(G^+)$, and so the equation (3.115) is Fredholm's equation on $C_{\Gamma'}(G^+)$, uniquely solvable and hence its solution we may represent by the formula

$$v(z) = \overline{\psi(z)} + R^*\bar{\psi}, \tag{3.117}$$

where R^* is an integral operator adjoint to R. Substituting (3.117) into (3.113) we obtain

$$\psi^+(t) - \overline{q_2(t)\,t'^2(s)}\,\overline{\psi^+(t)} + \bar{R}^*\psi - \overline{q_2(t)\,t'^2(s)}\,R^*\bar{\psi} = 0 \tag{3.118}$$

on Γ. This is the perturbed Schwartz condition on Γ

$$\operatorname{Re} i\,t'(s)\,(\psi^+ + \overline{R}^*\psi) = 0 \tag{3.118'}$$

in case $q_2 \equiv 1$, and the perturbed Riemann-Hilbert condition

$$\operatorname{Re}[\,t'(s)\,(1-q_2(t))\,(\psi^+ + \bar{R}^*\psi)\,] = 0 \tag{3.118''}$$

in case $q_2(t) \neq 0$ on Γ.

Since the conditions (3.118′), (3.118″) are homogeneous and adjoint to (3.110) and (3.111), then the theorem is proved in case $q_1 \equiv 0$, because of the theory of the Riemann-Hilbert problem for generalized analytic functions[3] (see [58]) if we note that in case $q_2(t) \neq 1$, $1 - q_2(t) = q_2(t)(\overline{q_2(t)} - 1)$ and so $\kappa_2 = \operatorname{Ind}_\Gamma(1 - q_2(t)) = 2\operatorname{Ind}_\Gamma q_2(t)$.

Now let us turn to general case of equation (3.85). From (3.86), (3.87) it follows that $1 - |q_1(z)|^2 \neq 0$ in $\bar{G}^+$ and hence we introduce the new function

$$u(z) = \frac{1-|q_1(z)|^2+|q_2(z)|^2+\sqrt{\sigma(z)}}{2(1-|q_1(z)|^2)}\,w(z) + \frac{q_2(z)}{1-|q_1(z)|^2}\,\overline{w(z)}, \tag{3.119}$$

where

[3] The perturbation of boundary conditions by means of the compact operator does not change result.

$$\sigma(z) = (1 - |q_1(z)|^2 + |q_2(z)|^2)^2 - 4|q_2(z)|^2 > 0, \quad z \in G^+.$$

As we have seen in Chapter 1 the function $u(z)$ defined by (3.119) in G^+ will be a solution of the equation

$$L_0 u \equiv u_{\bar{z}} - q(z)u_z$$

$$-\left[\frac{a_x^0 - \lambda a_y^0 - a}{(1-i\lambda)\Delta}\left(\frac{1-|q_1|^2+|q_2|^2+\sqrt{\sigma}}{2(1-|q_1|^2)}u - \frac{q_2}{1-|q_1|^2}\bar{u}\right)\right.$$

$$\left.+\frac{b_x^0 - \lambda b_y^0 - b}{(1-i\lambda)\Delta}\left(\frac{1-|q_1|^2+|q_2|^2+\sqrt{\sigma}}{2(1-|q_1|^2)}\bar{u} - \frac{\overline{q_2}}{1-|q_1|^2}u\right)\right]$$

$$= \frac{f}{1-i\lambda}, \tag{3.120}$$

where

$$q(z) = \frac{2\,\mathrm{Im}\,q_1(z) - i(\sqrt{\sigma(z)} - \delta(z))}{2\,\mathrm{Im}\,q_1(z) - i(\sqrt{\sigma(z)} + \delta(z))},$$

$$\delta(z) = |1 - q_1(z)|^2 - |q_2(z)|^2,$$

$$a^0(z) = \frac{1-|q_1(z)|^2+|q_2(z)|^2+\sqrt{\sigma(z)}}{2(1-|q_1(z)|^2)}, \quad b^0(z) = \frac{q_2(z)}{1-|q_1(z)|^2},$$

$$\lambda(z) = \frac{2\,\mathrm{Im}\,q_1(z)}{\delta(z)} + \frac{i\sqrt{\sigma(z)}}{\delta(z)},$$

$$\Delta(z) = \frac{\sqrt{\sigma(z)}(\sqrt{\sigma(z)}+1-|q_1(z)|^2+|q_2(z)|^2)}{2(1-|q_1(z)|^2)^2}.$$

From (3.86), (3.87) it follows that

$$|q(z)| < 1 \tag{3.121}$$

in G^+ and

$$|q(t)| \equiv 1, \ \sigma(t) = \Delta(t) = 0 \tag{3.122}$$

on Γ.

Since we are seeking a solution of (3.85) that is continuous in the closure $\bar{G}^+$, it follows from (3.110) that the function u satisfies the condition

$$\frac{1-|q_1(t)|^2+|q_2(t)|^2}{2} u^+(t) - q_2(t)\,\overline{u^+(t)} = 0 \tag{3.123}$$

on Γ.

Since $|q_2(t)| = 1 - |q_1(t)|$ on Γ, then (3.123) is

$$u^+(t) - G(t)\,\overline{u^+(t)} = 0, \tag{3.123'}$$

where

$$G(t) = \frac{q_2(t)}{1-|q_1(t)|}. \tag{3.124}$$

Now we simplify equation (3.120). Since

$$u_{\bar{z}} - qu_z = \frac{i\delta}{-2\,\mathrm{Im}\,q_1 + i\sqrt{\sigma} + \delta)}\left(u_x - \frac{2\,\mathrm{Im}\,q_1}{\delta}u_y + \frac{i\sqrt{\sigma}}{\delta}u_y\right), \tag{3.125}$$

then in case Γ is not characteristic we denote by $\xi(x, y)$ a solution of the characteristic equation

$$\frac{\partial \xi}{\partial x} - \frac{2\,\mathrm{Im}\,q_1}{\delta}\frac{\partial \xi}{\partial y} = 0$$

with $\xi_y \neq 0$ in $\bar{G}^+$, and change the variables x, y to ξ, η by the nondegenerate transformation

$$\xi = \xi(x, y), \ \eta = x. \tag{3.126}$$

Then (3.125) becomes

$$u_{\bar z} - q(z)u_z = \frac{\mathrm{i}\,\delta}{-2\,\mathrm{Im}\,q_1 + \mathrm{i}\sqrt{\sigma} + \delta)}\,(u_\eta + \frac{\mathrm{i}\sqrt{\sigma}}{\delta}\xi_y \cdot u_\xi).$$

Hence equation (3.120) reduces to

$$u_\eta + \frac{\mathrm{i}\sqrt{\sigma}}{\delta}\xi_y \cdot u_\xi$$

$$-\frac{2\,\mathrm{Im}\,q_1 - \mathrm{i}(\sqrt{\sigma}+\delta)}{\mathrm{i}\,\delta}\Big[\frac{(\frac{\mathrm{i}\sqrt{\sigma}}{\delta}\,\xi_y\,\overset{0}{a_\xi} + a)}{(1-\mathrm{i}\lambda)\Delta}\Big(\frac{1-|q_1|^2+|q_2|^2+\sqrt{\sigma}}{2(1-|q_1|^2)}u - \frac{q_2}{1-|q_1|^2}\bar{u}\Big)$$

$$+\frac{(\frac{\mathrm{i}\sqrt{\sigma}}{\delta}\,\xi_y\,\overset{0}{b_\xi} + b)}{(1-\mathrm{i}\lambda)\Delta}\Big(\frac{1-|q_1|^2+|q_2|^2+\sqrt{\sigma}}{2(1-|q_1|^2)}\bar{u} - \frac{\bar{q}_2}{1-|q_1|^2}u\Big)\Big]$$

$$= \frac{-2\,\mathrm{Im}\,q_1 + \mathrm{i}\sqrt{\sigma} + \delta)}{\mathrm{i}\,\delta(1-\mathrm{i}\lambda)}f. \tag{3.127}$$

If by $E(\zeta - z)$ we denote the fundamental solution (Cauchy kernel) of the Beltrami equation

$$u_x + \mathrm{i}\Big(1 + \int_0^\eta \Big(\frac{\sqrt{\sigma}}{\delta}\xi_y\Big)_\xi \,d\eta\Big)\,u_y = 0, \tag{3.128}$$

then evidently the function

$$\hat{E}(\zeta - z) = E(\xi - x + \mathrm{i}\Big(\int_0^\eta \Big(\frac{\sqrt{\sigma}}{\delta}\xi_y\Big)_\xi \,d\eta - \int_0^y \Big(\frac{\sqrt{\sigma}}{\delta}\xi_y\Big)_\xi \,d\eta\Big)\Big)$$

will be the fundamental solution for the equation

$$u_\eta + \frac{\mathrm{i}\sqrt{\sigma}}{\delta}\,\xi_y \cdot u_\xi = 0. \tag{3.129}$$

By means of $\hat{E}(\zeta - z)$ the equation (3.127) reduces to the equivalent integral equation

$$u(z) - \iint_{G^+} \hat{E}(\zeta - z)\,\hat{u}(\zeta)\,dG_\zeta^+ = \varphi(z) + F_0(z), \tag{3.130}$$

where $f(z)$ is a solution in G^+ of the Beltrami equation (3.128) and

$$\hat{u}(z) = \frac{2\operatorname{Im} q - i(\sqrt{\sigma}+\delta)}{i\delta}\Big[\frac{\frac{i\sqrt{\sigma}}{\delta}\xi_y a^0_{\bar\xi} + a}{(1-i\lambda)\Delta}\Big(\frac{1-|q_1|^2+|q_2|^2+\sqrt{\sigma}}{2(1-|q_1|^2)}u$$

$$-\frac{q_2}{1-|q_1|^2}\bar{u}\Big) + \frac{\frac{i\sqrt{\sigma}}{\delta}\xi_y b^0_{\bar\xi} + b}{(1-i\lambda)\Delta}\Big(\frac{1-|q_1|^2+|q_2|^2+\sqrt{\sigma}}{2(1-|q_1|^2)}\bar{u} - \frac{\overline{q_2}}{1-|q_1|^2}u\Big)\Big],$$

$$F_0(z) = -\iint_{G^+} \hat{E}(\zeta - z)\frac{2\operatorname{Im} q_1(\zeta) - i(\sqrt{\sigma(\zeta)}+\delta(\zeta))}{i\delta(\zeta)(1-i\lambda(\zeta))} f(\zeta)\, dG^+_\zeta .$$

The equation, as may be shown as above, is the Fredholm equation in the space $\hat{C}_\Gamma(G^+)$ of complex-valued functions, continuous in closure $\bar{G}^+$, satisfying the condition (3.123), and is uniquely solvable in this space. Representing its solution through the resolvent operator R in the form (3.108) with φ being a solution of Beltrami's equation (3.128) we reduce the boundary condition (3.123′) in case $G(t) \equiv 1$ to the Schwartz condition and in case $G(t) \neq 1$ to the Riemann–Hilbert condition for function φ perturbed by the compact operator

$$\operatorname{Re}\left[(1 - \overline{G(t)})(\varphi^+ + R\varphi)\right] = -\operatorname{Re}(1 - \overline{G(t)})\, Qf), \tag{3.131}$$

In the same way as above we reduce the adjoint problem (3.99), (3.100) to the homogeneous problem, adjoint to (3.131),

$$\operatorname{Re}\left[t'(s)(1 - G(t))(\psi^+ + R^*\psi)\right] = 0 \tag{3.131'}$$

for a solution ψ of the equation, adjoint to Beltrami's equation (3.128). But since on $\Gamma: |G(t)| \equiv 1$ and so

$$1 - G(t) = (\overline{G(t)} - 1)\, G(t),$$

then

$$\kappa_G = 2\operatorname{Ind}_\Gamma G(t) = 2\operatorname{Ind}_\Gamma q_2(t) = 2\kappa_2.$$

The conditions (3.98) are the consequences of the identity following by Green's formula

$$\operatorname{Re}\iint\limits_{G^+} f\bar{v}\,dx\,dy = \operatorname{Re}\frac{1}{2i}\int_\Gamma u(t)\left(\overline{v^+(t)} + \overline{q_1(t)t'^2(s)}\,\overline{v^+(t)}\right.$$

$$\left. - \overline{q_2(t)t'^2(s)}\,v^+(t)\right)dt = 0.$$

Thus the theorem is proved also in the general case.

Theorem 3.4. *If* $\kappa_2 < 0$, *then the homogeneous equation corresponding to (3.85)* $(f \equiv 0)$ *has no nontrivial solutions, but if* $\kappa_2 > m - 1$ *then it has exactly* $2\kappa_2 + 1 - m$ *linearly independent (over the field of real numbers) nontrivial solutions.*

Indeed, from boundary condition (3.123′) it follows that the numbers N_Γ, N_{G^+} of zeros of the solution of equation (3.127) on Γ and in G^+ (counting with their multiplicities) satisfy the equality

$$N_\Gamma + 2N_{G^+} = 2\kappa_2. \tag{3.132}$$

Hence if $\kappa_2 < 0$, then the homogeneous equation (3.85) has no nontrivial solutions, because it reduces to the problem (3.123′) for equation (3.127). The second assertion of Theorem 3.4, follows by the same argument applied to the adjoint problem.

(b) Now we make some remarks concerning the more general situation, when the equation (3.85) is elliptic in domain G^+, including some part Γ' of its boundary Γ, degenerate at the remaining part $\Gamma'' = \Gamma \setminus \Gamma'$ i.e. $q_1(t)$ and $q_2(t)$ satisfy inequality

$$|q_1(z)| + |q_2(z)| < 1 \tag{3.133}$$

for $z \in G^+ + \Gamma'$ and the identity

$$|q_1(t)| + |q_2(t)| \equiv 1 \tag{3.134}$$

for $t \in \Gamma''$.

In this case it is evident that for the solution of equation (3.85) to be determinate a boundary condition on Γ' is needed. We take the Riemann–Hilbert boundary condition

$$\operatorname{Re} G_0(t)\, w(t) = 0 \tag{3.135}$$

with given Hölder conditions function $G_0(t)$ such that $G_0(t) \neq 0$ on Γ'.

Let

$$G_1(t) = G_0(t)\Big\{\overline{\kappa(t)}\,\overline{q(t)}\,(1 - |q_1(t)|^2 + |q_2(t)|^2)\Big[(2(1 - \overline{q_1(t)})$$
$$- (1 - |q_1(t)|^2 + |q_2(t)|^2)\Big]$$

$$- q_2(t)\,\kappa(t)\Big[|q_2(t)|^2 - (1 - \overline{q_1(t)})(1 - |q_1(t)|^2 + |q_2(t)|^2)\Big\}$$

$$+ \overline{G_0(t)}\Big\{\overline{q_2^2(t)}\,\overline{\kappa(t)}\Big[2(1 - \overline{q_1(t)}) - (1 - |q_1(t)|^2 + |q_2(t)|^2)\Big]$$

$$- \kappa(t)\,(1 - |q_1(t)|^2 + |q_2(t)|^2)\Big[|q_2(t)|^2 - (1 - \overline{q_1(t)})(1 - |q_2(t)|^2)|q_2(t)|^2)\Big]\Big\}.$$

Using all constructions, which we did above we come finally to the same Riemann-Hilbert problem

$$\operatorname{Re} \tilde{G}(t)\, u = \tilde{h}(t)$$

for elliptic equation (3.120) with

$$\tilde{G}(t) = \begin{cases} G^0(t) & \text{on } \Gamma' \\ 1 - \overline{G(t)} & \text{on } \Gamma'' \end{cases}$$

$$G^0(t) = \frac{1}{2}(1 - |q_1(t)|^2 + |q_2(t)|^2)\, G_0(t) - \overline{q_2(t)}\,\overline{G_0(t)}$$

and hence in case $G(t)$ is continuous also at the common points of curves Γ' and Γ'' we obtain the following theorem.

Theorem 3.5. *If $q_2(t) \neq 0$ on Γ'' and Γ'' is not characteristic for (3.85), then the problem (3.135) in case (3.133), (3.134) possesses the Noetherian property: the inhomogeneous problem (3.85), (3.135) is solvable if and only if its right-hand side satisfies a finite number of orthogonality conditions*

$$\operatorname{Re} \iint_{G^+} f(z)\,\overline{v_j(z)}\,dx\,dy = 0, \quad j = 1,\ldots,\ell',$$

where $\{v_j(z)\}$ is a complete system of nontrivial solutions in G^+ of homogeneous adjoint equation (3.99), continuous in $\bar{G}^+$ satisfying the condition

$$\operatorname{Re} i\, G_1(t)\, v(t) = 0$$

on Γ' and the condition (3.100) on Γ''.

In case $\kappa_{\tilde{G}} = \operatorname{Ind}_\Gamma \tilde{G}(t) < 0$ *homogeneous problem (3.85), (3.135) has no nontrivial solutions and in case* $\kappa_{\tilde{G}} \geq 0$ *it has* $2\kappa_{\tilde{G}} + 1 - m$ *nontrivial solutions*.

The proof of this theorem is almost the same as in Theorems 3.2 and 3.3. We only note that by means of Green's formula we have for solutions $v(z)$ of the adjoint problem:

$$\operatorname{Re} \iint_{G^+} f(z)\,\overline{v(z)}\,dx\,dy = \operatorname{Re} \frac{1}{2i} \int_{\Gamma'} u(t)\,((t'(s) - q_1(t)\,\overline{t'(s)}\,\overline{v(t)}$$

$$- \overline{q_2(t)\,t'(s)}\,v(t))\,ds$$

$$= \operatorname{Re} \frac{1}{2i} \int_\Gamma G_0(t)\,u(t)\,\overline{G_0(t)}\left[(t'(s) - q_1(t)\,\overline{t'(s)})\,\overline{v(t)} - \overline{q_2(t)\,t'(s)}\,v(t)\right)ds$$

$$= \frac{1}{2} \int_{\Gamma'} (\operatorname{Re}\left[G(t)\,u(t)\right] \operatorname{Im}\left[\overline{G_0(t)}\left\{(t'(s) - q_1(t)\,\overline{t'(s)})\,\overline{v(t)} - \overline{q_2(t)\,t'(s)}\,v(t)\right\}\right]$$

$$+ \operatorname{Im}\left[G_0(t)\,u(t)\right] \operatorname{Re}\left[\overline{G_0(t)}\left\{(t'(s) - q_1(t)\,\overline{t'(s)})\,\overline{v(t)} - \overline{q_2(t)\,t'(s)}\,v(t)\right\}\right]\Big)ds =$$

0.

3.6 The boundary-value problem for a second-order elliptic system degenerate at the boundary

In this section we restrict our attention on the equation

$$a(z)\,u_{\bar{z}z} + b(z)\,\bar{u}_{\bar{z}^2} = f(z) \tag{3.136}$$

in a bounded multiple connected domain G^+. As we know from Chapter 1, if (3.136) is purely elliptic, i.e.

$$|\,a(z)\,| \neq |\,b(z)\,| \tag{3.137}$$

in the closure $\bar{G}^+ = G^+ + \Gamma$, then for determinacy of a solution of (3.136) it needs two real boundary conditions on Γ (Dirichlet, Neumann or condition A_n). We assume that the condition (3.137) holds inside the domain G^+ and is degenerate

$$|\,a(t)\,| \equiv |\,b(t)\,| \tag{3.138}$$

for $t \in \Gamma$.

In this case we show that for determinacy of a solution of equation (3.136) it needs only one real condition. Namely we consider the following boundary-value problem: find a solution u of (3.136) in G^+, continuous in $\bar{G}^+$, satisfying the boundary condition:

$$a(t)\,\overline{u^+(t)} - \overline{b(t)\,t'^2(s)}\,u^+(t) = 0 \qquad (3.139)$$

on Γ.

We assume $a(z), b(z)$ to be $H_p^1(G^+)$ functions and for $f(z)$ to be an $L_p(G^+)$ function with $p > 2$.

Theorem 3.6. *The problem (3.136), (3.138) with* $|\, a(z)\,| \neq |\, b(z)\,|$ *in* G^+ *and* $|\, a(t)\,| \equiv |\, b(t)\,|$ *on* Γ *is uniquely solvable for any right-hand side* $f \in L_p(G^+), p > 2$.

Proof. Denote

$$u_z = w. \qquad (3.140)$$

Then (3.136) is equation for unknown $w(z)$:

$$a\,w_{\bar{z}} + b\,\overline{w_z} = f(z). \qquad (3.141)$$

By Theorem 3.2, this equation is solvable if and only if its right-hand side f satisfies the conditions

$$\operatorname{Re} \iint_{G^+} f(z)\,\overline{v_j(z)}\,dx\,dy = 0,\ j,\ldots,k', \qquad (3.142)$$

where $\{\, v_j(z)\,\}$ form a complete system of nontrivial solutions of the equation

$$(\bar{a}\,u)_z + (\bar{b}\,u)_{\bar{z}} = 0 \qquad (3.143)$$

in G^+, continuous in $\bar{G}^+$, satisfying the condition

$$a(t)\,\overline{u^+(t)} - b(t)\,\overline{t'^2(s)}\,u^+(t) = 0 \qquad (3.144)$$

on Γ.

In case (3.141) is solvable its solution is expressed as

$$w(z) = \tilde{K} f + \sum_{j=1}^{\ell} c_j\, w_j(z), \qquad (3.145)$$

where $\tilde{K}$ is some integral operator compact in $L_p(G^+)$ and $\{w_j\}$ forms a solution of the homogeneous equation:

$$a\,w_{\bar{z}} + b\,\overline{w_z} = 0; \tag{3.141°}$$

c_j are arbitrary real constants.

Integrating (3.140), we obtain

$$u(z) = \overline{\psi(z)} + \bar{T}_{G^+} w,$$

where $\psi(z)$ is an arbitrary function, holomorphic in G^+. Hence

$$u(z) = \overline{\psi(z)} + \bar{T}_{G^+}\tilde{K}f + \sum_{j=1}^{\ell} c_j \bar{T}_{G^+} w_j\,. \tag{3.146}$$

Substituting this representation into the boundary condition (3.139), we obtain the following boundary condition for the holomorphic function ψ:

$$a(t)\,\psi^+(t) - \overline{b(t)}\,\overline{t'^2(s)}\,\overline{\psi^+(t)} = \sum_{j=1}^{\ell} c_j\left[\overline{b(t)}\,\overline{t'^2(s)}\;\bar{T}_{G^+} w_j - a(t)\,T_{G^+}\bar{w}_j\right]$$

$$+\, b(t)\,\overline{t'^2(s)}\,\bar{T}_{G^+}\tilde{K}f - a(t)\,T_{G^+}\overline{\tilde{K}f} = h(t). \tag{3.147}$$

This problem is solvable if and only if

$$\operatorname{Im}\int_\Gamma \frac{\overline{h(t)}\;a(t)\varphi_j(t)}{\overline{t'^2(s)}\;\overline{b(t)\,a(t)}}\,dt = 0,\; j = 1,\ldots,\ell, \tag{3.148}$$

where the $\varphi_j(z)$ are solutions of the homogeneous problem adjoint to problem (3.147), i.e. the problem

$$\overline{a(t)}\,\varphi^+(t) - b(t)\,\overline{\varphi^+(t)} = 0 \tag{3.149}$$

The number of linearly independent nontrivial solutions of this problem is the same as the number of nontrivial solutions of the homogeneous equation (3.141°).

The conditions (3.148) are

$$\sum_{j=1}^{\ell} c_j \operatorname{Im}\left[\int_\Gamma T_{G^+}\bar{w}_j\,\varphi_i(t)\,dt - \int_\Gamma \bar{T}_{G^+} w_j\,\varphi_i(t)\,d\bar{t}\right]$$

$$= \operatorname{Im}\left[\int_\Gamma T_{G^+}\tilde{K}f\ \varphi_i(t)\,dt - \int_\Gamma \bar{T}_{G^+}\tilde{K}f\ \overline{\varphi_i(t)}\,dt\right],$$

because $\overline{a(t)}\,\varphi^+(t) = b(t)\,\overline{\varphi^+(t)}$ and $\overline{t'^2(s)}\,dt = d\bar{t}$.

Hence by means of Green's formula, we obtain

$$\sum_{j=1}^{l} c_j \operatorname{Re} \iint_{G^+} w_j(z)\,\overline{\varphi_i(t)}\,dx\,dy = \operatorname{Re} \iint_{G^+} \tilde{K}f\ \overline{\varphi_i(t)}\,dx\,dy. \tag{3.150}$$

Since $w_1,\dots,w_l$ are linearly independent as well as $\varphi_1,\dots,\varphi_l$, then the constants $c_1,\dots,c_l$ are uniquely determined by (3.150) through the right-hand side f, because

$$\det\left(\operatorname{Re} \iint_{G^+} w_j(z)\,\overline{\varphi_i(t)}\,dx\,dy\right) \neq 0. \tag{3.151}$$

It means that in the right-hand sides (3.146), (3.147) all quantities are known and expressed through f and hence the function ψ and then u uniquely expressed through the right-hand side f of the equation (3.136).

Now considering the more general equation

$$a(z)\,u_{\bar{z}z} + b(z)\,\bar{u}_{\bar{z}^2} + F(z, u, u_z, u_{\bar{z}}) = 0 \tag{3.152}$$

with the linear function $F(z, u, v, w)$, we prove the following:

Theorem 3.7. *The problem (3.152), (3.138), with* $|a(z)| \neq |b(z)|$ *in* G^+ *and* $|a(z)| \equiv |b(z)|$ *on* Γ *possesses Fredholm's property.*

3.7 Nonelliptic degenerations

In this section we consider nonelliptic first-order equations in $\mathbb{R}^3$ with variable complex coefficients with properties which we noted above as degenerations. First consider the equation

$$u_{\bar{z}} + i\,\bar{z}\,u_{x_1} = f(x_1, z), \quad z = x_2 + i\,x_3. \tag{3.153}$$

This equation is not elliptic. If we assume f to have compact support in $\mathbb{R}^3$ and to vanish at point $z = 0$, then this equation has a solution for any such function and the corresponding homogeneous equation ($f \equiv 0$) has infinitely many nontrivial solutions bounded and continuous on the whole space $\mathbb{R}^3$:

$$u_k(x_1, z) = \frac{(x_1 - 2x_2 x_3)^k}{(1+(x_1 - 2x_2 x_3)^2)^k}, \quad k = 0,1,\ldots . \tag{3.154}$$

Indeed, after transformation $\zeta = \xi + i\,\eta = z^2/2$ (3.153) becomes

$$u_{\bar\zeta} + i\, u_{x_1} = \frac{f(x_1, \sqrt{(2\zeta)})}{\sqrt{(2\bar\zeta)}} \tag{3.155}$$

and after the transformation

$$\xi' = x_1 - 2\eta, \quad \eta' = x_1/2, \tag{3.156}$$

(3.155) becomes the inhomogeneous Cauchy–Riemann equation with parameter ξ' :

$$u_{\bar\zeta'} = \frac{f(2\eta', \sqrt{[2(\xi + i(\eta' - \xi'/2)]})}{\sqrt{[2(\xi - i(\eta' - \xi'/2)]}}, \tag{3.157}$$

where $u_{\bar\zeta'} = 1/2(u_\xi + i\, u_{\eta'})$. Integrating (3.157) we obtain its solution in the form

$$u(\xi', \zeta') = -\frac{1}{\pi} \iint \frac{f(2y', \sqrt{[2(x' + i(y' - \xi'/2)]})\, dx'dy'}{x' + iy' - (\xi + i\eta')}$$

apart from the solution of the homogeneous equation.

Let us now consider the following equation, obtained from (3.153) by replacing $\bar z$ by $z/2$ in the second term of the left-hand side:

$$u_{\bar z} + \frac{iz}{2}\, u_{x_1} = f(x_1, z), \tag{3.158}$$

which is the famous Lewy [46] equation without solution. We assume for $f(x_1, z)$ to be $L_2(\mathbb{R}^3)$ function, having compact support in z-plane $\mathbb{C}$. In contrast to equation (3.153) this equation is not solvable for any such right-hand sides. Here we describe the subspace of right-hand sides for which the equation has solution. We use the Fourier transform $\tilde u(\xi_1, z)$ with respect to x_1:

$$\tilde u(\xi_1, z) = \int_{-\infty}^{\infty} u(x_1, z)\, e^{-i\xi_1 x_1}\, dx_1.$$

We obtain instead of (3.158) the elliptic equation

$$\tilde{u}_{\bar{z}} - \frac{\xi_1}{2} z \tilde{u} = \tilde{f}(\xi_1, z). \tag{3.159}$$

Hence

$$\tilde{u}(\xi_1, z) = e^{(\xi_1/2)|z|^2}\left(\Phi_0(\xi_1, z) - \frac{1}{\pi}\iint_{\mathbb{C}} \frac{e^{-(\xi_1/2)|\zeta|^2}\tilde{f}(\xi_1, \zeta)\, d\xi\, d\eta}{\zeta - z}\right),$$

where $\Phi_0(\xi_1, z)$ is an L_2-function in ξ_1 and holomorphic in z. Since $\tilde{f}(\xi_1, z)$ has compact support in the z-plane, then for $|z|$ large enough we obtain

$$-\frac{1}{\pi}\iint_{\mathbb{C}} \frac{e^{-(\xi_1/2)|\zeta|^2}\tilde{f}(\xi_1, \zeta)\, d\xi\, d\eta}{\zeta - z}$$

$$= \sum_{k=0}^{\infty}\left(\frac{1}{\pi}\iint_{\mathbb{C}} \zeta^k e^{-(\xi_1/2|\zeta|^2}\tilde{f}(\xi_1, \zeta)\, d\xi\, d\eta\right)\frac{1}{z^{k+1}},$$

i.e. the necessary and sufficient conditions for $\tilde{u}(\xi_1, z)$ to be in L_2 in case $\xi_1 > 0$ are $\Phi_0(\xi_1, z) = 0$ and

$$\iint_{\mathbb{C}} \zeta^k\, e^{-(\xi_1/2|\zeta|^2}\tilde{f}(\xi_1, \zeta)\, d\xi\, d\eta = 0, \quad k = 0,1,..., \tag{3.160}$$

because $e^{\xi_1|z|^2/2}$ is growing in case $\xi_1 > 0$. But (3.160) means that $\tilde{f}(\xi_1, z)$ is orthogonal to functions $\nu_k : (\tilde{f}, \nu_k) = 0$,

$$\nu_k(\xi_1, z) = \bar{z}^k e^{-(\xi_1/2)(|z|^2+1)}, \quad k = 0,1,...,$$

which are the solutions of the homogeneous equation

$$\tilde{v}_z + \frac{\xi_1}{2}\bar{z}\,\tilde{v}(\xi_1, z) = 0, \tag{3.161}$$

obtained by means of the Fourier transformation from the homogeneous equation

$$-v_z + \frac{i\bar{z}}{2} v_{x_1} = 0, \tag{3.161'}$$

adjoint to (3.158).

Now we consider instead (3.159) the equation

$$\tilde{u}_{\bar{z}} - \frac{\xi_1}{2} z\, \tilde{u}(\xi_1, z) = \tilde{F}(\xi_1, z) \tag{3.159'}$$

with right-hand side equal to

$$\tilde{F}(\xi_1, z) = \tilde{f}(\xi_1, z) - H(\xi_1) \sum_{\ell=0}^{\infty} a_\ell(\xi_1)\, \bar{z}^\ell\, e^{-(\xi_1/2)(|z|^2+1)}, \tag{3.162}$$

where $H(\xi_1)$ is the Heaviside function

$$H(\xi_1) = \begin{cases} 1, & \xi_1 > 0 \\ 0, & \xi_1 < 0 \end{cases}.$$

Substituting (3.162) into (3.166), we obtain for $\xi_1 > 0$

$$\iint \tilde{F}(\xi_1, z) z^k e^{-(\xi_1/2)(|z|^2+1)}\, dx\, dy = \iint \tilde{f}(\xi_1, z)\, z^k e^{-(\xi_1/2)(|z|^2+1)}\, dx\, dy$$

$$- H(\xi_1) \sum_{\ell=0}^{\infty} a_\ell(\xi_1) \iint z^k \bar{z}^\ell e^{-\xi_1(|z|^2+1)}\, dx\, dy = 0.$$

But since

$$\iint z^k \bar{z}^\ell e^{-\xi_1(|z|^2+1)}\, dx\, dy = \begin{cases} \dfrac{\pi k!}{\xi_1^{k+1}}\, e^{-\xi_1}, & \ell = k, \\ 0, & \ell \neq k, \end{cases}$$

because for $\xi_1 > 0$:

$$\iint |z|^{2k} e^{-\xi_1(|z|^2+1)}\, dx\, dy = 2\pi \int_0^\infty \rho^{2k+1} e^{-\xi_1(\rho^2+1)}\, d\rho$$

$$= \pi \int_0^\infty \rho_1^k e^{-\xi_1 \rho_1}\, d\rho_1\, e^{-\xi_1} = \frac{\pi k!}{\xi^{k+1}}\, e^{-\xi_1},$$

then for $\xi_1 > 0$:

$$a_k(\xi_1) = \frac{\xi_1^{k+1}}{\pi k!} \iint \zeta^k e^{-(\xi_1/2)(|\zeta|^2-1)} \tilde{f}(\xi_1, \zeta)\, d\xi\, d\eta.$$

Consequently

$$\tilde{F}(\xi_1, z) = \tilde{f}(\xi_1, z) - \iint \tilde{K}(\xi_1, \zeta, z) \tilde{f}(\xi_1, \zeta)\, d\xi\, d\eta,$$

where

$$\tilde{K}(\xi_1, \zeta, z) = \frac{\xi_1 H(\xi_1)}{\pi} \sum_{k=0}^{\infty} \frac{(\xi_1 \zeta \bar{z})^k}{k!} e^{-(\xi_1/2(|\zeta|^2+|z|^2)}$$

$$= \frac{\xi_1 H(\xi_1)}{\pi} e^{(\xi_1/2)(\zeta\bar{z} - (|\zeta|^2+|z|^2))}. \tag{3.163}$$

Taking the inverse Fourier transformation we obtain

$$F(x_1, z) = f(x_1, z) - \frac{1}{\pi^2} \iint d\xi\, d\eta \int_0^{\infty} \xi_1 e^{\xi_1(\zeta\bar{z} - \frac{1}{2}|\zeta|^2+|z|^2)) + i\xi_1 x_1} \tilde{f}(\xi_1, z)\, d\xi_1 \tag{3.164}$$

Thus we have proved the following:

Lemma 3.9. *The equation (3.15') is solvable in* $L_2(\mathbb{R}^3)$ *if and only if its right-hand side* F *has the form (3.164) for any* $f \in L_2$ *with compact support in the* z*-plane.*

Denote by $\bar{L}$ the subspace of the space $L_2(\mathbb{R}^3)$, consisting solutions of adjoint homogeneous equation (3.161′). If $f \in \bar{L}$, then $\tilde{f}(\xi_1, z)$ will be a solution of equation (3.161), i.e.

$$\tilde{f}(\xi_1, z) = \sum_{k=0}^{\infty} \tilde{a}_k \bar{z}^k e^{-(\xi_1/2(|z|^2+1)}.$$

Then we obtain

$$\tilde{K}\tilde{f} = \frac{1}{\pi} \iint \xi_1 H(\xi_1) e^{\xi_1(\zeta\bar{z} - \frac{1}{2}|\zeta|^2+|z|^2))} \tilde{f}(\xi_1, \zeta)\, d\xi\, d\eta$$

$$= \sum_{k,\ell=0}^{\infty} \frac{1}{\pi} \iint \xi_1 H(\xi_1) \frac{(\xi_1 \zeta \bar{z})^\ell}{\ell!} \tilde{a}_k \bar{\zeta}^k e^{-\xi_1(|\zeta|^2+1)}\, d\xi\, d\eta\; e^{-(\xi_1/2)(|z|^2-1)}$$

$$= \begin{cases} \sum_{k=0}^{\infty} a_k \bar{z}^k e^{-(\xi_1/2(|z|^2+1)}, & \xi_1 > 0, \\ 0, & \xi_1 < 0, \end{cases}$$

i.e.

$$\tilde{K}\tilde{f} = \begin{cases} \tilde{f}(\xi_1, z), & \xi_1 > 0, \\ 0, & \xi_1 < 0. \end{cases} \tag{3.165}$$

Taking inverse Fourier transformation of function

$$H(\xi_1) \sum_{k=0}^{\infty} \tilde{a}_k \xi^{k+1} \bar{z}^k e^{-\xi_1 (|z|^2+1)}$$

we obtain

$$\frac{1}{2\pi} \int_0^{\infty} \sum_{k=0}^{\infty} \tilde{a}_k \bar{z}^k \xi_1^{k+1} e^{-(\xi_1/2)(|z|^2+1) + i\xi_1 x_1} d\xi_1$$

$$= \sum_{k=0}^{\infty} \tilde{a}_k \bar{z}^k \frac{1}{2\pi} \int_0^{\infty} \xi_1^{k+1} e^{i\xi_1 (x_1 + \frac{i}{2}(|z|^2+1))} d\xi_1$$

$$= \sum_{k=0}^{\infty} \frac{\tilde{a}_k \bar{z}^k (-1)^{k+2} (k+1)!}{(x_1 + (i/2)(|z|^2+1))^{k+2} i^{k+2}} = \sum_{k=0}^{\infty} \tilde{a}_k i^{k+2} \frac{\bar{z}^k}{(x_1 + (i/2)(|z|^2+1))^{k+2}}.$$

The functions

$$v_k(x_1, z) = \frac{\bar{z}^k}{(x_1 + (i/2)(|z|^2+1))^{k+2}} \tag{3.166}$$

are solutions of homogeneous equation (3.161′). The relation (3.164) can be written also as

$$F(x_1, z) = f(x_1, z) = \frac{1}{2\pi} \iint d\xi \, d\eta \int_{-\infty}^{\infty} d\sigma \int_0^{\infty} \xi_1 e^{i\xi_1 (x_1 - \sigma - i(\zeta\bar{z} - \frac{1}{2}(|\zeta|^2 + |z|^2))} d\xi_1 f(\sigma, \zeta).$$

But

$$\int_0^\infty \xi_1\, e^{i\xi_1(x_1-\sigma - i(\zeta\bar{z} - \frac{1}{2}(|\zeta|^2+|z|^2))} = \frac{-1}{(\sigma - x_1 + i(\zeta\bar{z} - \frac{1}{2}(|\zeta|^2+|z|^2)))^2}.$$

Consequently

$$F(x_1, z) = f(x_1, z) - \iiint_{\mathbb{R}^3} N(\xi_1, \zeta; x_1, z) f(\xi_1, \zeta)\, d\xi_1\, d\xi\, d\eta \tag{3.167}$$

where

$$N(\zeta_1, \zeta; x_1, z) = \frac{-1}{(\xi_1 - x_1 + i(\zeta\bar{z} - \frac{1}{2}(|\zeta|^2+|z|^2)))^2}. \tag{3.168}$$

Let

$$N_1(\xi_1, \zeta; x_1, z) = \frac{1}{i(\xi_1 - x_1 + i(\zeta\bar{z} - \frac{1}{2}(|\zeta|^2+|z|^2))}.$$

Then

$$-\frac{\partial N_1}{\partial \zeta} + \frac{i}{2}\bar{\zeta}\frac{\partial N_1}{\partial \xi_1} = (\bar{\zeta} - \bar{z}) N,$$

i.e.

$$N = \frac{\left(-\frac{\partial N_1}{\partial \zeta} + \frac{i\bar{\zeta}}{2}\frac{\partial N_1}{\partial \xi_1}\right)}{(\bar{\zeta} - \bar{z})}$$

and for any function f, satisfying the equation (3.161′) we obtain

$$\frac{1}{2\pi^2}\iiint_{\mathbb{R}^3} N(\xi_1, \zeta; x_1, z) f(\xi_1, \zeta)\, d\xi_1\, d\xi\, d\eta$$

$$= \frac{1}{2\pi} \iint d\xi \, d\eta \int_{-\infty}^{\infty} \frac{\left(-\frac{\partial}{\partial \zeta} + \frac{i\zeta}{2} \frac{\partial}{\partial \xi_1}\right)(N_1 f)}{\bar{\zeta} - \bar{z}} \, d\xi_1.$$

But

$$-\frac{1}{\pi} \iint \frac{\frac{\partial N_1 f}{\partial \zeta} \, d\xi \, d\eta}{\bar{\zeta} - \bar{z}} = (N_1 f)_{\zeta = z},$$

$$\int_{-\infty}^{\infty} \frac{\partial}{\partial \xi_1} N_1 f \, d\xi_1 = \lim_{R \to \infty} \int_{-R}^{R} \frac{\partial N_1 f}{\partial \xi_1} \, d\xi_1 = \lim_{R \to \infty} \left\{ (N_1 f)_{\xi_1 = R} - (N_1 f)_{\xi_1 = -R} \right\} = 0,$$

because the function $N_1 f$ vanishes at infinity.

Hence

$$\frac{1}{2\pi^2} \iiint_{\mathbb{R}^3} N(\xi_1, \zeta; x_1, z) f(\xi_1, \zeta) \, d\xi_1 \, d\xi \, d\eta = \frac{1}{2\pi i} \int_{-\infty}^{\infty} \frac{f(\xi_1, z) \, d\xi_1}{\xi_1 - x_1},$$

because

$$N_1 |_{\zeta = z} = \left. \frac{1}{i(\xi_1 - x_1 + i(\zeta\bar{z} - \frac{1}{2}(|\zeta|^2 + |z|^2)))} \right|_{\zeta = z} = \frac{1}{i(\xi_1 - x_1)}.$$

Since

$$f(x_1, z) = \sum_{k=0}^{\infty} \tilde{a}_k \frac{\bar{z}^k}{(x_1 + (i/2)(|z|^2 + 1))^{k+2}},$$

then

$$\frac{1}{2\pi i} \int_{-\infty}^{\infty} \frac{f(\xi_1, z) \, d\xi_1}{\xi_1 - x_1} = \sum_{k=0}^{\infty} \frac{\tilde{a}_k \bar{z}^k}{2\pi i} \int_{-\infty}^{\infty} \frac{d\xi_1}{(\xi_1 + (i/2)(|z|^2 + 1))^{k+2} (\xi_1 - x_1)}.$$

Now taking into account the identity

$$\frac{1}{(\xi_1-\zeta)^m(\xi_1-x_1)}$$

$$=\frac{1}{x_1-\xi}\left(\frac{1}{\xi_1-x_1}-\frac{1}{\xi_1-\xi}\right)-\frac{1}{(x_1-\xi)^{m-1}(\xi_1-\xi)^2}-\dots-\frac{1}{(x_1-\xi)(\xi_1-\xi)^m},$$

we obtain

$$\frac{1}{2\pi i}\int_{-\infty}^{\infty}\frac{d\xi}{(\xi_1+(i/2)(|z|^2+1))^{k+2}(\xi_1-x_1)}=\frac{1}{(x_1+(i/2)(|z|^2+1))^{k+2}},$$

because

$$\frac{1}{2\pi i}\int_{-\infty}^{\infty}\frac{d\xi_1}{\xi_1-x_1}=\frac{1}{2\pi i}\lim_{R\to\infty}\int_{-R}^{R}\frac{d\xi_1}{\xi_1-x_1}=\frac{1}{2},$$

$$\frac{1}{2\pi i}\int_{-\infty}^{\infty}\frac{d\xi_1}{\xi_1+(i/2)(|z|^2+1)}=\frac{1}{2\pi i}\lim_{R\to\infty}\log(\xi_1+(i/2)|z|^2+1))\Big|_{-R}^{R}$$

$$=\frac{1}{2\pi}\{\arg(\infty+(i/2)(|z|^2+1))-\arg(-\infty+(i/2)(|z|^2+1))\}=-1/2$$

and hence

$$\frac{1}{2\pi i}\int_{-\infty}^{\infty}\frac{d\xi_1}{\xi_1-x_1}-\frac{1}{2\pi i}\int_{-\infty}^{\infty}\frac{d\xi_1}{\xi_1+(i/2)(|z|^2+1)}=1,$$

and the remaining integrals are equal to zero.

Thus, we have

$$\frac{1}{2\pi i}\int_{-\infty}^{\infty}\frac{f(\xi_1,z)\,d\xi_1}{\xi_1-x_1}=\sum_{k=0}^{\infty}\frac{\tilde{a}_k\,\bar{z}^k}{(\xi_1+(i/2)(|z|^2+1))^{k+2}}=f(x_1,z).$$

This means that for any $f\in\bar{L}$,

$$f(x_1, z) = \frac{1}{2\pi^2} \int_{\mathbb{R}^3} \frac{f(\xi_1, \zeta)\, d\xi_1\, d\xi\, d\eta}{((\xi_1 - x_1 - \operatorname{Im} \zeta \bar{z} + (i/2)|\zeta - z|^2)^2} \tag{3.169}$$

i.e. the kernel

$$K(\xi, x) = \frac{1}{2\pi^2 (\xi_1 - x_1 - \operatorname{Im} \zeta \bar{z} + (i/2)|\zeta - z|^2)^2}$$

reproduces $\bar{L}$ and integral operator Kf with this kernel is the orthogonal projection in $L_2(\mathbb{R}^3)$.

3.8 Elliptic systems in a plane, degenerate on a line

(a) Consider Beltrami's equation on plane:

$$w_{\bar{z}} - q(z)\, w_z = 0 \tag{3.170}$$

with ellipticity condition

$$|q(z)| \neq 1 \tag{3.171}$$

everywhere on $\mathbb{C}$, except simple analytic curve γ, along which this condition is violated:

$$|q(z)| \equiv 1, \quad z \in \gamma. \tag{3.172}$$

We assume $|q(z)|$ to be an analytic function of the variables x, y in some neighbourhood D of γ. If all the partial derivatives of the function $|q(z)|^2 - 1$ up to the $(n-1)$th order (inclusively) are equal to zero along γ and at least one of the derivatives of nth order is nonzero, then in a certain neighbourhood $\Delta \subset D$ of the curve γ the following representation holds:

$$|q(z)|^2 - 1 = \eta^n(x, y) \cdot a(x, y) \tag{3.173}$$

where $\eta(x, y) = 0$ is the equation of γ, and $a(x, y) \neq 0$.

First we assume that the direction of the characteristic of the equation (3.170) at the points of γ do not coincide with the direction of the tangent to γ, i.e. that along γ

$$(1 - \operatorname{Re} q(z))\eta_x - \operatorname{Im} q(z)\,\eta_y \neq 0. \tag{3.174}$$

Let the function $\xi(x, y)$ be a solution of

$$(1 - \operatorname{Re} q(z))\,\xi_x - \operatorname{Im} q(z)\,\xi_y = 0. \tag{3.175}$$

Evidently one can find a subdomain $\Delta' \subset \Delta$ such that it contains γ and such that the function $b(x, y)$, satisfying the equations

$$\xi_x = b(x, y) \operatorname{Im} q(z), \xi_y = b(x, y)\,(1 - \operatorname{Re} q(z)) \tag{3.176}$$

does not vanish in it.

Since

$$J = \begin{vmatrix} \xi_x & \xi_y \\ \eta_x & \eta_y \end{vmatrix} = - b(x, y)\,\{(1 - \operatorname{Re} q)\eta_x - \operatorname{Im} q \cdot \eta_y\} \neq 0$$

in Δ', the mapping $\xi = \xi(x, y)$, $\eta = \eta(x, y)$ is a homeomorphism of the neighbourhood Δ' onto some domain $\tilde{\Delta}'$ in the plane of the variables ξ, η. As a result of this mapping, the equation (1) takes the form

$$(\xi_{\bar{z}} - q(z)\xi_z)w_\xi + (\eta_{\bar{z}} - q(z)\eta_z)\,w_\eta = 0. \tag{3.177}$$

Since, in virtue of (3.173) and (3.176),

$$\xi_{\bar{z}} - q(z)\xi_z = -\frac{i\,b(x,y)}{2}(|q(z)|^2 - 1) = \frac{-i\,a(x,y)\,b(x,y)}{2}\eta^n$$

and in virtue of the inequality (3.174),

$$\eta^*(z) = \eta_{\bar{z}} - q(z)\,\eta_z \neq 0$$

in Δ', it follows that, dividing (3.177) by $\eta^*(z)$, we obtain

$$\eta^n k(\zeta)\,w_\xi + i\,w_\eta = 0, \tag{3.178}$$

where

$$k(\zeta) = a\,b/2\eta^*. \tag{3.179}$$

Note that in virtue of (3.174)

$$\operatorname{Re} k(\zeta) = \frac{a\,b}{|\eta^*|^2}\left[(1 - \operatorname{Re} q)\eta_x - \operatorname{Im} q\, \eta_y\right] \neq 0 \tag{3.180}$$

in $\tilde{\Delta}'$.

Now we consider the case when the equation

$$(1 - \operatorname{Re} q)\eta_x - \operatorname{Im} q\, \eta_y = 0 \tag{3.181}$$

holds along γ, i.e. that γ is characteristic of (3.170). In this case one can find functions $c(x, y)$ and $d(x, y)$, for which

$$(1 - \operatorname{Re} q)c - \operatorname{Im} q\, d \neq 0. \tag{3.182}$$

Let $\xi(x, y)$ be a solution of (3.175) and $\tilde{\eta}(x, y)$ satisfy the equation

$$d(x, y)\, \tilde{\eta}_x - c(x, y)\, \tilde{\eta}_y = 0. \tag{3.183}$$

Obviously, one can find a section γ_1 of the curve γ and domain Δ' containing the arc γ_1, in which inequality (3.182) holds and the functions $a_1(x, y)$, $b_1(x, y)$, which satisfy the relations

$$\xi_x = a_1(x, y) \operatorname{Im} q(z), \quad \xi_y = a_1(x, y)\,(1 - \operatorname{Re} q(z)),$$

$$\tilde{\eta}_x = b_1(x, y)\, c(x, y), \quad \tilde{\eta}_y = b_1(x, y)\, d(x, y) \tag{3.184}$$

do not vanish.

It follows from these relations and the equality (3.183) that the Jacobian of the transformation $\xi = \xi(x, y)$, $\tilde{\eta} = \tilde{\eta}(x, y)$ is nonzero in Δ'':

$$J = a_1 a_2\,(\operatorname{Im} q(z)\, d(x, y) - (1 - \operatorname{Re} q(z))\, c(x, y) \neq 0.$$

Let $\eta^0(\xi, \tilde{\eta}) = \eta[x(\xi, \tilde{\eta}), y(\xi, \tilde{\eta})]$. Since the directions of the tangent curves $\eta(x, y) =$ constant, $\xi(x, y) =$ constant coincide along γ_1 by virtue of (3.174) and (3.181), it follows that $\eta^0(0, \tilde{\eta}) = 0$. Moreover, it is easy to see that

$$\eta^0_\xi = \frac{b_1}{J}\,(d(x, y)\eta_x - c(x, y)\eta_y) \neq 0. \tag{3.185}$$

and since

$$\xi_{\bar z} - q(z)\xi_z = \frac{a_1}{2\mathrm{i}}(|q(z)|^2 - 1) = [\eta^0(\xi, \hat\eta)]^n a(x, y)$$

we see that as a result of the nondegenerate transformation $\xi = \xi(x, y)$, $\hat\eta = \hat\eta(x, y)$ equation (3.170) takes the form

$$\xi^n \tilde k(\zeta)\, w_\xi + \mathrm{i}\, w_{\hat\eta} = 0 \tag{3.186}$$

with

$$\tilde k(\zeta) = \frac{a_1 b_1 \eta_0(\xi, \hat\eta)}{2(\hat\eta_{\bar z} - q(z)\hat\eta_z)_{z = z(\zeta)}}.$$

Note that by virtue of inequality (3.182) it is easy to see that the function $\operatorname{Re} \tilde k(\zeta)$ is nonzero in the domain $\tilde\Delta_1$, the image of the domain Δ_1 in the mapping $\xi = \xi(x, y)$, $\hat\eta = \hat\eta(x, y)$. Thus we have proved the following.

Lemma 3.10. *In a neighbourhood of the line of degeneracy γ equation (3.170) can be reduced to the form (3.178) if γ is not characteristic and to the form (3.186) if γ is characteristic of (3.170).*

This means that the line of degeneracy γ locally may be straightened into the line $y = 0$ around which equation (3.170) may be taken in the form:

$$y^n k\, w_x + \mathrm{i}\, w_y = 0, \quad \operatorname{Re} k \neq 0 \tag{3.178'}$$

in case γ is not characteristic and into the line $y = 0$ around which equation (3.170) may be taken in the form

$$w + \mathrm{i}\, y^n k\, w_y = 0 \tag{3.186'}$$

in case γ is characteristic.

If $k \equiv 1$, then for n even the equation (3.178′) reduces to the Cauchy-Riemann equation by the transformation $y \to y^{n+1}/(n+1)$. But for n odd the equation

$$y^n w_x + \mathrm{i}\, w_y = f(z) \tag{3.187}$$

possesses the same property in $\mathbb{R}^2$ as the Lewy equation (3.158) in $\mathbb{R}^3$. To show this we make the following transformation:

$$\eta = \begin{cases} \dfrac{y^{n+1}}{n+1}, & y > 0, \\ -\dfrac{y^{n+1}}{n+1}, & y < 0, \end{cases} \tag{3.188}$$

and as a result we obtain instead of (3.187) the following equation in the plane $x + i\eta$:

$$w_x + \text{i sign}\,\eta\, u_\eta = f_1(x, \eta), \tag{3.189}$$

where

$$f_1(x, \eta) = \left\{ \begin{array}{ll} \dfrac{1}{[(n+1)\eta]^{\frac{n}{n+1}}} f(x + \text{i}[(n+1)\eta]^{\frac{1}{n+1}}), & \eta > 0, \\ \dfrac{1}{[-(n+1)\eta]^{\frac{n}{n+1}}} f(x + \text{i}[-(u+1)\eta]^{\frac{1}{n+1}}), & \eta < 0. \end{array} \right\} \tag{3.190}$$

The further transformation

$$x + \text{i}\eta = \frac{\omega - \text{i}}{\text{i}(\omega + \text{i})}, \tag{3.191}$$

which maps the upper half-plane $\eta > 0$ onto the interior of the disc $|\omega| < 1$ and the lower half-plane onto the exterior of this disc such that the degeneracy line $\eta = 0$ maps onto the circle $|\omega| = 1$, transforms equation (3.189) into the following:

$$(1 + \text{sign}\,(1 - |\omega|))\, W_{\bar{\omega}} + (1 - \text{sign}\,(1 - |\omega|))\, w_{\bar{\omega}} = F(\omega),$$

$$F(\omega) = \begin{cases} \dfrac{f_1}{(\bar{\omega} - \text{i})^2}, & |\omega| < 1, \\ \dfrac{f_1}{(\omega + \text{i})^2}, & |\omega| > 1, \end{cases} \tag{3.192}$$

which we considered in Section 3 (Chapter 2). Hence by virtue of Lemma 3.1, we come to the following.

Lemma 3.11. *The equation (3.187) with* $f \in L_p^{\text{loc}}(\mathbb{C}), p > 2,$ having compact support in $\mathbb{C}$, is solvable in $L_2(\mathbb{C})$ if and only if

$$\iint\limits_{y>0} f(z)\left(\frac{t}{\tau}\right)^k \frac{\bar{\tau}}{\tau}\,dx\,dy + \iint\limits_{y<0} f(z)\left(\frac{t}{\tau}\right)^{k+2} \frac{\bar{t}}{t}\,dx\,dy = 0, k = 0,1,... \quad (3.193)$$

where

$$t = x + i(y^{n+1/n+1} - 1), \; \tau = x + i(y^{n+1/n+1} + 1).$$

3.9 Elliptic systems in $\mathbb{R}^3$ degenerate on a line and on a plane

The system of equations with respect to complex-valued functions u, v:

$$z u_{x_1} + 2v_z = f, \quad -2u_{\bar{z}} + \bar{z}\, v_{x_1} = g \quad (3.194)$$

has characteristic form $\chi = |z|^2\xi_1 + |\zeta|^2$, and hence is elliptic everywhere in $\mathbb{R}^3$, except on the line $z = x_2 + ix_3 = 0$. By means of the transformation

$$\zeta = z^2/2 \quad (3.195)$$

as a result (3.194) reduces to the elliptic system with constant coefficients

$$u_{x_1} + 2v_\zeta = \frac{f}{\sqrt{(2\zeta)}}, \quad -2u_{\bar{\zeta}} + v_{x_1} = \frac{g}{\sqrt{(2\bar{\zeta})}}, \quad (3.196)$$

which was considered in Chapter 2.

In contrast to (3.197) the system

$$\bar{z}\, u_{x_1} + 2\, v_z = f, \quad -2\, u_{\bar{z}} + z\, v_{x_1} = g, \quad (3.197)$$

with the same characteristic form as for (3.194), is not reducible to the system with constant coefficients. We consider this system in the ball $\Omega : x_1^2 + |z|^2 \le R$. Let $f, g \in H_p^1(\Omega)$, $p > 3$.

First we consider the following problem: find a solution of (3.197) in $C^2(\Omega)$, continuous in the closure $\bar{\Omega} = \Omega + \Gamma$, satisfying the boundary condition

$$v = 0 \quad (3.198)$$

on $\Gamma = \partial\Omega$ and the condition

$$\text{Re}\,(\bar{\lambda}\, u) = 0 \quad (3.199)$$

on the circle $\gamma: x_2^2 + x_3^2 = R^2$, where $\lambda(z)$ is a given Hölder-continuous function and $\lambda(z) \neq 0$ on γ.

Theorem 3.8. *If* $\kappa_\lambda = \mathrm{Ind}_\gamma \lambda \geq 0$, *then the problem (3.197) - (3.199) is solvable for any right-hand sides* $f, g \in H_p^1(\Omega), p > 3$ *and the corresponding homogeneous problem has* $2\kappa_\lambda + 1$ *nontrivial solutions, but if* $\kappa_\lambda < 0$, *then this problem is solvable if and only if its right-hand side satisfies* $-2\kappa_\lambda - 1$ *conditions and the corresponding homogeneous problem has no nontrivial solutions.*

Proof. By differentiation of the second equation with respect to x_1 and by means of the first equation, we obtain

$$|z|^2 v_{x_1x_1} + 4 v_{z\bar{z}} = 2f_{\bar{z}} + \bar{z}\, g_{x_1} \tag{3.200}$$

Now using the transformation (3.195), we see that v satisfies the inhomogeneous Laplace equation

$$v_{x_1x_1} + 4v_{\zeta\bar{\zeta}} = (2\zeta)^{-1/2}(2f_{\bar{\zeta}} + g_{x_1}). \tag{3.201}$$

Hence the function v is determined through the right-hand side of (3.1907) as a solution of the Dirichlet problem (3.201), (3.198). Then the function u is determined as a solution of an overdetermined system (3.197) uniquely through f and g apart from the holomorphic function φ of z :

$$u = \varphi(z) + \tilde{K}(f, g), \tag{3.202}$$

where $\tilde{K}(f, g)$ is some integral operator, compact in $L_p(\Omega)$, $p > 3$. Substituting (3.202) into (3.199) we obtain for the holomorphic function φ the Riemann-Hilbert problem in the disc $x_2^2 + x_3^2 \leq R^2$ with the condition $\mathrm{Re}\, \bar{\lambda}\, \varphi = \tilde{\gamma}$ on γ and thus we complete the proof of the theorem.

Let Ω_+ be the upper half-ball $\Omega \cap (x_3 > 0)$ and D_+ be the half disc obtained by crossing Ω_+ with plane $x_2 - x_3 = 0$.

Problem R_H^+. Find a solution of (3.197) in $C^2(\Omega_+)$ continuous in the closure $\bar{\Omega}$, satisfying the conditions

$$\mathrm{Im}\, u = 0, \;\; \mathrm{Im}\, v = 0 \tag{3.203}$$

on $\Gamma_+ = \partial\Omega_+$ and the condition

$$\alpha \operatorname{Re} u + \beta \operatorname{Re} v = 0 \tag{3.204}$$

on $\gamma_+ = \partial D_+$ with given real Hölder-continuous functions such that $\alpha^2 + \beta^2 \neq 0$ on γ_+.

Theorem 3.9. *If* $\kappa_+ = \operatorname{Ind}_{\gamma_+}(\alpha + i\beta) \geq 0$, *then the problem* $\overset{+}{R_H}$ *is solvable for any right-hand sides* $f, g \in H_p^1(\Omega_+), p > 3$, *and the corresponding homogeneous problem has* $2\kappa_+ + 1$ *nontrivial solutions, but if* $\kappa_+ < 0$, *then this problem is solvable if and only if its right-hand sides satisfy* $-2\kappa_+ - 1$ *conditions and the corresponding homogeneous problem has no nontrivial solution.*

Proof. By differentiation of the first equation of (3.197) with respect to x_1 and by means of the second equation we obtain

$$|z|^2 u_{x_1x_1} + 4u_{z\bar{z}} = z f_{x_1} - 2g_z .$$

Using transformation (3.195), we obtain

$$u_{x_1x_1} + 4u_{\zeta\bar{\zeta}} = \frac{f_{x_1} - 2g_\zeta}{\sqrt{(2\bar{\zeta})}}, \tag{3.205}$$

i.e. the function $\operatorname{Im} u$ as well as the function $\operatorname{Im} v$ can be determined uniquely through the right-hand sides f and g as a solution of the Dirichlet problems (3.203) for the inhomogeneous Laplace equations

$$(\operatorname{Im} u)_{x_1x_1} + 4(\operatorname{Im} u)_{\zeta\bar{\zeta}} = \operatorname{Im}\frac{f_{x_1} - 2g_\zeta}{\sqrt{(2\bar{\zeta})}},$$

$$(\operatorname{Im} v)_{x_1x_1} + 4(\operatorname{Im} v)_{\zeta\bar{\zeta}} = \operatorname{Im}\frac{2f_{\bar{\zeta}} + g_{x_1}}{\sqrt{(2\zeta)}}.$$

Then (3.197) becomes an overdetermined system for the unknown function $w = \operatorname{Re} u + i \operatorname{Re} v$:

$$w_{x_2} + i x_2 w_{x_1} = f_2, \quad w_{x_2} + i x_3 w_{x_1} = f_3 \tag{3.206}$$

with right-hand sides expressed in terms of f and g; i.e. in the variables (x_1, x_2) and (x_1, x_3) these equations are elliptic, degenerate on the lines $x_2 = 0$ and $x_3 = 0$ respectively, as considered in the previous section. Now after the transformation

$$\xi_2 = \begin{cases} x_2^2/2, & x_2 > 0, \\ -x_2^2/2, & x_2 < 0, \end{cases}$$

$$\xi_3 = x_3^2/2,$$

(3.206) reduces to

$$w_{\xi_2} + \mathrm{i}\,\mathrm{sign}\,\xi_2\, w_{x_1} = \tilde{f}_2, \quad w_{\xi_3} + \mathrm{i}\, w_{x_1} = \tilde{f}_3 \tag{3.207}$$

and then after transformation

$$\eta = \xi_2 + \xi_3, \quad \xi = -\xi_2 + \xi_3 \tag{3.208}$$

we obtain

$$w_\xi = f_1(x_1, \xi, \eta), \quad \frac{1}{2}(w_\eta + \mathrm{i}\,\mathrm{sign}\,\eta\, w_{x_1}) = f_0(x_1, \xi, \eta). \tag{3.209}$$

Since domain D_+ is located in the quarter $x_2 > 0, x_3 > 0$, then its D_+^* in variables x_1, ξ, η is the domain in the plane $\xi = 0$ bounded in the upper half-plane $\eta > 0$ by the parabola

$$x_1^2 + 2\eta = R^2 \tag{3.210}$$

and on the x_1-axis by the segment $[-R, R]$. Integrating the first equation (3.209) we have

$$w = \omega(x_1, \eta) + \int_0^\xi f_1(x_1, \xi, \eta)\, d\xi, \tag{3.211}$$

where ω is an arbitrary complex valued C^2-function, depending only on two variables (x_1, η). Substituting (3.211) into the second equation (3.209) we obtain in the domain D_+^* the inhomogeneous Cauchy-Riemann equation for ω:

$$\omega_{\bar{\zeta}^*} = F(\zeta^*), \quad \zeta^* = x_1 + \mathrm{i}\,\eta. \tag{3.212}$$

Thus from (3.211) and (3.212) we obtain

$$w = \operatorname{Re} u + i \operatorname{Re} v = \varphi(\zeta^*) + K^*(f, g), \tag{3.213}$$

where $\varphi(\zeta^*)$ is an arbitrary function, holomorphic on D_+^* and continuous in $\overline{D}_+^*$, and $K^*(f, g)$ is some an integral operator, depending on the right-hand sides f and g, compact in $L_p(\Omega_+)$, $p > 3$. The boundary condition (3.203) transforms to the condition

$$\operatorname{Re}(\alpha - i\beta)w = 0 \tag{3.203'}$$

on the curve $\Gamma_+^* = \partial D_+^*$ consisting of the parabola (3.210) and the segment $[-R, R]$. Substituting (3.213) into (3.203′) we obtain the Riemann-Hilbert problem for the holomorphic function $\varphi(\zeta^*)$ in the plane domain D_+^* and hence the theorem follows from the results on the Riemann-Hilbert problem.

It is obvious how to formulate the above problems in the case of a more general domain Ω, Ω_+ than a ball or half-ball, and also how to investigate them for systems with lower terms.

The following system

$$u_{x_1} + 2x_1 v_z = f, \quad -2x_1 u_{\bar{z}} + v_{x_1} = g, \tag{3.214}$$

having characteristic form $\chi = \xi_1^2 + x_1^2 |\zeta|^2$ and hence an elliptic system for $x_1 \neq 0$ degenerate at the plane $x_1 = 0$, reduces to the system with constant coefficients

$$u_{x_1} + 2 v_z = \frac{f}{\sqrt{(2x_1)}}, \quad -2 u_{\bar{z}} + v_{x_1} = \frac{g}{\sqrt{(2x_1)}}$$

if we replace x_1 by $x_1^2/2$.

Hence a set of solutions of (3.214) is reach enough and so we may consider the boundary-value problem, for instance, in the half-space $x_1 > 0$. In contrast to (3.217) the system

$$x_1 u_{x_1} + 2 v_z = f, \quad -2 u_{\bar{z}} + x_1 v_{x_1} = g, \tag{3.215}$$

having characteristic form $\chi = x_1^2 \xi_1^2 + |\zeta|^2$ and thus also an elliptic system degenerate on the plane $x_1 = 0$ has a quite different property:

Lemma 3.12. *For any* $f, g \in L_p^{\text{loc}}(\mathbb{R}_+^3)$, $p > 3$ *decreasing fast enough at infinity and* *at the plane* $x_1 = 0$, *the system has a unique solution, bounded in the half-space* $\mathbb{R}^3_+ = \{x_1, z): x_1 \geq 0, z \in \mathbb{C}\}$ *apart from constants, and the corresponding homogeneous system has no nontrivial solution bounded in the half-space* $\overline{\mathbb{R}^3_+}$, *except constants.*

Proof. Evidently the mapping $(x_1, z) \to (\log x_1, z)$ reduces (3.215) given on the half-space to the system with constant coefficients

$$u_{x_1} + 2\, v_z = f, \quad -2\, u_{\bar{z}} + v_{x_1} = g \tag{3.215}$$

on the whole space $\mathbb{R}^3$, which is solvable for any right-hand sides from $L_p^{\mathrm{loc}}(\mathbb{R}^3)$, $p > 3$, decreasing fast enough at infinity, and the corresponding homogeneous system has the unique solution $u = \text{const}$, $v = \text{const}$.

It would also be interesting to investigate the properties of the following elliptic system degenerate on the lines:

$$u_{x_1} + 2\, \bar{z}\, v_z = f, \quad -2\, z\, u_{\bar{z}} + v_{x_1} = g$$

with characteristic form $\chi = \xi_1^2 + |z|^2 |\zeta|^2$.

4. Nonstationary systems

4.1 First-order hyperbolic systems in $\mathbb{R}^3$ and $\mathbb{R}^4$

(a) In this section we consider Cauchy problems for first-order hyperbolic systems, and in a later section we will investigate initial-boundary-value problems for them. Let $z = x + \mathrm{i}\, y$ be a complex variable on the complex plane $\mathbb{C}$ and let t denote a real variable on the half-line $t \geq 0$. We shall consider hyperbolic systems of first-order equations in the half-space $(t, z) \in \mathbb{R}^3_+$. First we consider the system of two real equations, which may be put into the following single complex equation

$$\bar{u}_t - Lu = f(t, z), \tag{4.1}$$

where L is an elliptic operator

$$Lu \equiv a(z)\, u_{\bar{z}} + b(z)\, u_z + c(z)u + d(z)\bar{u}, \;\; 2u_{\bar{z}} = u_x + \mathrm{i}\, u_y, \;\; 2u_z = u_x - \mathrm{i}\, u_y$$

and the bar over a function, as usual, denotes the complex conjugate. We assume that $a(z)$, $b(z)$ are functions vanishing at infinity with bounded first-order derivatives $|a_{\bar{z}}| \leq k, |b_z| \leq \ell$, that $c(z), d(z)$ are bounded in $\mathbb{C} : |c(z)| \leq \alpha, |d(z) \leq \beta$ and that $f(t, z)$ is the following function, having bounded $L_2(\mathbb{C})$-norm:

$$\|f(t)\|_{L_2(\mathbb{C})} = \left(\iint_{\mathbb{C}} |f(t, z)|^2 \,\mathrm{d}x\, \mathrm{d}y \right)^{1/2} < \infty$$

for any $t \geq 0$.

Let $u(t, z)$ be a solution of (4.1) continuously differentiable on t with finite $L_2(\mathbb{C})$-norm:

$$\| u(t) \|_{L_2(\mathbb{C})} = \left(\iint_{\mathbb{C}} | u(t, z)|^2 \,\mathrm{d}x\, \mathrm{d}y \right)^{1/2} < \infty.$$

Then we obtain, by (4.1),

$$\frac{d\|u(t)\|^2}{dt} = 2\,\mathrm{Re}\iint_{\mathbb{C}} u\,\bar{u}_t\,dx\,dy = \mathrm{Re}\iint_{\mathbb{C}}\Big[a(z)\,u_{\bar{z}}^2 + b(z)\,u_z^2$$

$$+ 2c(z)u^2 + 2d(z)\,|\,u\,|^2\Big]\,dx\,dy + 2\,\mathrm{Re}\iint_{\mathbb{C}} u(t,z)f(t,z)\,dx\,dy$$

$$= \mathrm{Re}\iint_{\mathbb{C}}\Big[(2c(z) - a_{\bar{z}} - b_z)u^2 + 2d(z)\,|\,u\,|^2\Big]\,dx\,dy + 2\,\mathrm{Re}\iint_{\mathbb{C}} u f\,dx\,dy, \tag{4.2}$$

because by Green's formula we obtain

$$\iint_{\mathbb{C}}\Big[(au^2)_{\bar{z}} + (bu^2)_z\Big]\,dx\,dy = \lim_{R\to\infty}\iint_{|z|>R}\Big[(au^2)_{\bar{z}} + (bu^2)_z\Big]\,dx\,dy$$

$$= \lim_{R\to\infty}\frac{R}{2}\int_0^{2\pi}\Big[a\,\mathrm{Re}^{i\theta})\,e^{i\theta} - b(\mathrm{Re}^{i\theta})e^{-i\theta}\Big]u(t,\mathrm{Re}^{i\theta})\,d\theta = 0$$

since a, b vanish at infinity and $u \in L_2(\mathbb{C})$ for $t \geq 0$. From (4.2), by means of the Schwartz inequality it follows that

$$\frac{d\|u(t)\|^2}{dt} \leq 2\gamma\,\|\,u(t)\,\|^2 + 2\,\|\,u(t)\,\|\;\|\,f(t)\,\|,$$

or

$$\frac{d\|u(t)\|}{dt} - \chi\,\|\,u(t)\,\| \leq \|\,f(t)\,\|, \tag{4.3}$$

where

$$\gamma = \alpha + \beta + \frac{k+\ell}{2}.$$

Integrating (4.3) from 0 to t we obtain the inequality

$$\|\,u(t)\,\| \leq e^{\gamma t}\,\|\,u(0)\,\| + \int_0^t e^{\gamma(t-s)}\,\|\,f(s)\,\|\,ds. \tag{4.4}$$

This means that the following assertion holds:

Lemma 4.1. *The Cauchy problem for equation (4.1) with initial condition* $u(0, z) = f_0(z) \in L_2(\mathbb{C})$ *is uniquely solvable and there is an estimate*

$$\| u(t) \| \le e^{\gamma t} \| f_0 \|_{L_2(\mathbb{C})} + \int_0^t e^{\gamma(t-s)} \| f(s) \| \, ds. \tag{4.4'}$$

The second hyperbolic system is four real systems of equations, which may be put in the form of two complex equations with respect to two complex-valued functions $u(t, z)$, $v(t, z)$ as follows:

$$\begin{aligned} &u_t + 2v_z + a_{11}\, u + a_{12} v + b_{11}\, \bar{u} + b_{12}\, \bar{v} = f, \\ &v_t + 2u_{\bar{z}} + a_{21}\, u + a_{22} v + b_{21}\, \bar{u} + b_{22}\, \bar{v} = g. \end{aligned} \tag{4.5}$$

We assume that the coefficients of this system are complex-valued functions, given on the whole plane $\mathbb{C}$, bounded on $\mathbb{C}$, and the right-hand sides $f(t, z)$, $g(t, z)$ are functions with a bounded $L_2(\mathbb{C})$ norm. If $u = (u, v)$ is a solution of (4.5) from C^1 in $t \ge 0$ with finite $L_2(\mathbb{C})$-norm:

$$\| u(t) \|^2 = \| u(t) \|^2 + \| v(t) \|^2 = \Big(\iint_{\mathbb{C}} | u(t, z) |^2 \, dx\, dy \Big)^{1/2} + \Big(\iint_{\mathbb{C}} | v(t, z) |^2 \, dx\, dy \Big)^{1/2} < \infty$$

then we have by means of Green formula and Schwartz inequality:

$$\begin{aligned} \frac{d \| u(t) \|^2}{dt} &= 2 \operatorname{Re} \iint_{\mathbb{C}} \big[\bar{u}\, u_t + \bar{v}\, v_t \big] dx\, dy - 4 \operatorname{Re} \iint_{\mathbb{C}} (\bar{u}\, v)_z \, dx\, dy \\ &- 2 \operatorname{Re} \iint_{\mathbb{C}} \big[(a_{11} + a_{21}) | u |^2 + (a_{12} + a_{22})\, \bar{u}\, v + (b_{11} + b_{12})\, \bar{u}^2 \\ &+ (b_{12} + b_{22})\, \bar{u}\, \bar{v} \big] dx\, dy + 2 \operatorname{Re} \iint_{\mathbb{C}} (\bar{u} f + \bar{v} g)\, dx\, dy \\ &\le \tilde{\gamma} \| u(t) \|^2 + 2 \| u(t) \| \| F(t) \|, \quad F = (f, g) \end{aligned}$$

and hence

$$\| u(t) \| \le e^{\tilde{\gamma} t} \| u(0) \| + \int_0^t e^{\tilde{\gamma}(t-s)} \| F(s) \| \, ds, \tag{4.6}$$

where $\tilde{\gamma}$ is some positive constant.

From (4.6) as above we obtain the following assertion:

Lemma 4.2. *The Cauchy problem for system (4.5) with initial conditions* $u(0, z) = f_0(z), v(0, z) = g_0(z)$ *from* $L_2(\mathbb{C})$ *is uniquely solvable and there is an estimate*

$$\left(\| u(t) \|^2 + \| v(t) \|^2\right)^{1/2} \leq e^{\tilde{\gamma} t}\Big[(\| f_0 \|_{L_2(\mathbb{C})} + \| g_0 \|_{L_2(\mathbb{C})})^{1/2} + \int_0^t e^{\tilde{\gamma}(t-s)} \left(\| f(s) \|^2 + \| g(s) \|^2\right)^{1/2} ds \Big]. \tag{4.7}$$

The final first-order hyperbolic system is the following system in $\mathbb{R}^4_+ = \{(t, x) : t \geq 0, x = (x_1, x_2, x_3) \in \mathbb{R}^3\}$ for the complex-valued vector function $u(t, x) = (u_1, u_2)$:

$$(I \frac{\partial}{\partial t} - \hat{\partial}_x)\, u + A\, u = F, \tag{4.8}$$

where

$$\hat{\partial}_x = \begin{pmatrix} \dfrac{\partial}{\partial x_3} & 2\dfrac{\partial}{\partial z} \\ 2\dfrac{\partial}{\partial \bar{z}} & -\dfrac{\partial}{\partial x_3} \end{pmatrix}, \quad z = x_1 + i\, x_2, \quad \frac{\partial}{\partial \bar{z}} = \frac{1}{2}\left(\frac{\partial}{\partial x_1} + i \frac{\partial}{\partial x_2} \right),$$

A is some square matrix, depending on x, I the identity operator. The homogeneous system (4.8) as well as the system

$$(I \frac{\partial}{\partial t} + \hat{\partial}_x)\, u + A\, u = 0 \tag{4.8'}$$

may be obtained formally from the Dirac system, which in the general case of a relativistic particle of mass m_0 with a half-spin has the form (see [12])

$$\left(\frac{1}{c} \gamma^0 \frac{\partial}{\partial t} + i \sum_{K=1}^{3} \gamma^k \frac{\partial}{\partial x_k} - m_0 I \right) \psi = 0,$$

where $\psi(t, x) = (\psi_1, \psi_2, \psi_3, \psi_4)$ is the wave vector function, c is the speed of light and γ^k are Dirac's matrices:

$$\gamma^0 = \begin{pmatrix} 1 & 0 & 0 & 0 \\ 0 & 1 & 0 & 0 \\ 0 & 0 & -1 & 0 \\ 0 & 0 & 0 & -1 \end{pmatrix}, \quad \gamma^1 = \begin{pmatrix} 0 & 0 & 0 & 1 \\ 0 & 0 & 1 & 0 \\ 0 & -1 & 0 & 0 \\ -1 & 0 & 0 & 0 \end{pmatrix},$$

$$\gamma^2 = \begin{pmatrix} 0 & 0 & 0 & -i \\ 0 & 0 & i & 0 \\ 0 & i & 0 & 0 \\ -i & 0 & 0 & 0 \end{pmatrix}, \quad \gamma^3 = \begin{pmatrix} 0 & 0 & 1 & 0 \\ 0 & 0 & 0 & -1 \\ -1 & 0 & 0 & 0 \\ 0 & 1 & 0 & 0 \end{pmatrix}.$$

In case $c = 1$, multiplying the first and the second lines by $-i$ and the third and fourth lines by i, we see that the system may be written as

$$\frac{\partial \psi'}{\partial t} + \hat{\partial}_x \psi'' = 0, \quad \frac{\partial \psi''}{\partial t} + \hat{\partial}_x \psi' \tag{4.9}$$

for

$$\psi' = (\psi_1, \psi_2), \quad \psi'' = (\psi_3, \psi_4).$$

Now it is obvious that (4.8) and (4.8′) are obtained from (4.9) by addition and by subtraction.

The Cauchy problem $u(0, x) = F_0(x)$ for the system (4.8) (or (4.8′)) is also uniquely solvable and an estimate of the $L_2(\mathbb{R}^3)$-norm of the solution through L_2-norms of initial conditions and the right-hand side also can be obtained.

4.2 Initial-boundary-value problems for first-order hyperbolic systems

(a) Let G be bounded region in the z-plane with boundary Γ a simple smooth closed curve, and let

$$\Omega = \mathbb{R}_+ \times G = \{(t, z) : t \geq 0, z \in G\}$$

be a cylindrical region. We shall consider the following:

Problem M: Find in Ω a solution of equation (4.1), continuous in $\bar{\Omega}$, satisfying the initial condition

$$u(0, z) = 0, \quad z \in G \tag{4.10}$$

and the boundary condition

$$u(t, z) + \sigma(z)\, \overline{u(t, z)} = 0, \quad (t, z) \in \mathbb{R}_+ \times \Gamma, \tag{4.11}$$

where

$$\sigma(z) = \frac{\overline{\theta(z)}}{|\theta(z)|}, \; \theta(z) = a(z)z'(s) - b(z)\,\overline{z'(s)}, \; z'(s) = \frac{dz(s)}{ds},$$

$z = z(s)$ is parametric equation of curve Γ.

Lemma 4.3. *If the functions* $a(z)$, $b(z)$ *have bounded first-order derivatives in* $\bar{G} = G + \Gamma$, $|a(z)| \le k$, $|b(z)| \le \ell$, *the functions* $c(z)$, $d(z)$ *are bounded in* $\bar{G}$: $|c(z)| \le \alpha$, $|d(z)| \le \beta$ *and the function* $f(t, z)$ *belongs to the space* $L_2(\bar{G})$ *for any* $t \ge 0$, *then for any solution of equation (4.1), which is continuously differentiable with respect to* t, *with finite norm*

$$\| u(t) \|_G = \Big(\iint_G | u(t, z) |^2 \, dx\, dy \Big)^{1/2},$$

the inequality

$$\| u(t) \|_G \le e^{\gamma t} \| u(0) \|_G + \int_0^t e^{\gamma(t-s)} \| f(s) \|_G \, ds \tag{4.12}$$

holds with $\gamma = \alpha + \beta + (k + \ell)/2$.

Proof. Using (4.1) and Green's formula, we have

$$\frac{d\| u(t) \|_G^2}{dt} = \iint_G \Big(u(t, z)\, \overline{u(t, z)} \Big)_t \, dx\, dy = 2\,\mathrm{Re} \iint_G u\, \bar{u}_t \, dx\, dy$$

$$= \mathrm{Re} \iint_G \{ a\, u_{\bar{z}}^2 + b\, u_z^2 + 2\,(cu^2 + d\,| u |^2 + uf) \} \, dx\, dy$$

$$= \mathrm{Re} \iint_G [(2c - a_{\bar{z}} - b_z)\, u^2 + 2d\,| u |^2 + 2\, uf] \, dx\, dy$$

$$+ \mathrm{Re} \frac{1}{2i} \int_\Gamma \theta(z)\, u^2(t, z) \, ds. \tag{4.13}$$

Now by virtue of boundary condition (4.11) we have

$$\int_\Gamma \theta(z)u^2(t,z)\,ds \equiv \int_\Gamma \theta(z)\,u(t,z)\left(-\frac{\overline{\theta(z)}}{|\theta(z)|}\;\overline{u(t,z)}\right)ds$$

$$= -\int_\Gamma |\theta(z)\,u(t,z)|^2\,ds,$$

and last integral in (4.13) disappears and thus by means of the Schwartz inequality we obtain

$$\frac{d\|u(t)\|_G}{dt} - \gamma\|u(t)\|_G \le \|f(t)\|_G.$$

Integrating this inequality, we obtain (4.12).

Lemma 4.3 proves the uniqueness of Problem M. To prove the existence we use the Laplace transformation with respect to the variable t. Equation (4.1) reduces to

$$a(z)\hat{u}_{\bar{z}} + b(z)\hat{u}_z + c(z,\lambda)\hat{u}(z,\lambda) + d(z,\lambda)\,\overline{\hat{u}(z,\bar{\lambda})} + \hat{f}(z,\lambda) \tag{4.14}$$

and the boundary condition

$$\hat{u}(z,\lambda) + \sigma(z)\,\overline{\hat{u}(z,\bar{\lambda})} = 0, \tag{4.15}$$

where

$$\hat{u}(z,\lambda) = \int_0^\infty u(t,z)e^{-\lambda t}\,dt,\quad d(z,\lambda) = d(z) - \lambda. \tag{4.16}$$

Henceforth, we shall assume that the functions $a(z)$, $b(z)$ and $f(t,z)$ for $t \ge 0$ belong to the space $L_p(\bar{G})$, $p > 2$. If $|a(z)| > |b(z)|$, then denoting by $\zeta(z)$ the homeomorphism satisfying the Beltrami equation $\zeta_{\bar{z}} - q(z)\,\zeta_z = 0$ with $q(z) = -b(z)/a(z)$, which maps the region G to the unit disc $K = \{|\zeta| < 1\}$ centred at the origin, equation (4.14) reduces to the form (see [58])

$$\hat{u}_{\bar{\zeta}} + A(\zeta)\,\hat{u}(\xi,\lambda) + B(\zeta,\lambda)\,\overline{\hat{u}(\zeta,\bar{\lambda})} = C(\zeta,\lambda) \tag{4.17}$$

for $|\zeta| < 1$, and the boundary condition to the form

$$\hat{u}(\zeta,\lambda) - i\,\bar{\zeta}\,\sigma_0(\zeta)\,\overline{\hat{u}(\zeta,\bar{\lambda})} = 0 \tag{4.18}$$

or $|\zeta| = 1$, where $\sigma_0(\zeta)$ is a function such that $|\sigma_0(\zeta)| \equiv 1$ for $|\zeta| = 1$ and has zero index: $\text{Ind}_\gamma\, \sigma_0(\zeta) = 0$.

The problem (4.17), (4.18) is equivalent to an integral equation

$$\hat{u}(\zeta,\lambda) - \frac{1}{\pi}\iint_{|\zeta|<1}\left[\frac{\omega(\tau,\lambda)}{\tau-\zeta} - i\,\frac{\overline{\omega(\tau,\overline{\lambda})}}{1-\overline{\tau}\zeta}\right] dK_\tau$$

$$= \frac{1}{\pi}\iint_{|\zeta|<1}\left[\frac{C(\tau,\lambda)}{\tau-\zeta} - \frac{i\,\overline{C(\tau,\overline{\lambda})}}{1-\overline{\tau}\zeta}\right] dK_\tau,$$

$$\omega(\zeta,\lambda) = A(\zeta)\,\hat{u}(\zeta,\lambda) + B(\tau,\lambda)\,\overline{\hat{u}(\zeta,\overline{\lambda})}\,,$$

whose unconditional solvability in the class of the functions continuous in K is established as in the theory of generalized analytical functions for the Riemann-Hilbert problem in the unit disc. We assume that the function $f(t,z)$ has a sufficient number of derivatives with respect to t for it to be arranged that

$$|c(z,\lambda)| \le \text{const.}\, |\lambda|^{-s},$$

which will guarantee the required convergence of the inverse Laplace transform of $\hat{u}(z,\lambda)$. The case $|b(z)| > (a(z)|$ reduces to the preceding one on transforming $\check{z} \to \bar{z}$. Thus we have proved the following:

Theorem 4.1. *If $|a(z)| \neq |b(z)|$ and the function $f(t,z)$ satisfies the concordance conditions*

$$\overline{f(0,z)} + \overline{\sigma(z)}f(0,z),\;\; \overline{f_t(0,z)} + \bar{L}f(0,z) + \overline{\sigma(z)}\left(f_t(0,z) + L\overline{f_t(0,z)}\right) = 0,$$

for $z \in \Gamma$, then there exists a unique classical solution of Problem M.

Problem N. Find in Ω a solution of the system of equations (4.5), continuous in $\bar{\Omega}$, satisfying the initial conditions

$$u(0,z) = v(0,z) = 0, \quad z \in G \tag{4.19}$$

and the boundary condition

$$u(t,z) - \overline{z'(s)}\,v(t,z) = 0, \quad (t,z) \in \mathbb{R}_+ \times \Gamma. \tag{4.20}$$

Lemma 4.4. *If the coefficients of the system (4.5) are bounded in $\bar{G}$ and $f(t,z)$, $g(t,z)$ are in $L_2(\bar{G})$ for any $t \ge 0$, then for any solution $u(t,z)$, $v(t,z)$ of this* *system, which is continuously differentiable with respect to t, having finite norm*

$$\| v(t) \|_G^2 + \| u(t) \|_G^2 = \iint_G | u(t,z) |^2 \, dx\, dy + \iint_G | v(t,z) |^2 \, dx\, dy,$$

the inequality

$$I(t) = \left(\| u(t) \|_G^2 + \| v(t) \|_G^2 \right)^{1/2} \le e^{\tilde{\gamma} t} \Big[\left(\| u(0) \|_G^2 + \| v(0) \|_G^2 \right)^{1/2}$$
$$+ \int_0^t e^{\tilde{\gamma}(t-s)} \left(\| f(s) \|_G^2 + \| g(s) \|_G^2 \right)^{1/2} ds \Big] \qquad (4.21)$$

holds.

Proof. Since by (4.5) and Green's formula

$$\frac{dI^2(t)}{dt} = 2 \operatorname{Re} \iint_G (\bar{u} u_t + \bar{v} v_t)\, dx\, dy = 2 \operatorname{Re} \iint_G \{ \bar{u} f + \bar{v} g$$
$$- 2(a_{11} + a_{21}) | u |^2 + (a_{12} + a_{22}) \bar{u} v + (b_{11} + b_{12}) \bar{u}^2$$
$$+ (b_{12} + b_{22}) \bar{u} \bar{v} \}\, dx\, dy + 2 \operatorname{Re} \frac{1}{i} \int_\Gamma \overline{u(t,z)}\, v(t,z)\, d\bar{z},$$

then by means of the boundary condition (4.20) and the Schwartz inequality we have the inequality from which (4.21) follows.

Note that we obtain the same inequality if instead of boundary condition (4.20) we take the following boundary conditions:

$$u(t,z) = 0, \quad (t,z) \in \mathbb{R}_+ \times \Gamma, \qquad (N_1)$$

$$v(t,z) = 0, \quad (t,z) \in \mathbb{R}_+ \times \Gamma, \qquad (N_2)$$

$$\operatorname{Re} u(t,z) = 0, \ \operatorname{Re} \overline{z'(s)}\, v(t,z) = 0, \ (t,z) \in \mathbb{R}_+ \times \Gamma, \qquad (N_3)$$

$$\operatorname{Im} u(t,z) = 0, \ \operatorname{Im} \overline{z'(s)}\, v(t,z) = 0, \ (t,z) \in \mathbb{R}_+ \times \Gamma, \qquad (N_4)$$

$$\operatorname{Re} z'(s)\, u(t,z) = 0, \ \operatorname{Re} v(t,z) = 0, \ (t,z) \in \mathbb{R}_+ \times \Gamma, \qquad (N_5)$$

$$\operatorname{Im} z'(s)\, u(t,z) = 0, \ \operatorname{Im} v(t,z) = 0, \ (t,z) \in \mathbb{R}_+ \times \Gamma. \qquad (N_6)$$

This means that if a solution exists for the initial-boundary-value problems for system (4.5) with initial conditions (4.19) and correspondingly boundary conditions (4.20), (N_j), $j = 1,...,6$, that solution is unique. To prove the existence of the solution we may, as above, use the Laplace transform with respect to variable t. Alternatively we may prove it directly. For instance, consider the problem with initial conditions (4.19) and boundary conditions (N_1) and let us take only the principal part of the system (4.5):

$$u_t + 2v_z = f, \quad v_t + 2\, u_{\bar{z}} = g. \tag{4.22}$$

Differentiating first equation with respect to t and using the second equation we see that the function $u(t, z)$ satisfies the inhomogeneous d'Alembert equation

$$u_{tt} - 4\, u_{z\bar{z}} = f_t - 2\, g_z. \tag{4.23}$$

From (4.19) by means of the first equation (4.22) we obtain initial conditions for $u(t, z)$:

$$u(0, z) = 0, \quad u_t(0, z) = f(0, z) \tag{4.24}$$

Hence in the cylinder Ω we have initial-boundary-value problem (4.23), (N_1) for the function $u(t, z)$ which is uniquely solvable. Substituting the value of $u(t, z)$ expressed in terms of the right-hand side (f, g) into the second equation (4.22) and then integrating from 0 to t we obtain:

$$v(t, z) = -2 \int_0^t u_{\bar{z}}(\tau, z)\, dt + \int_0^t g(\tau, z)\, dt \tag{4.25}$$

The functions $u(t, z)$, $v(t, z)$ obtained in this way gives the solution of the problem (4.22), (4.19), (N_1). Indeed, the first condition (4.24) and (N_1) shows that $u(t, z)$ satisfies (4.19) and (N_1). Also from (4.25) it follows that $v(0, z) = 0$, i.e. they satisfy the initial and boundary conditions. Now from (4.25), by means of the second condition (4.24), we have

$$2v_z = -4 \int_0^t u_{z\bar{z}}(\tau, z)\, d\tau + 2 \int_0^t g_z(\tau, z)\, d\tau = -\int_0^t u_{tt}(\tau, z)\, d\tau$$

$$+ \int_0^t (f_t(\tau, z) - 2\, g_z(\tau, z))\, d\tau + 2 \int_0^t g_z(\tau, z)\, d\tau = -u_t(t, z)$$

$$+ u_t(0, z) + f(t, z) - f(0, z) = -u_t(t, z) + f(t, z),$$

i.e. the first equation (4.22) is satisfied and from (4.25) it follows that the second equation is also satisfied.

4.3 Modified initial-boundary-value problem for first-order hyperbolic systems

(a) Let G be a bounded region in x-space $\mathbb{R}^3$ with boundary $\Gamma = \partial G$, a smooth closed Liapunov surface and let $\Omega_t = \mathbb{R}_+ \times G = \{(t, x) : t \in \mathbb{R}_+, x \in G\}$ be a cylindrical region. The usual initial-boundary-value problem for the system (4.8) in Ω_t with given initial conditions

$$u(0, x) = h(x), \quad x \in G \tag{4.26}$$

and with given boundary conditions on the surface $S_t = \{(t, x) : t \in \mathbb{R}_+, \ x \in \Gamma\}$ of the cylinder in general is not well posed. We state the following modified initial-boundary-value problem for the system (4.8). Let α, β be nonzero complex numbers, which we assume to have unit modulus[1]: $|\alpha| = |\beta| = 1$. Denote by G' the plane region, which is the intersection of G by the plane $\mathrm{Re}(\alpha\bar{\beta})x_1 - \mathrm{Im}(\alpha\bar{\beta})x_2 = 0$, i.e.

$$G' = G \cap (\mathrm{Re}\,(\alpha\bar{\beta})x_1 - \mathrm{Im}\,(\alpha\bar{\beta})x_2 = 0)$$

and suppose that the boundary of this region $\Gamma' = \partial G'$ is a simple smooth closed curve. Let $S_t' = \{(t, x) : t \in \mathbb{R}_+, x \in \Gamma'\}$ be a two-dimensional subset of the three-dimensional set S_t.

Problem M'. *Find in* Ω_t a solution $u(t, x) = (u_1, u_2)$ of the system (4.8), continuous in $\bar{\Omega}_t$, satisfying the initial conditions (4.26) on the base of Ω_t, the boundary conditions

$$\mathrm{Re}\,(\alpha u_1) = 0, \quad \mathrm{Re}\,(\beta u_2) = 0 \tag{4.27}$$

on the surface S_t of Ω_t and the boundary condition

$$[\,\mathrm{Re}\,(\alpha\bar{\beta})x_1'(s) + \mathrm{Im}\,(\alpha\bar{\beta})x_2'(s)\,]\,\mathrm{Im}\,(\alpha u_1) + (1 - x_3'(s))\,\mathrm{Im}\,(\beta u_2) = 0 \tag{4.28}$$

on the curve $\Gamma' = \partial G'$, where $x_1 = x_1(s), x_2 = x_2(s), x_3 = x_3(s)$ is a parametric equation of the curve Γ' and $x_j'(s) = \mathrm{d}x_j/\mathrm{d}s$.

[1] As is clear from the context there is no loss of generality.

Let $A \equiv 0$ and

$$\varphi = \alpha u_1, \quad \psi = \beta u_2;$$

then from (4.8) it follows that

$$\beta\varphi_t = \beta\varphi_{x_3} + 2\alpha\psi_z + \alpha\beta f_1,$$

$$\alpha\psi_t = 2\beta\,\varphi_{\bar{z}} - \alpha\psi_{x_3} + \alpha\beta f_2. \tag{4.29}$$

Since $\hat{\partial}_x^2 \equiv I\cdot\Delta$, then we have from (4.8) that

$$\frac{\partial^2 u}{\partial t^2} = \Delta u + \hat{\partial}_x F$$

and so

$$\frac{\partial^2 \varphi}{\partial t^2} - \Delta\varphi = \alpha\left(\frac{\partial f_1}{\partial t} + \frac{\partial f_1}{\partial x_3} + 2\frac{\partial f_2}{\partial z}\right), \tag{4.30}$$

$$\frac{\partial^2 \psi}{\partial t^2} - \Delta\psi = \beta\left(2\frac{\partial f_1}{\partial \bar{z}} + \frac{\partial f_2}{\partial t} - \frac{\partial f_2}{\partial x_3}\right). \tag{4.31}$$

Moreover, from (4.26) we have initial conditions

$$\varphi(0, x) = \alpha h_1, \quad \psi(0, x) = \beta h_2 \tag{4.26'}$$

$$\operatorname{Re}\varphi = 0, \ \operatorname{Re}\psi = 0 \tag{4.32}$$

on its surface S_t. Hence the real parts of the functions φ, ψ is determined uniquely by the right-hand sides f_1 and f_2, f_3 as a solution of the first initial-boundary-value problem for the D'Alembert equation. Then for imaginary parts of φ and ψ from (4.29) we obtain an overdetermined system of four real equations, which may be put into the following complex form with respect to the unknown complex-valued function $w(t, x) = \operatorname{Im}\varphi + i\operatorname{Im}\psi$:

$$\operatorname{Re}(\alpha\bar{\beta})\,\bar{w}_t = -\mathrm{i}\left(\frac{\partial}{\partial x_1} + \mathrm{i}\operatorname{Re}(\alpha\bar{\beta})\frac{\partial}{\partial x_3}\right)w + g_1(t,x),$$

$$\operatorname{Im}(\alpha\bar{\beta})\,\bar{w}_t = -\mathrm{i}\left(\frac{\partial}{\partial x_2} + \mathrm{i}\operatorname{Im}(\alpha\bar{\beta})\frac{\partial}{\partial x_3}\right)w + g_2(t,x) \qquad (4.33)$$

where

$$g_1(t,x) = \operatorname{Im}(\alpha\bar{\beta})f_t - \mathrm{i}\left(\frac{\partial}{\partial x_1} - \mathrm{i}\operatorname{Im}(\alpha\bar{\beta})\frac{\partial}{\partial x_3}\right)\bar{f} + \mathrm{i}\,(\operatorname{Im}\alpha f_2 + \mathrm{i}\operatorname{Im}\beta f_2),$$

$$g_2(t,x) = \operatorname{Re}(\alpha\bar{\beta})f_t - \mathrm{i}\left(\frac{\partial}{\partial x_2} - \mathrm{i}\operatorname{Re}(\alpha\bar{\beta})\frac{\partial}{\partial x_3}\right)\bar{f} + \mathrm{i}\,(\operatorname{Re}\alpha f_2 - \mathrm{i}\operatorname{Im}\beta f_1),$$

$$f(t,x) = \operatorname{Re}\varphi + \mathrm{i}\operatorname{Re}\psi.$$

Taking into account (4.30), (4.31) it is easy to see that the right-hand sides in (4.33) satisfy the compatibility condition of this system:

$$\mathrm{i}\operatorname{Im}(\alpha\bar{\beta})\frac{\partial g_1}{\partial t} - \left(\frac{\partial}{\partial x_2} + \mathrm{i}\operatorname{Im}(\alpha\bar{\beta})\frac{\partial}{\partial x_3}\right)g_1$$

$$= \mathrm{i}\operatorname{Re}(\alpha\bar{\beta})\frac{\partial g_2}{\partial x_1} - \left(\frac{\partial}{\partial x_1} + \mathrm{i}\operatorname{Re}(\alpha\bar{\beta})\frac{\partial}{\partial x_3}\right)g_2.$$

According to (4.26′) on the basis of Ω_t the function $w(t,x)$ satisfies the initial condition

$$w(0,x) = \operatorname{Im}\alpha h_1(x) + \mathrm{i}\operatorname{Im}\beta h_2(x) \qquad (4.34)$$

and according to (4.28) this function satisfies the following boundary condition on the set S_t':

$$\operatorname{Im}[(1+\mathrm{i}\,G(s))\,w(t,x)]\,_{S_t'} = 0, \qquad (4.35)$$

where

$$G(s) = \operatorname{Re}(\alpha\bar{\beta})x_1'(s) + \operatorname{Im}(\alpha\bar{\beta})\,x_2'(s) + \mathrm{i}\,x_3'(s). \qquad (4.36)$$

Making the nondegenerate linear transformation

$$\eta = \mathrm{Re}\,(\alpha\bar{\beta})x_1 + \mathrm{Im}\,(\alpha\bar{\beta})x_2,$$

$$\xi = -\,\mathrm{Im}\,(\alpha\bar{\beta})x_1 + \mathrm{Re}\,(\alpha\bar{\beta})x_2, \tag{4.37}$$

we transform (4.33) into the following equations:

$$w_\xi = \mathrm{Im}\,(\alpha\bar{\beta})g_1 - \mathrm{Re}\,(\alpha\bar{\beta})g_2, \tag{4.38}$$

$$\bar{w}_t = -\,2\mathrm{i}\; w_{\bar{\zeta}} - \mathrm{i}\,\mathrm{Re}\,(\alpha\bar{\beta})g_1 + \mathrm{Im}\,(\alpha\bar{\beta})g_2, \tag{4.39}$$

where

$$w_{\bar{\zeta}} = \frac{1}{2}(w_\eta + \mathrm{i}\; w_{x_3}).$$

Integrating the first equation, we have

$$w(t, x) = \omega(t, \zeta) + w_*\,, \tag{4.40}$$

where $\bar{w}_*$ is a known function and $\omega(t, \zeta)$ is an arbitrary complex-valued continuous function of the variables t, $\zeta = \eta + \mathrm{i}\,x_3$. Substituting (4.40) into equation (4.39), we obtain the following hyperbolic equation with respect to $\omega(t, \zeta)$:

$$\bar{\omega}_t + 2\mathrm{i}\;\omega_{\bar{\zeta}} = f_*(t, \zeta), \quad (t, \zeta) \in \tilde{\Omega}_t' \tag{4.41}$$

in the three-dimensional cylindrical region $\tilde{\Omega}_t' = \{(t, \eta, x_3) : t \in \mathbb{R}_+,\ (\eta, x_3) \in \tilde{G}'\}$, where $\tilde{G}'$ is the image of G' of the mapping (4.37), and $f_*(t, \zeta)$ is a known function. The equation (4.41) evidently is a particular case of equation (4.1). Note that on mapping (4.37) the set S_t' becomes the surface $\tilde{S}_t' = \{(t, \eta, x_3),\ t \in \mathbb{R}^+,\ (\eta, x_3) \in \tilde{\Gamma}'\}$ of the cylinder $\tilde{\Omega}_t'$, where $\tilde{\Gamma}'$ is the image of $\Gamma' = \partial G'$, which is a boundary of the plane region $\tilde{G}'$obtained from the region G by intersection with the plane $\xi = 0$ in the space of variables (ξ_1, η, x_3). The boundary condition (4.35) by means of (4.40) becomes

$$\omega(t, \zeta) + \mathrm{i}\,\overline{\zeta'(s)}\;\overline{\omega(t, \zeta)} = \gamma(t, \zeta), \quad (t, \zeta) \in S_t' \tag{4.42}$$

on $\tilde{S}_t'$.

The initial-boundary-value problem (4.41), (4.34), (4.42) is a particular case of problem M, considered in the previous section.

(b) The usual initial-boundary-value problem in Ω_t with initial conditions

$$\psi'(0,x) = H^1(x), \quad \psi''(0,x) = H^2(x) \tag{4.43}$$

on the base of Ω_t and boundary conditions

$$\text{Re}\,\psi' = 0, \quad \text{Re}\,\psi'' = 0 \tag{4.44}$$

on its surface S_t is also not well posed for the system (4.9). We state the following modified initial-boundary value problem:

Problem N'**.** Find in Ω_t a solution $\psi' = (\psi_1, \psi_2), \psi'' = (\psi_3, \psi_4)$ of the system (4.9), continuous in $\bar{\Omega}_t$, satisfying the initial conditions (4.43) on the basis of Ω_t, the boundary conditions (4.44) on its surface S_t and the boundary condition

$$\text{Im}\,\psi_1 + \text{i}\,\text{Im}\,\psi_2 = \text{i}\,\overline{z'(s)}\,(\text{Im}\,\psi_3 + \text{i}\,\text{Im}\,\psi_4) \tag{4.45}$$

on the two-dimensional subset $S_t' = \{(t,x) : t \in \mathbb{R}_+, x \in \Gamma'\}$ of S_t, where $z(s) = x_1(s) + \text{i}\,x_3(s)$ is a parametrical equation of the curve $\Gamma' = \partial G'$, $G' = G \cap (x_2 = 0)$, which we suppose to be simple, closed and smooth.

Theorem 4.2. *If the right-hand sides of (4.9) satisfy some smoothness and compatibility conditions, then there exists a unique classical solution of Problem* N'.

The proof of this theorem is the same as in Problem M', but in this case instead of (4.33) we obtain the following kind of overdetermined system for the imaginary parts of the functions ψ_j in the region G':

$$w^{(1)}_{x_2} = f^{(1)}, \quad \bar{w}^{(2)}_t - 2\text{i}\,w^{(1)}_z = g^{(1)}, \tag{4.46}$$

$$w^{(2)}_{x_3} = f^{(2)}, \quad \bar{w}^{(1)}_t - 2\text{i}\,w^{(2)}_z = g^{(2)}, \tag{4.46}$$

where

$$w_{\bar{z}} = w_{x_1} + \text{i}\,w_{x_3}, w^{(1)} = \text{Im}\,\psi_1 + \text{i}\,\text{Im}\,\psi_2, \quad w^{(2)} = \text{Im}\,\psi_3 + \text{i}\,\text{Im}\,\psi_4.$$

Moreover, in this case the uniqueness theorem needs to be proved not for the scalar equation (4.1), but for the system of two equations

$$\bar{\omega}_t^{(1)} = 2\mathrm{i}\,\omega_z^{(2)} + F^{(1)},\ \bar{\omega}_t^{(2)} = 2\mathrm{i}\,\omega_z^{(1)} + F^{(2)}. \tag{4.47}$$

Considering the $L_2(G')$-norm of the vector function $\omega(t, \zeta) = (\omega^{(1)}, \omega^{(2)})$:

$$I^2(t) = \|\omega\|_{G'}^2 = \iint_{G'} |\omega^{(1)}(t,z)|^2\,\mathrm{d}x\,\mathrm{d}y + \iint_{G'} |\omega^{(2)}(t,z)|^2\,\mathrm{d}x\,\mathrm{d}y, \tag{4.48}$$

we obtain, by means of (4.47), (4.45) and Green's formula,

$$\frac{\mathrm{d}I^2(t)}{\mathrm{d}t} = \mathrm{Re}\iint_{G'}\left[\omega^{(1)}\,\omega_t^{(1)} + \omega^{(2)}\,\bar{\omega}_t^{(2)}\right]\mathrm{d}x_1\,\mathrm{d}x_3$$

$$= \mathrm{Re}\,2\mathrm{i}\iint_{G'}\frac{\partial}{\partial\bar{z}}(\omega^{(1)}\,\omega^{(2)})\,\mathrm{d}x_1\,\mathrm{d}x_3 + \mathrm{Re}\iint_{G'}\left[\omega^{(1)}\,F^{(1)} + \omega^{(2)}\,F^{(2)}\right]\mathrm{d}x_1\,\mathrm{d}x_3$$

$$= \mathrm{Re}\int_{\Gamma'}\omega^{(1)}(t,z)\,\omega^{(2)}(t,z)\,\mathrm{d}z + \iint_{G'}\left[\omega^{(1)}\,F^{(1)} + \omega^{(2)}\,F^{(2)}\right]\mathrm{d}x_1\,\mathrm{d}x_3$$

$$= \mathrm{Re}\,\mathrm{i}\int_{\Gamma'}|\omega^{(1)}(t,z)|^2\,\mathrm{d}s + \mathrm{Re}\iint_{G'}\left[\omega^{(1)}(t,z)\,F^{(1)}(t,z) + \omega^{(2)}(t,z)\,F^{(2)}(t,z)\right]\mathrm{d}x_1\,\mathrm{d}x_3$$

$$\leq I(t)\,\|F\|_{G'}$$

and hence

$$I(t) \leq I(0) + \|F\|_{G'}.$$

4.4 Initial-boundary-value problem for second-order hyperbolic systems

As in Section 4.2, let G be a bounded region in the z-plane with boundary Γ a simple smooth closed curve and let

$$\Omega = \mathbb{R}_+ \times G = \{(t,z) : t \geq 0,\ z \in G\}$$

be a cylindrical region. For the inhomogeneous D'Alembert equation with respect to the complex-valued function $u(t, z)$:

$$u_{tt} - 4\,u_{z\bar{z}} = f(t,z) \tag{4.49}$$

we consider the following.

Problem M_z. Find in Ω a solution of equation (4.49) in $C^1(\bar{\Omega})$, satisfying the initial conditions

$$u(0, z) = 0, \quad u_t(0, z) = 0, \quad z \in G \tag{4.50}$$

and the boundary conditions

$$\mathrm{Re}\,[\lambda(z)u_z] = 0, \quad \mathrm{Re}\,[\lambda(z)\,\overline{z'(s)}\,u(z)] = 0, \quad (t, z) \in \mathbb{R}_+ \times \Gamma \tag{4.51}$$

Problem $M_{\bar{z}}$. Find in Ω a solution of equation (4.49) in $C^1(\bar{\Omega})$, satisfying the initial condition (4.50) and the boundary conditions

$$\mathrm{Re}\,[\lambda(z)u_{\bar{z}}] = 0, \quad \mathrm{Re}\,p\,\lambda(z)\,z'(s)u\,] = 0, \quad (t, z) \in \mathbb{R}_+ \times \Gamma \tag{4.52}$$

where $\lambda(z)$ is given on Γ Hölder continuous function such that $\lambda(z) \neq 0$, $z \in \Gamma$ and $z = z(s)$ is parametrical equation of the curve Γ. Without loss of generality we assume that $|\lambda(z)| \equiv 1$.

Lemma 4.5. *For any solution in C^2 of Problem M_z with finite $\| u(t) \|_z$ norm:*

$$\| u(t) \|_z = \Big(\iint_G \big[|u_t|^2 + 4|u_z|^2 \big] \mathrm{d}x\, \mathrm{d}y \Big)^{1/2}, \tag{4.53}$$

there is an estimate with respect to t

$$\| u(t) \|_z \le \| u(0) \|_z + \| f(t) \|, \tag{4.54}$$

and for any solution in C^2 of the problem $M_{\bar{z}}$ with finite $\| u(t))\|_{\bar{z}}$-norm:

$$\| u(t) \|_{\bar{z}} = \Big(\iint_G \big[|u_t|^2 + 4|u_{\bar{z}}|^2 \big] \mathrm{d}x\, \mathrm{d}y \Big)^{1/2} \tag{4.55}$$

there is an estimate with respect to t

$$\| u(t) \|_{\bar{z}} \le \| u(0) \|_{\bar{z}} + \| f(t) \|. \tag{4.56}$$

Proof. Differentiating (4.53) with respect to t and taking into account (4.49) we have

$$\frac{\mathrm{d}\| u(t) \|_z^2}{\mathrm{d}t} = \mathrm{Re} \iint_G (\bar{u}_t\, u_{tt} + 4u_z\, \bar{u}_{t\bar{z}})\, \mathrm{d}x\, \mathrm{d}y = 4\,\mathrm{Re} \iint_G \bar{u}_t\, u_{z\bar{z}}\, \mathrm{d}x\, \mathrm{d}y$$

$$+\operatorname{Re}\iint_G \bar{u}_t f(t,z)\,dx\,dy + 4\operatorname{Re}\iint_G u_z \cdot \bar{u}_{t\bar{z}}\,dx\,dy.$$

But by Green's formula we have

$$\operatorname{Re}\iint_G u_z\,\bar{u}_{t\bar{z}}\,dx\,dy = -\operatorname{Re}\iint_G \bar{u}_t \cdot u_{z\bar{z}}\,dx\,dy + \operatorname{Re}\iint_G (u_z \cdot \bar{u}_t)\,dx\,dy$$

$$= -\operatorname{Re}\iint_G \bar{u}_t \cdot u_{z\bar{z}}\,dx\,dy + \operatorname{Re}\frac{1}{2i}\int_\Gamma u_z \cdot \bar{u}_t\,dz = -\operatorname{Re}\iint_G \bar{u}_t \cdot u_{z\bar{z}}\,dx\,dy,$$

because by (4.51)

$$\operatorname{Re}\frac{1}{2i}\int_\Gamma u_z \cdot \bar{u}_t\,dz = \operatorname{Re}\frac{1}{2i}\int_\Gamma \lambda(z)u_z\,\overline{\lambda(z)}\,z'(s)\,\bar{u}_t\,ds$$

$$= \frac{1}{2}\int_\Gamma \operatorname{Re}[\lambda(z)u_z]\operatorname{Im}[\overline{\lambda(z)}\,z'(s)\,\bar{u}_t]\,ds$$

$$+\frac{1}{2}\int_\Gamma \operatorname{Im}[\lambda(z)u_z]\operatorname{Re}[\overline{\lambda(z)}\,z'(s)\,\bar{u}_t]\,ds = 0.$$

Hence by the Schwartz inequality

$$\frac{d\|u(t)\|_z^2}{dt} = \operatorname{Re}\iint_G \bar{u}_t f\,dx\,dy$$

$$\leq \Big(\iint_G |u_t|^2\,dx\,dy\Big)^{1/2}\Big(\iint_G |f(t,z)|^2\,dx\,dy\Big)^{1/2} \leq \|u(t)\|_z\,\|f\|,$$

i.e.

$$\frac{d\|u(t)\|_z}{dt} \leq \|f\|.$$

Integrating this inequality from 0 up to t gives (4.54). The proof of (4.55) is analogous. From this inequality the uniqueness of the solution of problem M_z and $M_{\bar{z}}$ follows. Indeed, let $u(t,z)$ be a solution of the homogeneous Problem M_z. Then

$$\| u(t) \|_z = \| u(0) \|_z = 0,$$

i.e.

$$u_t = 0, \; u_z = 0 \text{ in } \Omega$$

and so

$$u(t, z) = \overline{\psi(z)},$$

where $\psi(z)$ is holomorphic in G. But since $u(0, z) = 0$, then $u(t, z) \equiv 0$. Analogously with Problem $M_{\bar{z}}$. To prove existence theorems for solutions of this problem, we consider eigenvalue problems. For instance, let $v(z)$ be a solution in G of the equation

$$Lv - \lambda v = 0, \tag{4.57}$$

satisfying the boundary conditions

$$\operatorname{Re}[\lambda(z) v] = 0, \;\; \operatorname{Re}[\lambda(z)\overline{z'(s)}\, v(z)] = 0 \tag{4.58}$$

on Γ, where L is an elliptic operator

$$Lv \equiv 4v_{z\bar{z}} + a(z)v + b(z)\bar{v}. \tag{4.59}$$

Since the operator L with boundary conditions, in case $a(z)$ is real, determines a self-adjoint operator, since for solutions u, v of this problem

$$\operatorname{Re} \iint_G \bar{v}\, Lu \, dx\, dy = \operatorname{Re} \iint_G \bar{u}\, Lv \, dx\, dy,$$

then in case the problem (4.58) is uniquely solvable for the equation $Lu = F$ it follows that there exist positive eigenvalues $\{\lambda_k\}$ and a corresponding complete system of eigenfunctions $\{v_k(z)\}$, by which in the usual way we construct a solution of the initial-value problem (4.50), (4.51) for the equation

$$u_{tt} - 4u_{z\bar{z}} - a(z)u - b(z)\bar{u} = f(t, z) \tag{4.60}$$

with $\overline{a(z)} = a(z)$, $z \in G$ in the form of a Fourier series. We do the same for Problem $M_{\bar{z}}$ and thus we come to the following (see [28].)

Theorem 4.3. *If $f(t,z) \in L_2(G)$ for any $t \geq 0$ and satisfies some concordance conditions, then there exists a unique solution of the problem (4.60), (4.50), (4.51) as well as of the problem (4.60), (4.50), (4.52), provided the function $\lambda(z)$ is such that the problem (4.58) is uniquely solvable for the equation $Lu = F$.*

4.5 Degenerate hyperbolic systems

The equation

$$\bar{z}\,\bar{u}_t + a(z)u_{\bar{z}} + b(z)u_z + c(z)u + d(z)\bar{u} = f(t,z) \tag{4.61}$$

with $|a(z)| \neq |b(z)|$, $z \in G$, is hyperbolic in Ω and degenerate on the line $z = 0$. If we put $\zeta = z^2/2$, then we obtain the hyperbolic equation

$$\bar{u}_t + au_{\bar{\zeta}} + b^* u_\zeta + c^* u + d^* \bar{u} = f^* \tag{4.62}$$

with

$$b^* = b\left(\frac{\zeta}{\bar{\zeta}}\right)^{1/2},\ c^* = \frac{c}{(2\bar{\zeta})^{1/2}},\ d^* = \frac{d}{(2\bar{\zeta})^{1/2}},\ f^* = \frac{f}{(2\bar{\zeta})^{1/2}}.$$

If coefficients c, d and right-hand side of (4.61) vanishes at the line $z = 0$, then we may consider for equation (4.62) the initial-boundary-value problem in Ω with initial condition

$$u(0,z) = 0 \tag{4.63}$$

and boundary condition

$$u(t,z) + \sigma(z)\,\overline{u(t,z)} = 0 \tag{4.64}$$

on the surface of the cylinder Ω, where

$$\sigma(z) = \frac{\overline{\theta(z)}}{|\theta(z)|},\ \theta(z) = a(z)\,z'(s) - b(z)\,\overline{z'(s)}.$$

The initial condition (4.63) for the transformed equation will be

$$u(0,(2\zeta)^{1/2}) = 0, \tag{4.63'}$$

and the boundary condition (4.64) will be

$$u(t,(2\zeta)^{1/2})+\sigma(\zeta)\,\overline{u(t,(2\zeta)^{1/2})}=0, \tag{4.64'}$$

where

$$\sigma(\zeta)=\frac{a\bar{\zeta}\,\zeta'(s)-b\zeta\overline{\zeta'}}{|a\bar{\zeta}\zeta'(s)-b\zeta\overline{\zeta'(s)}|}, \tag{4.65}$$

i.e. we have Problem M. Since $|a|\neq|b^*|=|b|$, then we may apply results from Section 4.2.

The next hyperbolic system is

$$\begin{aligned}&zu_t+2v_z+a_{11}u+a_{12}v+b_{11}\bar{u}+b_{12}\bar{v}=f,\\ &\bar{z}v_t+2u_{\bar{z}}+a_{21}u+a_{22}v+b_{21}\bar{u}+b_{22}\bar{v}=g,\end{aligned} \tag{4.66}$$

which is degenerate on the line $z=0$. By the same transformation $\zeta=z^2/2$ this system reduces to the system (4.5), so we may apply the results of Section 4.4.

It would be interesting also to investigate the properties of the following hyperbolic system, degenerate on line $z=0$:

$$u_t+2zv_z+a_{11}u+a_{12}v+b_{11}\bar{u}+b_{12}\bar{v}=f,$$

$$v_t+2\bar{z}u_{\bar{z}}+a_{21}u+a_{22}v+b_{21}\bar{u}+b_{22}\bar{v}=g.$$

Considering the second-order hyperbolic equation

$$u_{tt}-4|z|^2u_{z\bar{z}}=f(t,z), \tag{4.67}$$

degenerate on line $z=0$, we may reduce it to the d'Alembert equation by the same transformation $\zeta=z^2/2$ and use the results of the previous section. But what about equation

$$|z|^2u_{tt}-4u_{z\bar{z}}=f(t,z) \tag{4.68}$$

which is also hyperbolic in Ω, except line $z=0$, where it is degenerate?

4.6 Mixed type first-order systems of equations

(a) Let G be the region in x,y-plane bounded by a single smooth curve σ in upper halfplane with ends at point A with coordinates $x=y=0$ and at point B with coordinates $x = 1,\ y = 0$ and bounded by curves $AC : x - 2/3\,(-y)^{3/2} = 0$ and $BC : x + 2/3\,(-y)^{3/2} = 1$ in lower halfplane $y < 0$, where C is a point with coordinates $x = 1/2,\ y = -(3/4)^{2/3}$.

In this region we consider a system of two real first-order equations written in the form of the following single complex equation

$$y^{1/2} w_x + \mathrm{i}\, w_y + a(z)w + b(z)\bar{w} = 0 \tag{4.69}$$

with complex-valued given functions $a(z), b(z)$ bounded in $\bar{G}$.

Evidently this equation is elliptic in the upper half-plane $y > 0$, parabolic in the lower half-plane $y < 0$ and degenerate on line $y = 0$.

First we investigate the following:

Problem μ**.** Find in G a solution of (4.69), continuous in $\bar{G}$, which satisfies the condition

$$\operatorname{Re} \zeta'(s) w = g \tag{4.70}$$

on the curve σ and the condition

$$\operatorname{Im} w = h \tag{4.71}$$

on the curve AC.

Theorem 4.4. *In case* $a = b = 0$ *the problem* μ *is uniquely solvable for any Hölder-continuous right-hand sides* g, h. *In the general case this problem is solvable if and only if its right-hand sides satisfy a finite number of conditions*

$$\int_\sigma g(t) v_j(t)\, \mathrm{d}s + \int_0^1 h\left(\frac{x}{2}\right) \nu_j(x)\, \mathrm{d}x = 0, \tag{4.72}$$

where $v_j(t), \nu_j(x)$ are completely determined functions.

Proof. Making transformation

$$\xi = x,\ \eta = \begin{cases} \dfrac{2}{3} y^{3/2}, & y > 0, \\ -\dfrac{2}{3}(-y)^{3/2}, & y < 0 \end{cases}$$

the equation (4.69) reduces to

$$\mathfrak{L}\, w \equiv w_{\bar{\zeta}} + a_0(\zeta)w + b_0(\zeta)\,\bar{w} = 0, \quad \zeta = \xi + i\,\eta \tag{4.73}$$

in D_+-image of $G \cap (y > 0)$ and to

$$w_\xi + w_\eta - a_1(\zeta)w - b_1(\zeta)\bar{w} = 0 \tag{4.74}$$

in D_--image of $G \cap (y < 0)$, where

$$2a_0(\zeta) = \left(\frac{4\,i}{3(\zeta - \bar{\zeta})}\right)^{1/3} a, \quad 2b_0(\zeta) = \left(\frac{4\,i}{3(\zeta - \bar{\zeta})}\right)^{1/3} b,$$

$$a_1(\zeta) = i\left(\frac{2}{3\eta}\right)^{1/3} a, \quad b_1(\zeta) = i\left(\frac{2}{3\eta}\right)^{1/3} b.$$

Since, obviously, $a_0(\zeta), b_0(\zeta) \in L_p(D_+), p > 2$, then (4.73) means that $w(\zeta)$ is a generalized analytic function. Any such function is generated by a holomorphic function. More precisely the equation (4.73) is equivalent to a uniquely solvable Fredholm integral equation

$$w(\zeta) - \frac{1}{\pi}\iint_{D_+} \frac{a_0(\tau)\,w(\tau) + b_0(\tau)\overline{w(\tau)}}{\tau - \zeta}\, dD_{+\tau} = \varphi(\zeta), \tag{4.75}$$

where $\varphi(\zeta)$ is an arbitrary function, holomorphic in D_+, all solutions of which are expressed by the formula (see [58]):

$$w(\zeta) = \varphi(\zeta) + \iint_{D_+} \left[\Gamma_1(\tau, \zeta)\,\varphi(\tau) + \Gamma_2(\tau, \zeta)\,\overline{\varphi(\tau)}\right] dD_{+\tau}\,, \tag{4.76}$$

where $\Gamma_j(\tau, \tau)$ are resolvent kernels for (4.75). Considering the equation adjoint to (4.73),

$$\mathfrak{L}^* v \equiv -\,v_{\bar{\zeta}} + a_0(\zeta)v + \overline{b_0(\zeta)}\,\bar{v} = 0, \tag{4.77}$$

we also express all of its solutions in the form

$$v(\zeta) = \psi(\zeta) + \iint_{D_+} \left[\Gamma_1'(\zeta, \tau)\psi(\tau) + \Gamma_2'(\zeta, \tau)\,\overline{\psi(\tau)}\right] dD_{+\tau}\,. \tag{4.78}$$

If ξ_0 is any inner point in the segment AB, then we may consider a special fundamental solution of (4.77):

$$v^0(\zeta, \xi_0) = \frac{1}{\zeta - \xi_0} + \iint_{D_+} \left[\frac{\Gamma_1'(\zeta, \tau)}{\tau - \xi_0} + \frac{\Gamma_2'(\zeta, \tau)}{\bar{\tau} - \zeta_0}\right] dD_{+\tau}. \tag{4.79}$$

Evidently the integral term $\tilde{v}^0(\zeta, \xi_0)$ of $v^0(\zeta, \xi_0)$ may have only a weak singularity at the point ξ_0:

$$|\tilde{v}^0(\zeta, \xi_0)| < \frac{\text{const.}}{|\zeta - \xi_0|^{\varepsilon}}, \quad \varepsilon < 1. \tag{4.80}$$

Denote by $h(\zeta, \xi_0)$ the solution of equation (4.77) in D_+ which satisfies the conditions

$$\operatorname{Re} h(\zeta, \xi_0)|_{\zeta \in \sigma} = -\operatorname{Re} v^0(\zeta, \xi_0)|_{\zeta \in \sigma} \tag{4.81}$$

$$\operatorname{Im} h(\zeta, \zeta_0)|_{\zeta \in AB} = 0. \tag{4.82}$$

The existence of a unique function of this kind follows from the generalized Riemann-Schwartz symmetry principle (see [58]). Indeed, the function

$$h^*(\zeta, \xi_0) = \begin{cases} h(\zeta, \xi_0), & \zeta \in D_+, \\ -\overline{h(\bar{\zeta}, \xi_0)}, & \zeta \in D_+^* \end{cases} \tag{4.83}$$

where D_+^*, a mirror image of D_+, is a solution of the equation in $D_+ \cup D_+^* \cup AB$:

$$-h_{\bar{\zeta}} + a_0^* h + \overline{b_0^*}\, \bar{h} = 0 \tag{4.84}$$

with coefficients a_0^*, b_0^* obtained from a_0, b_0 by the same rule (4.83) and on $\Gamma = \sigma + \sigma^*$:

$$\operatorname{Re} h^* (\zeta, \xi_0)|_{\zeta \in \Gamma} = \tilde{g}(\zeta, \xi_0), \tag{4.85}$$

where σ^* is a mirror image of σ and

$$\tilde{g}(\zeta, \xi_0) = \begin{cases} -\operatorname{Re} v^0(\zeta, \xi_0), & \zeta \in \sigma, \\ -\operatorname{Re} v^0(\bar{\zeta}, \xi_0); & \zeta \in \sigma_*. \end{cases}$$

Moreover, since by (4.82) the function $\operatorname{Im} h^*(\tau, \xi_0)$ vanishes at points A and B, then $h^*(\zeta, \xi_0)$ determines uniquely as a solution of Schwartz problem (4.84), (4.85). Now let us consider the function

$$\hat{V}(\zeta, \xi_0) = V^0(\zeta, \xi_0) + h(\zeta, \zeta, \xi_0), \tag{4.86}$$

which, obviously, is a solution of equation (4.77), satisfying the conditions

$$\operatorname{Re} \hat{V}(\zeta, \xi_0)\,|_{\zeta \in \sigma} = 0. \tag{4.87}$$

$$\operatorname{Im} \hat{V}\,|_{AB} = \operatorname{Im} \hat{V}\,|_{AB}.$$

Let σ_ε be upper half circle $|\zeta - \xi_0| = \varepsilon$, $\operatorname{Im} \zeta > 0$. We apply Green's formula to

region $D_+^\varepsilon = D_+ - |\zeta - \xi_0| < \varepsilon$, $\operatorname{Im} \zeta > 0$ and obtain for any solution $w(\zeta)$ of the problem μ:

$$0 = \operatorname{Re} \iint_{D_+^\varepsilon} (\hat{V}\,\mathfrak{L}\, w - w\, \mathfrak{L}^*\, \hat{v})\, d\xi\, d\eta = \operatorname{Re} \iint_{D_+^\varepsilon} \Big[(w\,\hat{v})_{\bar{\zeta}}$$

$$+ 2i \operatorname{Im} (b_0\, \bar{w}\, \hat{v}) \Big] d\xi\, d\eta = \operatorname{Re} \iint_{D_+^\varepsilon} (w\,\hat{v})_{\bar{\zeta}}\, d\xi\, d\eta$$

$$= \operatorname{Re} \frac{1}{2i} \int_\sigma w(\zeta)\, \hat{v}(\zeta, \xi_0)\, d\zeta + \operatorname{Re} \frac{1}{2i} \int_0^{\xi_0 - \varepsilon} w(\xi)\, \hat{V}(\xi, \xi_0)\, d\xi$$

$$+ \operatorname{Re} \frac{1}{2i} \int_{\sigma_\varepsilon} w(\zeta)\, \hat{V}(\zeta, \xi_0)\, d\zeta + \operatorname{Re} \frac{1}{2i} \int_{\xi_0 + \varepsilon}^{1} w(\xi)\, \hat{v}(\xi, \xi_0)\, d\xi.$$

According to (4.86), (4.79), (4.80), we have

$$\operatorname{Re} \frac{1}{2i} \int_{\sigma_\varepsilon} w(\zeta)\, \hat{V}(\zeta, \xi_0)\, d\zeta = \frac{1}{2} \operatorname{Re} \int_0^\pi w(\xi_0 + \varepsilon e^{i\theta})\, d\theta$$

$$+ \operatorname{Re} \frac{1}{2i} \int_{\sigma_\varepsilon} w(\zeta)\, \hat{V}^0(\zeta, \xi_0)\, d\zeta \to -\frac{\pi}{2} \operatorname{Re} w(\xi_0),$$

$$\operatorname{Re}\frac{1}{2i}\int_0^{\xi_0-\varepsilon}\hat{V}(\xi,\xi_0)\,w(\xi)\,d\xi+\operatorname{Re}\frac{1}{2i}\int_{\xi_0+\varepsilon}^{1}\hat{V}(\xi,\xi_0)\,w(\xi)\,d\xi$$

$$=\operatorname{Re}\frac{1}{2i}\left(\int_0^{\xi_0-\varepsilon}+\int_{\xi_0+\varepsilon}^{1}\right)\frac{w(\xi)\,d\xi}{\xi-\xi_0}+\operatorname{Re}\frac{1}{2i}\left(\int_0^{\xi_0-\varepsilon}+\int_{\xi_0+\varepsilon}^{1}\right)\tilde{V}^0(\xi,\xi_0)\,w(\zeta)\,d\xi$$

$$\to\frac{1}{2}\int_0^1\frac{\operatorname{Im}w(\xi)\,d\xi}{\xi-\xi_0}+\int_0^1 k_1(\xi,\xi_0)\operatorname{Re}w(\xi)\,d\xi$$

$$+\int_0^1 \ell_1(\xi,\xi_0)\operatorname{Im}w(\xi)\,d\xi,$$

where $k_1(\xi,\xi_0)$, $\ell_1(\xi,\xi_0)$ may have only a weak singularity in a neighbourhood of ξ_0.

Thus we have

$$\operatorname{Re}w(\xi_0)-\frac{1}{\pi}\int_0^1\frac{\operatorname{Im}w(\xi)\,d\xi}{\xi-\xi_0}-\frac{1}{\pi}\int_0^1 k_1(\xi,\xi_0)\operatorname{Re}w(\xi)\,d\xi$$

$$-\frac{1}{\pi}\int_0^1 \ell_1(\xi,\xi_0)\operatorname{Im}w(\xi)\,d\xi=\operatorname{Re}\frac{1}{2\pi i}\int_\sigma\hat{V}(\zeta,\xi_0)\,w(\zeta)\,d\zeta.$$

But since by (4.70) and (4.87)

$$\operatorname{Re}\frac{1}{2\pi i}\int_\sigma\hat{V}(\zeta,\xi_0)\,w(\zeta)\,d\zeta=\frac{1}{2\pi}\int_\sigma\operatorname{Re}\tilde{V}^0(\zeta,\xi_0)\operatorname{Im}\left[\zeta'(s)\,w(\zeta)\right]ds$$

$$+\frac{1}{2\pi}\int_\sigma\operatorname{Im}\tilde{V}^0(\zeta,\zeta_0)\operatorname{Re}\left[\zeta'(s)\,w(\zeta)\right]ds=\frac{1}{2\pi}\int_\sigma\operatorname{Im}\tilde{V}^0(\zeta,\zeta_0)\,g(\zeta)\,d\zeta,$$

then we obtain the following relation between real and imaginary parts of an unknown function $w(\zeta)$ on a segment AB :

$$\operatorname{Re}w(\xi)-\frac{1}{\pi}\int_0^1\frac{\operatorname{Im}w(\xi')\,d\xi'}{\xi'-\xi}-\frac{1}{\pi}\int_0^1\left[k_1(\xi',\xi)\operatorname{Re}w(\xi')\right.$$

$$\left.+\ell_1(\xi',\xi)\operatorname{Im}w(\xi')\right]d\xi'=\frac{1}{2\pi i}\int_\sigma\tilde{V}^0(\xi',\xi_0)\,g(\xi')\,d\xi'. \qquad (4.88)$$

All solutions of equation (4.74) are expressed by the formula (see [29])

$$w(\zeta) = \Gamma_1(\zeta)\,\omega(\xi - \eta) + \Gamma_2(\zeta)\,\overline{\omega(\zeta - \eta)}, \tag{4.89}$$

where ω is an arbitrary complex valued function and $\Gamma_j(\zeta)$ completely determined through a_1, b_1 complex valued functions such that $|\Gamma_1(\zeta)|^2 - |\Gamma_2(\zeta)|^2 \neq 0$ for all $\zeta = \xi + i\eta,\ \eta < 0$ and $\Gamma_1|_{\xi+\eta=0} = 1,\ \Gamma_2|_{\xi+\eta=0} = 0.$

From (4.71) and (4.89) it follows that

$$\operatorname{Im}\omega(\xi) = h(\xi/2). \tag{4.90}$$

But from (4.89) we have

$$\omega(\xi) = (|\Gamma_1(\xi)|^2 - |\Gamma_2(\xi)|^2)^{-1}(\overline{\Gamma_1(\xi)}\,w(\xi) - \Gamma_2(\xi)\,\overline{w(\xi)})$$

and thus

$$\operatorname{Im}(\overline{\Gamma_1(\xi)}\,w(\xi) - \Gamma_2(\xi)\,\overline{w(\xi)}) = (|\Gamma_1(\xi)|^2 - |\Gamma_2(\xi)|^2)\,h(\xi/2),$$

or we obtained another one relation between real and imaginary parts of unknown function w on segment AB:

$$\operatorname{Re}(\Gamma_1(\xi) + \Gamma_2(\xi)\operatorname{Im} w(\xi) - \operatorname{Im}(\Gamma_1(\xi) + \Gamma_2(\xi)\operatorname{Re} w(\xi)$$

$$= (|\Gamma_1(\xi)|^2 - |\Gamma_2(\xi)|^2)\,h(\xi/2). \tag{4.91}$$

By virtue of this relation, the relation (4.88) reduces to singular integral equation on segment AB with respect to $\operatorname{Re} w(\xi)$

$$\operatorname{Re} w(\xi) - \frac{\lambda(\xi)}{\pi}\int_0^1 \frac{\operatorname{Re} w(\xi')\,d\xi'}{\xi' - \xi} + \int_0^1 k_0(\xi, \xi')\operatorname{Re} w(\xi')\,d\xi' = G(\xi), \tag{4.92}$$

where $k_0(\xi, \xi')$ is the kernel with weak singularity and

$$\lambda(\xi) = \frac{\operatorname{Im}[\Gamma_1(\xi) + \Gamma_2(\xi)]}{\operatorname{Re}[\Gamma_1(\xi) + \Gamma_2(\xi)]}, \tag{4.93}$$

$$G(\xi) = \frac{1}{2\pi}\int_\sigma \operatorname{Im} \tilde{V}^0(\xi, \xi')\, g(\xi')\, d\xi' + \int_0^1 \frac{|\Gamma_1(\xi')|^2 - |\Gamma_2(\xi')|^2}{\operatorname{Re}(\Gamma_1(\xi)+\Gamma_2(\xi'))} \frac{h(\xi'/2)\, d\xi'}{\xi'-\xi}.$$

If in neighbourhood of points A and B the curve σ becomes as normal curve, i.e. as $|2\zeta - 1| = 1$, $\zeta = x + (2i/3)\, y^{3/2}$, $y > 0$ then we can apply the Carlemann's approach to the equation (4.92) (see [29]) and we prove that (4.92) has zero index. From this equation we determine the real part of w on AB, and then by (4.91) we determine also its imaginary part on AB. Hence a solution of the problem μ is determined as a solution of the Schwartz problem in D_+ and as a solution of the Cauchy problem in D_-. In case $a = b \equiv 0$ we have $\Gamma_1(\zeta) \equiv 1$, $\Gamma_2(\zeta) \equiv 0$ and everything is determined uniquely without solving the integral equation (4.92).

In the same way we investigate the following similar boundary-value problem:

Problem ν. Find in G a solution of (4.69), continuous in $\bar{G}$, which satisfies the condition

$$\operatorname{Im} \zeta'(s)\, w(\zeta) = g(\zeta) \tag{4.94}$$

on curve σ and the condition

$$\operatorname{Re} w = h \tag{4.95}$$

on curve AC.

Theorem 4.5. *The results of Theorem 4.4 on problem* μ *are valid also on problem* ν.

(b) The following mixed type system of two real equations with respect to real functions

$$u_y - y\, v_x = f_1, \quad v_y + u_x = f_2 \tag{4.96}$$

is elliptic in $G \cap (y > 0)$, hyperbolic in $G \cap (y < 0)$ and degenerate on the segment AB.

It is interesting to note that a solution of this system is determined in the region G described above, by given information on the curve σ:

Problem E. Find a solution (u, v) of system (4.96), continuous in $\bar{G}$, satisfying the boundary condition:

$$v = 0 \tag{4.97}$$

on σ.

Theorem 4.6. *Problem E is uniquely solvable for any right-hand sides continuous in $\bar{G}$.*

Proof. In the elliptic region $G \cap (y > 0)$ the system (4.96) may be put into complex form

$$w_y + \frac{i(1+y)}{2} w_x + \frac{i(1-y)}{2} \bar{w}_x = f \tag{4.98}$$

with respect to $w = u + iv$.

Introducing the new unknown function

$$\nu(z) = (1 + \sqrt{y})^2 w + (1 - y)\bar{w} \tag{4.99}$$

we obtain

$$\nu_y + i\sqrt{y}\,\nu_x = \frac{1+\sqrt{y}}{\sqrt{y}} \frac{(1+\sqrt{y})^2 \nu - (1+y)\bar{\nu}}{4\sqrt{y}\,(1+\sqrt{y})^2}$$

$$- \frac{\overline{(1+\sqrt{y})\nu - (1+y)\bar{\nu}}}{4\sqrt{y}\,(1+\sqrt{y})^2} + (1+\sqrt{y})f + (1-y)\bar{f}. \tag{4.100}$$

Since from (4.99) it follows that

$$(1+\sqrt{y})^2 \nu(z) - (1-y)\overline{\nu(z)} = 4\sqrt{y}\,(1+\sqrt{y})^2 w(z),$$

then it follows that the left-hand side of this relation vanishes on the line $y = 0$. Therefore, making the transformation

$$\xi = x, \ \eta = \frac{2}{3} y^{3/2}, \ y > 0,$$

we obtain the equation

$$\nu_{\bar\zeta} = \frac{1}{6\eta[1+\left(\frac{3}{2}\eta\right)^{1/3}]}\left(\left[1+\left(\frac{3}{2}\eta\right)^{1/3}\right]\nu-\left[1-\left(\frac{3}{2}\eta\right)^{2/3}\right]\bar\nu\right)$$

$$-\frac{1}{4\left(\frac{3}{2}\eta\right)^{2/3}\left[1+\left(\frac{3}{2}\eta\right)^{1/3}\right]^2}\overline{\left[1+\left(\frac{3}{2}\eta\right)^{1/3}\right]\nu-\left[1-\left(\frac{3}{2}\eta\right)^{2/3}\right]\bar\nu}$$

$$+\left(1+\left(\frac{3}{2}\eta\right)^{1/3}\right)f+\left(1-\left(\frac{3}{2}\eta\right)^{2/3}\right)\bar f \tag{4.101}$$

with right-hand side bounded on the line $\eta = 0$. The condition (4.97) on the σ^*-image of σ is $\operatorname{Im}\nu = 0$ and from (4.99) the condition $\operatorname{Im}\nu = 0$ on line $y = 0$ follows automatically. Thus in the region D_+ we obtain the Schwartz problem for equation (4.101). To find (u, v) in the region $G \cap (y < 0)$ we introduce instead of (u, v) the new unknown functions φ and ψ by the equalities

$$\varphi = \sqrt{(-y)}\, u - y v, \quad \psi = \sqrt{(-y)}\, u + y v. \tag{4.102}$$

Then we obtain the following equations for φ, ψ:

$$\varphi_y + \sqrt{(-y)}\,\varphi_x = -\frac{1}{2\sqrt{(-y)}}\,\frac{\varphi+\psi}{2\sqrt{(-y)}} - \frac{\varphi-\psi}{2y} + \sqrt{(-y)} f_1 - y f_2,$$

$$\psi_y - \sqrt{(-y)}\,\psi_x = -\frac{1}{2\sqrt{(-y)}}\,\frac{\varphi+\psi}{2\sqrt{(-y)}} + \frac{\psi-\varphi}{2y} + \sqrt{(-y)} f_1 + y f_2. \tag{4.103}$$

As follows form (4.102)

$$\varphi+\psi = 2\sqrt{(-y)}u, \quad \psi-\varphi = 2y\,v,$$

the left-hand sides of these relations, and hence also the functions φ and ψ, vanish on the line $y = 0$. Making the transformation

$$\xi = x, \quad \eta = -2/3\,(-y)^{3/2}, \; y<0$$

we see that equations (4.103) reduce to the equations:

$$\varphi_\eta + \varphi_\xi = \frac{\varphi+\psi}{6\eta} - \frac{\psi-\varphi}{3\eta} + \left(-\frac{3}{2}\eta\right)^{1/3} f_1 + \left(-\frac{3}{2}\eta\right)^{2/3} f_2,$$

$$\psi_\eta - \psi_\xi = \frac{\varphi+\psi}{6\eta} + \frac{\psi-\varphi}{3\eta} + \left(-\frac{3}{2}\eta\right)^{1/3} f_1 - \left(-\frac{3}{2}\eta\right)^{2/3} f_2. \tag{4.104}$$

A solution of these equations is uniquely determined in the D_--image of $G \cap (y<0)$ by means of the initial conditions $\varphi|_{\eta=0} = \psi|_{\eta=0} = 0$.

(c) As an example of multidimensional mixed type we consider the following system in $\mathbb{R}^3$:

$$u_{x_1} + 2x_1^{1/2} v_z = 0, \quad -2x_1^{1/2} u_{\bar{z}} + v_{x_1} = 0 \tag{4.105}$$

with respect to complex-valued functions u, v. This system is elliptic in the half-space

$$\mathbb{R}_+ = \{(x_1, z) : x_1 > 0, \ z = x_2 + i\, x_3 \in \mathbb{C} \sim \mathbb{R}^2\}$$

and hyperbolic in the half-space

$$\mathbb{R}_- = \{(x_1, z) : x_1 < 0, \ z \in \mathbb{C} \sim \mathbb{R}^2\}.$$

The simplest problem for system (4.105) is to find a solution of (4.105), continuous in $\mathbb{R}^3$, satisfying the condition

$$u(0, z) = \varphi(z), \ z \in \mathbb{C} \tag{4.106}$$

or the condition

$$v(0, z) = \psi(z), \ z \in \mathbb{C}. \tag{4.107}$$

Theorem 4.7. *The problem (4.105), (4.106) as well as the problem (4.105), (4.107) is uniquely solvable for any* $\varphi \in L_2(\mathbb{C})$ $(\psi \in L_2(\mathbb{C}))$, *apart from constants.*

Proof. Making the transformation

$$\xi_1 = \begin{cases} \frac{2}{3} x_1^{3/2}, & x_1 > 0, \\ -\frac{2}{3}(-x_1)^{3/2}, & x_1 < 0, \quad z \equiv z, \end{cases} \tag{4.108}$$

we find that $U = (u, v)$ is a solution of the elliptic system

$$\bar{\partial}_x u = 0 \tag{4.109}$$

in the half-space $\xi_1 > 0$, and of the hyperbolic system

$$u_{\xi_1} + 2i\, v_z = 0, \ -2i\, u_{\bar{z}} + v_{\xi_1} = 0 \tag{4.110}$$

in $\xi_1 < 0$. From (4.109) it follows that $u(\xi_1, z)$, form the solution of the Laplace equation in the half-space $\xi_1 > 0$ and hence in the case of both (4.106) and (4.107) this function is determined in this half-space uniquely apart from constants. Moreover, since on the plane $\xi_1 = 0$ the values of the functions u, v satisfying (4.109) are connected by the formulae (see [25]

$$u(0, z) = \frac{1}{2\pi} \iint_{\mathbb{C}} \frac{\bar{\zeta} - \bar{z}}{|\zeta - z|^3} v(0, \zeta)\, d\xi_2\, d\xi_3, \tag{4.111}$$

$$v(0, z) = \frac{1}{2\pi} \iint_{\mathbb{C}} \frac{\zeta - z}{|\zeta - z|^3} u(0, \zeta)\, d\xi_2\, d\xi_3, \tag{4.112}$$

then in the case of (4.106) we also have

$$v(0, z) = \frac{1}{2\pi} \iint_{\mathbb{C}} \frac{\zeta - z}{|\zeta - z|^3} \varphi(\zeta)\, d\xi_2\, d\xi_3 \tag{4.113}$$

and in the case of (4.107)

$$u(0, z) = \frac{1}{2\pi} \iint_{\mathbb{C}} \frac{\bar{\zeta} - \bar{z}}{|\zeta - z|^3} \psi(\zeta)\, d\xi_2\, d\xi_3; \tag{4.114}$$

i.e. in both cases the value of a solution of the hyperbolic system is given on the plane $\xi_1 = 0$ and hence its solution is determined uniquely in the half-space $\xi_1 < 0$ as a solution of the Cauchy problem. A more general problem for the system (4.105) is to find a solution of this system, continuous in $\mathbb{R}^3$, satisfying the conditions

$$\mathrm{Re}[G_{11} u + G_{12} v] = 0, \ \mathrm{Re}[G_{21} u + G_{22} v] = 0 \tag{4.115}$$

on the plane $x_1 = 0$ with given continuous functions $G_{ij}(z)$.

Theorem 4.8. *If G_{ij} are constants such that* $\det(G_{ij}) \neq 0$, *then the problem (4.105), (4.115) is uniquely solvable for any $\varphi, \psi \in L_2(\mathbb{R}^2)$ apart from constants.*

Proof. Making the same transformation (4.108) and the further transformation

$$u' = G_{11}u + G_{12}v, \quad v' = G_{21}u + G_{22}v, \tag{4.116}$$

we obtain the Dirichlet problem in the half-space $\xi_1 > 0$ for real functions $\operatorname{Re} u'$, $\operatorname{Re} v'$, satisfying the Laplace equation and then as in Chapter 2 we obtain an overdetermined elliptic system for $\operatorname{Im} u'$, $\operatorname{Im} v'$, from which they are determined for $\xi_1 \geq 0$ in terms of $\operatorname{Re} u'$, $\operatorname{Re} v'$ uniquely apart from constants. Thus in particular we find the values of u, v on the plane $\xi_1 = 0$. In the half-space $\xi_1 < 0$ the solution is determined as a solution of the Cauchy problem for the first-order hyperbolic system (4.110).

5. First-order overdetermined systems

5.1 Introduction

As we have seen in Chapter 2 and 4 some types of overdetermined first-order systems of partial differential equations arise. In this Chapter we shall investigate more general types of such overdetermined systems. Any such a linear system in $\mathbb{R}^m$ may be written as

$$Au := \sum_{j=1}^{m} A_j(x)u_{x_j} + A_0(x)u = F(x), \tag{5.1}$$

where $u(x) = (u_1,\dots,u_n)$ is an unknown real vector function, $F(x) = (F_1,\dots,F_n)$ is a given real vector function and $A_j(x)$ are given real matrices with r rows and n columns $r > n$. Here we give a classification of (5.1).

Let $\Delta_j(x, \xi)$, $j = 1,\dots,r!/[\,n!\,(r-n)!\,]$ be nth order minors of the matrix

$$A(x, \xi) := \sum_{j=1}^{m} A_j(x)\,\xi_j\,. \tag{5.2}$$

The overdetermined system (5.1) is called elliptic at the point x if there exists at least one value of j such that $\Delta_j(x, \xi) \neq 0$ for all $\xi := (\xi_1,\dots,\xi_m) \in \mathbb{R}^m\backslash 0$.

If we determine a characteristic set for the system (5.1) given in the domain $\Omega \subseteq \mathbb{R}^m$ as

$$\operatorname{char} A = \left\{ (x, \xi) : x \in \Omega, \xi \in \mathbb{R}^m\backslash 0 : \Delta_j(x, \xi) = 0, \forall_j \right\}, \tag{5.3}$$

then for elliptic systems this set is empty. The overdetermined system (5.1) we call the principal type at the point $x \in \Omega$ if there exists at least one value of j such that

$$\operatorname{grad}_\xi \Delta_j(x, \xi) \neq 0 \ \text{ for all } \xi \in \mathbb{R}^m\backslash 0. \tag{5.4}$$

The surface $\omega(x) = \text{const}$ with $\operatorname{grad} \omega \neq 0$, $x \in \Omega$, we call a characteristic if the real function $\omega(x)$ is a solution of the overdetermined system:

$$\Delta_j(x, \operatorname{grad} \omega) = 0 \ \text{ for all } j = 1,..,r!/[\,n!\,(r-n)!\,]. \tag{5.5}$$

Example 1. The overdetermined system in $\mathbb{R}^3$:

$$\operatorname{div} u = 0, \quad \operatorname{curl} u = 0 \tag{5.6}$$

is elliptic, because

$$\operatorname{rank} A(x, \xi) = \operatorname{rank} \begin{pmatrix} \xi_1 & \xi_2 & \xi_3 \\ 0 & -\xi_3 & \xi_2 \\ \xi_3 & 0 & -\xi_1 \\ -\xi_2 & \xi_1 & 0 \end{pmatrix} = 3, \ \forall \xi \in \mathbb{R}^3 \backslash 0.$$

For minors $\Delta_j(x, \xi) = \xi_j |\xi|^2$ of the matrix $A(x, \xi)$, evidently, $\operatorname{grad}_\xi \Delta_j(x, \xi) \neq 0$, $\forall \xi \in \mathbb{R}^3 \backslash 0$, because

$$\operatorname{grad}_\xi \Delta_1 = (3\xi_1^2 + \xi_2^2 + \xi_3^2, \ 2\xi_1 \xi_2, \ 2\xi_1 \xi_3),$$

$$\operatorname{grad}_\xi \Delta_2 = (2\xi_2 \xi_1, \ 3\xi_2^2 + \xi_1^2 + \xi_3^2, \ 2\xi_2 \xi_3),$$

$$\operatorname{grad}_\xi \Delta_3 = (2\xi_3 \xi_1, \ 2\xi_3 \xi_2, \ 3\xi_3^2 + \xi_1^2 + \xi_2^2);$$

hence the system (5.6) is of principal type.

Example 2. The Cauchy-Riemann system in $\mathbb{R}^{2m} \sim \mathbb{C}^m (w = u_1 + \mathrm{i}\, u_2)$:

$$\frac{\partial w}{\partial \bar{z}_j} = 0, \quad j = 1, \dots, m \tag{5.7}$$

is elliptic, because

$$\operatorname{rank} A(x, \xi) = \operatorname{rank} \begin{pmatrix} \xi_1 & -\xi_{m+1} \\ \xi_{m+1} & \xi_1 \\ \vdots & \\ \xi_m & -\xi_{2m} \\ \xi_{2m} & \xi_m \end{pmatrix} = 2.$$

For the minors $\Delta_j = \xi_j^2 + \xi_{m+j}^2$, $j = 1,\dots,m$ of the matrix $A(x, \xi)$:

$$\operatorname{grad}_\xi \Delta_1 = (2\xi_1, 0,\dots,0,\ 2\xi_{m+1}, 0,\dots,0),$$

$$\operatorname{grad}_\xi \Delta_2 = (0, 2\xi_2, 0,\dots,0,\ 2\xi_{m+2}, 0,\dots,0),$$

$$\vdots$$

$$\operatorname{grad}_\xi \Delta_m = (0,\dots,0,\ 2\xi_m, 0,\dots,0,\ 2\xi_{2m}),$$

then it follows that among $\operatorname{grad}_\xi \Delta_j$ at least one is nonzero for $\xi := (\xi_1,\dots,\xi_{2m}) \neq 0$, i.e. the system (5.7) is of principal type.

Example 3. The overdetermined system in $\mathbb{R}^{n+1}(w = u_1 + \mathrm{i}\, u_2)$:

$$w_{x_j} + \mathrm{i}\, w_t = 0, \quad j = 1,\dots,n$$

which we met in Chapter 2, is also elliptic and of principal type by the same reason as in the previous example.

Example 4. The overdetermined system in $\mathbb{R}^3$:

$$\frac{\partial u_1}{\partial x_1} + \frac{\partial u_2}{\partial x_2} + \frac{\partial u_3}{\partial x_3} = 0, \quad \frac{\partial u_3}{\partial x_2} - \frac{\partial u_2}{\partial x_3} = 0, \tag{5.8}$$

$$\frac{\partial u_1}{\partial x_3} + \frac{\partial u_3}{\partial x_1} = 0, \quad \frac{\partial u_2}{\partial x_1} + \frac{\partial u_1}{\partial x_2} = 0$$

is of principal type, but not elliptic. Indeed, since

$$A(x, \xi) = \begin{pmatrix} \xi_1 & \xi_2 & \xi_3 \\ 0 & -\xi_3 & \xi_2 \\ \xi_3 & 0 & \xi_1 \\ \xi_2 & \xi_1 & 0 \end{pmatrix},$$

then all third-order minors of $A(x, \xi)$ are

$$\Delta_0 = \begin{vmatrix} 0 & -\xi_3 & \xi_2 \\ \xi_3 & 0 & \xi_1 \\ \xi_2 & \xi_1 & 0 \end{vmatrix} \equiv 0, \ \Delta_1 = \begin{vmatrix} \xi_1 & \xi_2 & \xi_3 \\ \xi_2 & \xi_1 & 0 \\ \xi_3 & 0 & \xi_1 \end{vmatrix} = \xi_1(\xi_1^2 - \xi_2^2 - \xi_3^2),$$

$$\Delta_2 = \begin{vmatrix} \xi_1 & \xi_2 & \xi_3 \\ 0 & -\xi_3 & \xi_2 \\ \xi_2 & \xi_1 & 0 \end{vmatrix} = -\xi_2(\xi_1^2 - \xi_2^2 - \xi_3^2), \ \Delta_3 = \begin{vmatrix} \xi_1 & \xi_2 & \xi_3 \\ 0 & -\xi_3 & \xi_2 \\ \xi_3 & 0 & \xi_1 \end{vmatrix} = -\xi_3(\xi_1^2 - \xi_2^2 - \xi_3^2)$$

and hence (5.8) has nonempty characteristic set, but $\text{grad}_\xi \Delta_j \neq 0$ for $\xi \in \mathbb{R}^3\backslash 0$ and $j = 1,2,3$, because

$$\text{grad}_\xi \Delta_1 = (3\xi_1^2 - \xi_2^2 - \xi_3^2, \ -2\xi_1\xi_2, \ -2\xi_1\xi_3),$$

$$\text{grad}_\xi \Delta_2 = (-2\xi_2\xi_1, \ 3\xi_2^2 - \xi_1^2 + \xi_3^2, \ 2\xi_2\xi_3),$$

$$\text{grad}_\xi \Delta_3 = (-2\xi_3\xi_1, \ 2\xi_3\xi_2, \ 3\xi_3^2 - \xi_1^2 + \xi_2^2).$$

Equations (5.5) in this case reduce to one equation:

$$\left(\frac{\partial \omega}{\partial x_1}\right)^2 - \left(\frac{\partial \omega}{\partial x_2}\right)^2 - \left(\frac{\partial \omega}{\partial x_3}\right)^2 = 0$$

and hence the cones $x_1 - x_1^0 \pm \sqrt{[(x_2 - x_2^0)^2 + (x_3 - x_3^0)^2]} = \text{const.}$ are characteristic for (5.8).

Example 5. The overdetermined system in $\mathbb{R}^{n+1}$:

$$\frac{\partial u_0}{\partial x_0} + \frac{\partial u_1}{\partial x_1} + \dots + \frac{\partial u_n}{\partial x_n} = 0, \ \frac{\partial u_k}{\partial x_j} - \frac{\partial u_j}{\partial x_k} = 0, \ j < k,$$

$$\frac{\partial u_j}{\partial x_0} - \frac{\partial u_0}{\partial x_j} = 0, \ j = 1,\dots,n \tag{5.9}$$

is a multidimensional analogue of the preceding example.

Example 6. The overdetermined system in $\mathbb{R}^{n+1}(u = u_1 + \mathrm{i}\, u_2,\ \bar{u} = u_1 - \mathrm{i}\, u_2)$:

$$P_j\, u \equiv u_{x_j} + \lambda_j(x, t)u_t + \mu_j(x, t)\bar{u}_t = 0, \quad j = 1,\ldots,n \tag{5.10}$$

is elliptic at the point (x, t) if and only if at this point the functions $\lambda_j(x, t)$, $\mu_j(x, t)$ satisfy the inequality

$$[\operatorname{Im}\lambda_j(x, t]^2 > |\mu_j(x, t)|^2 \tag{5.11}$$

at least for one value of j. The system (5.10) is principal type if and only if at this point (x, t):

$$[\operatorname{Im}\lambda_j(x, t)]^2 \neq |\mu_j(x, t)|^2 \tag{5.12}$$

at least for one value of j. Indeed, write the system in the form (5.1) and take n second-order minors

$$\Delta_j(x, t; \xi, \tau) = \xi_j^2 + 2\operatorname{Re}\lambda_j(x, t)\,\xi_j\tau + (|\lambda_j(x, t)|^2 - |\mu_j(x, t)|^2\tau^2, \ j = 1,\ldots,n$$

of the matrix $A(x, t; \xi, \tau)$. Then it is easy to see that according to (5.11) at least for one value of j we have $\Delta_j(x, t; \xi, \tau) \neq 0$ for all $(\xi, \tau) \neq 0$. Further, if $\operatorname{grad}_\xi \Delta_1 = \cdots = \operatorname{grad}_\xi \Delta_n = 0$, then by (5.12) it follows that $(\xi, \tau) = 0$, $\xi := (\xi_1,\ldots,\xi_n)$. Note that if for $(x, t) \in \Omega$,

$$[\operatorname{Im}\lambda_j(x, t)]^2 \leq |\mu_j(x, t)|^2, \ \forall j, \tag{5.13}$$

then the system (5.10) has characteristic set

$$\operatorname{char} P = \{(x, t=; \xi, \tau) : (x, t) \in \Omega,\ (\xi, \tau) \in \mathbb{R}^{n+1}\backslash 0 : \Delta^{(j)}(x, t; \xi, \tau) = 0,\ \forall j\}.$$

The characteristics for the system (5.10) are surfaces $\omega(x, t) =$ const., satisfying the equations

$$\Delta^{(j)}(x, t\,; \operatorname{grad}\omega = 0 \tag{5.14}$$

for all second-order minors $\Delta^{(j)}$ of the matrix $A(x, t; \xi, \tau)$.

5.2 Elliptic overdetermined systems in $\mathbb{R}^{n+1}$

Let us consider the following overdetermined first-order system in Cartesian space $\mathbb{R}^{n+1}$ of points $(x, t) = (x_1,\ldots,x_n, t)$ with respect to the complex-valued function

$u(x, t)$:

$$P_j^0 u - a_j(x, t)u - b_j(x, t)\bar{u} = f_j(x, t), \quad j = 1, \ldots, n, \tag{5.15}$$

where

$$P_j^0 u :\equiv \frac{\partial u}{\partial x_j} + \lambda_j(x, t)\frac{\partial u}{\partial t} \tag{5.16}$$

and the bar denotes the complex conjugate. We assume that the system (5.15) is elliptic in $\mathbb{R}^{n+1}$, i.e. for $(x, t) \in \mathbb{R}^{n+1}$ and at least one value of j : $\operatorname{Im}\lambda_j(x, t) \neq 0$.

For determinancy we assume further that

$$\operatorname{Im}\lambda_n(x, t) \neq 0, \quad (x, t) \in \mathbb{R}^{n+1}. \tag{5.17}$$

We also assume that the functions $a_j(x, t)$, $b_j(x, t)$, $f_j(x, t)$ are continuously differentiable in $\mathbb{R}^{n+1}$ and vanish at infinity as $O((|x|^2 + t^2)^{-1/2-\varepsilon})$, $\varepsilon > 0$.

We refer to a function bounded in $\mathbb{R}^{n+1}$ as a generalized constant, if it is identically zero, provided it is zero only at one point.

Here we consider the system (5.15) whose principal coefficients satisfy the conditions

$$\frac{\partial\lambda_k}{\partial x_j} - \frac{\partial\lambda_j}{\partial x_k} = 0, \ j \neq k, \quad \frac{\partial\lambda_j}{\partial t} = 0, \ j = 1, \ldots, n. \tag{5.18}$$

Then the commutator $[P_j^0, P_k^0]u \equiv (P_j^0 P_k^0 - P_k^0 P_j^0)u$ is zero:

$$[P_j^0, P_k^0]u \equiv \left(\frac{\partial\lambda_k}{\partial x_j} - \frac{\partial\lambda_j}{\partial x_k}\right)\frac{\partial u}{\partial t} = 0, \ (x, t) \in \mathbb{R}^{n+1};$$

hence the compatibility conditions for (5.15) are

$$P_j^0 a_k - P_k^0 a_j = 0, \ (P_j^0 - a_j)b_k - (P_k^0 - a_k)b_j = 0,$$

$$(P_j^0 - a_j)f_k - (P_k^0 - a_k)f_j = 0, \quad j \neq k. \tag{5.19}$$

In addition to (5.18), we also assume that the real functions

$$\lambda_j^0 = \lambda_j - \frac{\lambda_n}{\operatorname{Im}\lambda_n}$$

atisfy the conditions

$$\frac{\partial \lambda_k^0}{\partial x_j} - \frac{\partial \lambda_j^0}{\partial x_k} = 0,\ k \neq j,\ \frac{\partial \lambda_j^0}{\partial x_n} = 0,\ j = 1,\dots,n-1. \tag{5.20}$$

Jow we prove the following theorem.

Theorem 5.1. *If the conditions (5.18), (5.19), (5.20) fulfilled, then the system (5.15) s solvable in* $\mathbb{R}^{n+1}$ *and the corresponding homogeneous system (i.e.* $f_j(x, t) \equiv 0$*) as no solution bounded in* $\mathbb{R}^{n+1}$*, except a generalized constant.*

Proof. According to (5.18) there exists a continuously differentiable function $\xi(x)$ uch that

$$\frac{\partial \xi}{\partial x_j} = \operatorname{Im} \lambda_j(x),\ j = 1,\dots,n \tag{5.21}$$

nd according to (5.20) there exists a continuously differentiable function $\varphi(x')$, $x' = (x_1,\dots,x_{n-1})$ such that

$$\frac{\partial \varphi}{\partial x_j} = \lambda_j^0(x'),\ \ j = 1,\dots,n-1. \tag{5.22}$$

For uniqueness of this function we assume also that $\xi(0) = 0,\ \varphi(0) = 0.$

Now making the transformation $(x, t) \to (\xi, \tau)$:

$$\xi_j = x_j,\ \ j = 1,\dots,n-1,\ \ \xi_n = \xi(x),\ \ \tau = t - \varphi(x'), \tag{5.23}$$

ve have

$$\frac{\partial u}{\partial x_j} = \frac{\partial u}{\partial \xi_j} + \frac{\partial u}{\partial \xi_n} \operatorname{Im} \lambda_j(x) - \frac{\partial u}{\partial \tau} \lambda_j^0(x'),\ \ j = 1,\dots,n-1,$$

$$\frac{\partial u}{\partial x_n} = \frac{\partial u}{\partial \xi_n} \operatorname{Im} \lambda_n(x),\ \ \frac{\partial u}{\partial t} = \frac{\partial u}{\partial \tau}$$

nd the system (5.15) takes the form

$$\frac{\partial u}{\partial \xi_j} - A_j(\xi, \tau)u - B_j(\xi, \tau)\bar{u} = F_j(\xi, \tau),\ \ j = 1,\dots,n-1,$$

$$\frac{\partial u}{\partial \xi_n} + \frac{\lambda_n}{\operatorname{Im}\lambda_n}\frac{\partial u}{\partial \tau} - \tilde{A}_n(\xi,\tau)u - \tilde{B}_n(\xi,\tau)\bar{u} = \tilde{F}_n(\xi,\tau), \tag{5.24}$$

where

$$A_j(\xi,\tau) = a_j - \frac{\operatorname{Im}\lambda_j}{\operatorname{Im}\lambda_n}a_n,\quad B_j(\xi,\tau) = b_j - \frac{\operatorname{Im}\lambda_j}{\operatorname{Im}\lambda_n}b_n, \tag{5.25}$$

$$F_j(\xi,\tau) = f_j - \frac{\operatorname{Im}\lambda_j}{\operatorname{Im}\lambda_n}f_n,\quad \tilde{A}_n = \frac{a_n}{\operatorname{Im}\lambda_n},\quad \tilde{B}_n = \frac{b_n}{\operatorname{Im}\lambda_n},\quad \tilde{F}_n = \frac{f_n}{\operatorname{Im}\lambda_n}.$$

According to (5.25) $A_j, B_j, \tilde{A}_n, \tilde{B}_n, F_j, \tilde{F}_n$ are continuously differentiable in $\mathbb{R}^{n+1}$ and vanishes as $O((|\zeta|^2+\tau^2)^{-1/2-\varepsilon})$, $\varepsilon > 0$ when $\sqrt{(|\xi|^2+\tau^2)} \to \infty$. Let $\zeta = \zeta(\xi_n\,\tau) = \xi_n^0 + i\,\tau^0$ be a complete homeomorphism of the plane (ξ_n, τ) onto the plane (ξ_n^0, τ^0), satisfying the Beltrami equation

$$\frac{\partial \zeta}{\partial \xi_n} + \frac{\lambda_n}{\operatorname{Im}\lambda_n}\frac{\partial \zeta}{\partial \tau} = 0. \tag{5.26}$$

Then the last equation of the system (5.24) takes the form

$$\frac{\partial u}{\partial \bar{\zeta}} - A_n(\xi',\zeta)u - B_n(\xi',\zeta)\bar{u} = F_n(\xi',\zeta), \tag{5.27}$$

where

$$\xi' := (\xi_1,\ldots,\xi_{n-1}),\quad \frac{\partial u}{\partial \bar{\zeta}} = \frac{1}{2}\left(\frac{\partial u}{\partial \xi_n^0} + i\frac{\partial u}{\partial \tau^0}\right)$$

and

$$A_n(\xi',\zeta) = (\bar{\zeta}_{\xi_n} + \frac{\lambda_n}{\operatorname{Im}\lambda_n}\bar{\zeta}_\tau)^{-1}\,\tilde{A}_n,$$

$$B_n(\xi',\zeta) = (\bar{\zeta}_{\xi_n} + \frac{\lambda_n}{\operatorname{Im}\lambda_n}\bar{\zeta}_\tau)^{-1}\,\tilde{B}_n,$$

$$F_n(\xi',\zeta) = (\bar{\zeta}_{\xi_n} + \frac{\lambda_n}{\operatorname{Im}\lambda_n}\bar{\zeta}_\tau)^{-1}\,\tilde{F}_n.$$

It can easily be verified that the conditions (5.19) lead to the following compatibility conditions for the system (5.24), (5.27):

$$\frac{\partial A_k}{\partial \xi_j} - \frac{\partial A_j}{\partial \xi_k} = 0, \quad \mathrm{Im}\,(B_j \bar{B}_k) = 0,$$

$$\frac{\partial B_k}{\partial \xi_j} - \frac{\partial B_j}{\partial \xi_k} = 0, \; A_j B_k + B_j \bar{A}_k - A_k B_j - B_k \bar{A}_j = 0, \tag{5.28}$$

$$\frac{\partial F_k}{\partial \xi_j} - \frac{\partial F_j}{\partial \xi_k} + A_k F_j + B_k \bar{F}_j - B_j \bar{F}_k = 0, \quad j \neq k,$$

$$\frac{\partial A_j}{\partial \bar{\zeta}} - \frac{\partial A_n}{\partial \xi_j} = 0, \; \frac{\partial B_j}{\partial \bar{\zeta}} - \frac{\partial B_n}{\partial \xi_j} = 0, \; \frac{\partial F_j}{\partial \bar{\zeta}} - \frac{\partial F_n}{\partial \xi_j} + A_j F_n - A_n F_j - B_n \bar{F}_j = 0,$$

$$A_j B_n - A_n B_j = 0,$$

$$\bar{B}_j \frac{\partial u}{\partial \zeta} - \bar{B}_n A_j u - \bar{B}_n B_j \bar{u} = 0, \quad j = 1,\dots,n-1, \tag{5.29}$$

where

$$\frac{\partial u}{\partial \zeta} = \frac{1}{2}\Big(\frac{\partial u}{\partial \xi_n^0} - \mathrm{i}\,\frac{\partial u}{\partial \tau^0}\Big).$$

If $A_j = B_j \equiv 0$ for all $j = 1,\dots,n-1$, then from (5.28) and (5.29) it follows that A_n, B_n, F_n do not depend from ξ'. Then from the first $n-1$ equations of the system (5.24) and from the equalities

$$\frac{\partial F_k}{\partial \xi_j} - \frac{\partial F_j}{\partial \xi_k} = 0, \quad j \neq k$$

we obtain

$$u(\xi, \tau) = u^0(\zeta) + \sum_{j=1}^{n-1} \int_\ell F_j(\xi', \zeta)\, d\xi_j\,, \tag{5.30}$$

where ℓ is an arbitrary curve, which connects point $\xi' = (\xi_1,\dots,\xi_{n-1})$ with origin $\xi' = (0,\dots,0)$ and $u^0(\zeta)$ is a complex-valued function of only the variable $\zeta = \xi_n^0 + \mathrm{i}\tau^0$.

Substituting (5.30) into (5.27), we obtain that $u^0(\zeta)$ is the solution of the equation

$$\frac{\partial u^0}{\partial \bar{\zeta}} - A_n u^0 - B_n \bar{u}^0 = F_n \tag{5.31}$$

on the whole plane ζ. Since by our assumptions $A_n(\zeta), B_n(\zeta), F_n(\zeta) \in L_{p,2}$ on the plane ζ with $p > 2$, then (5.31) is uniquely solvable and corresponding homogeneous equation (i.e. $F_n \equiv 0$) has no solution bounded in the plane ζ, except a generalized constant (see [58]). Hence the theorem is proved in case $A_j = B_j \equiv 0$, $j = 1,\dots,n-1$. If $B_{j_0}(\xi, \tau) \neq 0$ for some $j_0 : 1 \leq j_0 \leq n-1$, then the last equation (5.29) for $j = j_0$ gives

$$\frac{\partial u}{\partial \zeta} - \bar{B}_n \frac{A_{j_0}}{\bar{B}_{j_0}} u - \bar{B}_n \frac{B_{j_0}}{\bar{B}_{j_0}} \bar{u} = 0. \tag{5.32}$$

The first $n-1$ equations of the system (5.24) with compatibility conditions (5.28) are equivalent to the following Volterra integral equation:

$$u(\xi, \tau) - \sum_{j=1}^{n-1} \int_\ell \left[A_j(\xi', \zeta)\, u(\xi', \zeta) + B_j(\xi', \zeta)\, \overline{u(\xi', \zeta)}\right] d\xi_j$$

$$= u^0(\zeta) + u^*(\xi, \tau), \tag{5.33}$$

where

$$u^0(\zeta) = u\,|_{\xi' = 0}$$

and

$$u^*(\xi, \tau) = \sum_{j=1}^{n-1} \int_\ell F_j\,(\xi', \zeta)\, d\xi_j\,.$$

We may solve (5.33) by the successive-approximation method. If $F_j \equiv 0$ we then obtain (see [29])

$$u(\xi, \tau) = \Gamma_1(\xi, \tau)\, u^0(\zeta) + \Gamma_2(\xi, \tau)\, \overline{u^0(\zeta)},$$

where $\Gamma_j(\xi, \tau)$ are known functions such that

$$|\Gamma_1(\xi, \tau)|^2 - |\Gamma_2(\xi, \tau)|^2 \neq 0, \quad \Gamma_1|_{\xi' = 0} = 1, \quad \Gamma_2|_{\xi' = 0} = 0.$$

From (5.27) and (5.32) we can find the differential of the function $u^0(\xi) = u \mid_{\xi' = 0}$ through linear combination of u^0 and $\bar{u}^0$

$$du^0 = u^0_\zeta \, d\zeta + u^0_{\bar\zeta} \, d\bar\zeta$$

$$= \left(\overline{B_n(0,\zeta)} \, \frac{\overline{A_{j_0}(0,\zeta)}}{\overline{B_{j_0}(0,\zeta)}} \, u^0(\zeta) + \overline{B_n(0,\zeta)} \, \frac{\overline{B_{j_0}(0,\zeta)}}{\overline{B_{j_0}(0,\zeta)}} \, \overline{u^0(\zeta)} d\zeta \right. \tag{5.34}$$

$$+ (A_n(0,\zeta)\, u^0(\zeta) + B_n(0,\zeta)\, \overline{u^0(\zeta)} + F_n(0,\zeta))\, d\bar\zeta \, .$$

Integrating this equality, we obtain for u^0 a Volterra-type integral equation on the plane ζ, which can be solved by successive approximation. In particular if $F_n = 0$, then its solution reduces to a constant. The theorem is proved.

5.3 Degenerate overdetermined elliptic systems

It is interesting to note that the assertions of Theorem 5.1 hold also for some first-order overdetermined systems with degenerate ellipticity condition. For example, the system

$$\frac{\partial u}{\partial x_j} - \mathrm{i}\, x_j^m \frac{\partial u}{\partial t} = f_j(x,t), \quad j = 1,\ldots,n \tag{5.35}$$

is elliptic everywhere in $\mathbb{R}^{n+1}$ except axis t, along which it is degenerate. If m is even, then the transformation $y_j = x_j^{m+1}$ immediately reduces this system into a system elliptic everywhere in $\mathbb{R}^{n+1}$. Moreover, there exist degenerate elliptic overdetermined systems which possess a stronger solvability property. For example the system

$$x_j \frac{\partial u}{\partial x_j} - \mathrm{i} \frac{\partial u}{\partial t} = f_j(x,t), \quad j = 1,\ldots,n \tag{5.36}$$

is also elliptic everywhere in $\mathbb{R}^{n+1}$ and degenerate along the t-axis. This system is solvable even in the octant $x_j \geq 0$, $j = 1,\ldots,n$ for any right-hand side, vanishing at infinity and satisfying the compatibility conditions. Indeed, the system (5.36) considered in this octant reduces to the system

$$\frac{\partial u}{\partial y_j} - \mathrm{i} \frac{\partial u}{\partial t} = f_j\,, \quad j = 1,\ldots,n$$

on the whole CArtesian space $\mathbb{R}^{n+1}$ by means of the mapping

$$y_j = \ln x_j, \quad j = 1,\dots,n.$$

The homogeneous system (i.e. $f_j \equiv 0$) corresponding to (5.36) has no solution bounded in the octant $x_j \geq 0$, $j = 1,\dots,n$, except for a constant.

Another such example is the system

$$\frac{\partial u}{\partial x_j} - \mathrm{i}\, t \frac{\partial u}{\partial t} = f_j(x, t), \quad j = 1,\dots,n, \tag{5.37}$$

which is elliptic everywhere in $\mathbb{R}^{n+1}$, except on the n-dimensional hyperplane $t = 0$, on which it is degenerate. This system, considered on the half-space $t \geq 0$, possesses the same property as system (5.36). We could continue with more such examples, but there are degenerate first-order overdetermined elliptic systems for which the assertions of Theorem 5.1 drastically fail. Such a system is (5.35) for odd numbers m. Indeed, if we take for instance $m = 1$, then the inhomogeneous system (5.35) becomes 'almost unsolvable' and the corresponding homogeneous system (i.e. $f_j \equiv 0$) has an infinite number of linearly independent continuous solutions bounded in $\mathbb{R}^{n+1}$:

$$u_k(x, t) = \left[\frac{2t + \mathrm{i}(|x|^2 - 2)}{2t + \mathrm{i}(|x|^2 + 2)}\right]^k, \quad k = 1,2,\dots \tag{5.38}$$

with grad $u_k \neq 0$, where $|x| = \sqrt{(x_n^2 + \cdots + x_n^2)}$.

Note by the way that the functions (5.38) for $n = 2$ are bounded solutions continuous in $\mathbb{R}^3$, of the homogeneous equation

$$u_{x_1} + \mathrm{i}\, u_{x_2} - \mathrm{i}\,(x_1 + \mathrm{i}\, x_2) u_t = 0$$

corresponding to famous Lewy equation, considered above.

Analogously the system

$$t\frac{\partial u}{\partial x_j} - \mathrm{i}\, \frac{\partial u}{\partial t} = f_j(x, t), \quad j = 1,\dots,n \tag{5.39}$$

is also 'almost unsolvable' and the corresponding homogeneous system (i.e. $f_j \equiv 0$) has an infinite number of linearly independent solutions $u_k(x, t)$, bounded and continuous in $\mathbb{R}^{n+1}$, with grad $u_k \neq 0$:

$$u_k(x,t) = \left[\frac{2(x_1+\cdots+x_n)+i(t^2-2)}{2(x_1+\cdots+x_n)+i(t^2+2)}\right]^k, \quad k=1,2,\ldots. \tag{5.40}$$

The main reason for difference (5.35) and (5.36) is that the t-axis for (5.35) is not characteristic, but for (5.36) it is. For the same reason, the hyperplane $t = 0$ is not characteristic for (5.39), but for (5.37) it is. There are also some pure elliptic systems with singular coefficients which possess analogous properties. For example, the system

$$\frac{\partial u}{\partial x_j} + i\frac{\partial u}{\partial t} + 2\mu(x_j + i\,t)u = 0 \tag{5.41}$$

has an infinite number of linearly independent solutions, bounded and continuous in $\mathbb{R}^{n+1}$ ($\mathrm{Re}\,\mu \geq 0$):

$$u_k(x,t) = (x_1 + \cdots x_n + i\,t)^k e^{-\mu(|x|^2 + t^2)}, \quad k = 1,2,\ldots$$

and the system

$$\frac{\partial u}{\partial x_j} + i\frac{\partial u}{\partial t} + 2\mu(x_1 + \ldots + x_n + i\,t)u = 0 \tag{5.42}$$

has an infinite number of such solutions ($\mathrm{Re}\,\mu \geq 0$):

$$u_k(x,t) = (x_1 + \cdots x_n + i\,t)^k e^{-\mu[(x_1+\cdots+x_n)^2 + t^2]}, \quad k = 1,2,\ldots.$$

5.4 Boundary-value problems

(a) Let Ω be a bounded domain in $\mathbb{R}^{n+1}$ with smooth boundary $\Gamma = \partial\Omega$. For determinacy we assume that its intersection G with the two-dimensional plane $x_j = 0$, $j = 1,\ldots,n-1$ is a plane domain whose boundary γ is a closed curve with no crossing points. We consider the system (5.15) in domain Ω assuming its coefficients to be bounded in $\bar{\Omega} = \Omega + \Gamma$.

Theorem 5.2. *If the conditions (5.18), (5.19), (5.20) are fulfilled in domain Ω, then any solution of the homogeneous system corresponding to (5.15) vanishing on curve γ, vanishes identically in Ω.*

Let us assume $B_j(\xi,\tau) \equiv 0$, $1 \leq j \leq n$ in Ω^*, the image of Ω under the mapping

(5.23). Then from the first $n-1$ equations of the system (5.24), we obtain

$$u(\xi, \tau) = u^0(\xi_n, \tau)\, e^{\omega^0(\xi, \tau)} + u^1(\xi, \tau), \tag{5.43}$$

where

$$\omega^0(\xi, \tau) = \sum_{j=1}^{n-1} \int_\ell A_j(\xi', \xi_n, \tau)\, d\xi_j\,, \tag{5.44}$$

$$u^1(\xi, \tau) = e^{\omega^0(\xi, \tau)} \sum_{j=1}^{n-1} \int_\ell F_j(\xi', \xi_n, \rho)\, e^{-\omega^0(\xi', \xi_n, \tau)}\, d\xi_j\,,$$

where $u^0(\xi_n, \tau)$ is an arbitrary continuously differentiable function of only two variables ξ_n, τ.

Substituting (5.43) into the last equation of the system (5.24) we obtain in the domain $G^* = \Omega^* \cap (\xi_j = 0,\ j = 1,...,n-1)$ the following equation for $u^0(\xi_n, \tau) = u|_{\xi' = 0}$:

$$\frac{\partial u^0}{\partial \xi_n} + \frac{\lambda_n}{\operatorname{Im}\lambda_n} \frac{\partial u^0}{\partial \tau} - \tilde{A}_n(0, \xi_n, \tau) u^0 = \tilde{F}_n(0, \xi_n, \tau). \tag{5.45}$$

By the similarity principle (see [58]) from (5.45) we have

$$u^0(\xi_n, \tau) = \Phi(\xi_n, \tau)\, e^{\omega_1(\xi_n, \tau)} + u_1(\xi_n, \tau), \tag{5.46}$$

where $\omega_1(\xi_n, \tau)$, $u_1(\xi_n, \tau)$ are known functions and $\Phi(\xi_n, \tau)$ is an arbitrary solution of the Beltrami equation

$$\Phi_{\xi_n} + \frac{\lambda_n}{\operatorname{Im}\lambda_n} \Phi_\tau = 0.$$

Hence by (5.43) we obtain

$$u(\xi, \tau) = \Phi(\xi_n, \tau)\, e^{\omega^*(\xi, \tau)} + u^*(\xi, \tau), \tag{5.47}$$

where $\omega^*(\xi, \tau)$, $u^*(\xi, \tau)$ are known functions expressed in terms of the coefficients and right-hand sides of (5.24). In the case of the homogeneous system (5.24) we have

$$u(\xi, \tau) = \Phi(\xi_n, \tau)\, e^{\omega^*(\xi, \tau)}. \tag{5.48}$$

The assertion of Theorem 5.2 in case $B_j(\xi, \tau) \equiv 0$, $j = 1,\dots,n$, follows from (5.48).

In the general case we use almost the same arguments used in Section 5.2. In particular, from (5.48) it follows that all solutions of the homogeneous elliptic system

$$\frac{\partial u}{\partial x_j} + \lambda_j(x)\frac{\partial u}{\partial t} = 0, \; j = 1,\dots,n-1, \; \frac{\partial u}{\partial x_n} + \mathrm{i}\frac{\partial u}{\partial t} = 0 \tag{5.49}$$

with $C^1(\Omega)$-coefficients, satisfying the conditions

$$\frac{\partial \lambda_k}{\partial x_j} - \frac{\partial \lambda_j}{\partial x_k} = 0, \; j \neq k, \; \frac{\partial \lambda_j}{\partial x_n} = 0, \; j = 1,\dots,n-1$$

in Ω has the form

$$u(x, t) = \Phi(x_n + \mathrm{i}\, t - \mathrm{i}\textstyle\int_{\ell} (\lambda, \mathrm{d}x')), \tag{5.50}$$

where Φ is an arbitrary function, holomorphic on its argument, ℓ is an arbitrary curve, which connects the point $x' = (x_1,\dots,x_{n-1}) \in \Omega$ with the origin $x' = (0,..,0)$ and $\lambda(x') = (\lambda_1(x'),\dots,\lambda_{n-1}(x')$, $(\lambda, \mathrm{d}x') = \lambda_1(x')\mathrm{d}x_1 + \cdots + \lambda_{n-1}(x')\mathrm{d}x_{n-1}$.

The representation formula (5.47) for the solution of the system (5.24) in case $B_j(\xi, \tau) \equiv 0$, $1 \leq j \leq n$, can be used for the investigation of boundary-value problems. Consider, for instance, the following Riemann-Hilbert type of boundary-value problem:

Problem *RH*. Find a solution of the system (5.15) in domain Ω, continuous in $\bar{\Omega}$, satisfying the condition

$$\mathrm{Re}\,(\bar{\lambda} u) = h \tag{5.51}$$

on curve γ, where λ and h are given Hölder continuous functions on γ.

It is clear now that by means of (5.47) the boundary condition (5.51) reduces to the Riemann-Hilbert boundary condition for the solution $\Phi(\xi_n, \tau)$ of the Beltrami equation in G^*:

$$\mathrm{Re}\,(\lambda^* \Phi) = h^* \tag{5.51'}$$

on $\gamma^* = \partial G^*$.

Thus we come to the following.

Theorem 5.3. *If $\lambda \neq 0$ on γ and $\kappa = \mathrm{Ind}_\gamma \lambda$, then in case $\kappa \geq 0$ the problem RH is solvable and corresponding homogeneous problem has exactly $2\kappa + 1$ nontrivial* *linearly independent (over a field of real numbers) solutions; in case* $\kappa \geq 0$ *the problem RH is solvable if and only if its right hand sides satisfy* $-2\kappa - 1$ *orthogonality conditions, in addition to compatibility conditions and corresponding homogeneous problem has no non trivial solution.*

(b) Let us consider now the system (5.15) in the whole Cartesian space $\mathbb{R}^{n+1}$ with vanishing coefficients and right-hand sides at infinity as described in Section 5.1. We seek a solution of this system which is piecewise continuous in $\mathbb{R}^{n+1}$, i.e. it is continuous in the closure $\bar{\Omega} = \Omega + \Gamma$ and $\overline{\mathbb{R}^{n+1} \setminus \Omega}$, where Ω is a bounded domain as described above.

If we intersect the domain Ω with the plane $x_j = 0$, $j = 1,...,n-1$, then we get plane domains G and $\tilde{G} = (\mathbb{R}^{n+1} \setminus G) \cap (x_j = 0, j = 1,...,n-1)$ with common boundary γ, which we assume to be a closed smooth curve without crossing points. Now we formulate the following problem.

Find a piecewise continuous solution of system (5.15) in $\mathbb{R}^{n+1}$, which is continuous in $\bar{\Omega}$ and $\overline{\mathbb{R}^{n+1} \setminus \Omega}$ respectively and satisfies the Riemann-type boundary condition

$$u^+ = G^- + g \tag{5.52}$$

on the curve γ, where G and g are Hölder continuous functions, given on γ, and u^+, u^- are boundary values of u on γ, which are determined by

$$u^+ = \lim_{\Omega \ni (x,t) \to (x_n, t) \in \gamma} u(x, t), \qquad u^- = \lim_{(\mathbb{R}^{n+1} \Omega) \ni (x,t) \to (x_n, t) \in \gamma} u(x, t)$$

and (x, t) tends to $(x_n, t) \in \gamma$ nontangentially.

In the simplest case of system (5.15) with

$$\lambda_j(x, t) \equiv \mathrm{i}, \; a_j(x, t) = b_j(x, t) = f_j(x, t) \equiv 0$$

this problem reduces to the usual Riemann problem for a holomorphic function. Indeed in this case the solution is expressed by formula (5.50):

$$u(x, t) = \Phi(x_1 + \cdots + x_n + \mathrm{i}\, t)$$

with holomorphic function Φ, so substituting into (5.52) gives the Riemann problem on the plane (x_n, t):

$$\Phi^+(x_n + \mathrm{i}\, t) - G(x_n, t)\, \Phi^-(x_n + \mathrm{i}\, t) + g(x_n, t).$$

In the general case with $B_j(\xi, \tau) \equiv 0$, $1 \leq j \leq n$, we also obtain the usual Riemann problem for piecewise continuous solutions of the generalized Cauchy Riemann equation. Thus we obtain the following (in case $B_j(\xi, \tau) \equiv 0$, $1 \leq j \leq n$):

Theorem 5.4. *If $G \neq 0$ on γ and $\kappa = \mathrm{Ind}_\gamma G$, then in case $\kappa \geq 0$, Problem R is solvable and the corresponding homogeneous problem has exactly 2κ nontrivial linearly independent solutions; if $\kappa < 0$, then Problem R is solvable if and only if its right-hand side satisfies -2κ orthogonality conditions (with the exception of the compatibility conditions) and the corresponding homogeneous problem has no nontrivial solutions.*

5.5 Nonelliptic first-order overdetermined systems

Now we turn to the system

$$P_j u - a_j(x,t)u - b_j(x,t)\,\bar{u} = f_j(x,t), \quad j = 1,\dots,n \tag{5.53}$$

with P_j given by (5.10) and satisfying the condition (5.13). If λ_j, μ_j do not depend on t and λ_j are real, then it is easy to see that the commutator $[P_j, P_k]u \equiv (P_j P_k - P_k P_j)u$ is

$$[P_j, P_k]u \equiv \left(\frac{\partial \lambda_k}{\partial x_j} - \frac{\partial \lambda_j}{\partial x_k}\right)\frac{\partial u}{\partial t}$$

$$+\left(\frac{\partial \mu_k}{\partial x_j} - \frac{\partial \mu_j}{\partial x_k}\right)\frac{\partial \bar{u}}{\partial t} + 2\mathrm{i}\,\mathrm{Im}\,(\mu_j \bar{u}_k)\cdot\frac{\partial^2 u}{\partial t^2}$$

and hence is equal to zero if

$$\frac{\partial \lambda_k}{\partial x_j} - \frac{\partial \lambda_j}{\partial x_k} = 0, \quad \frac{\partial \mu_k}{\partial x_j} - \frac{\partial \mu_j}{\partial x_k} = 0, \quad \mathrm{Im}\,\mu_j \bar{\mu}_k = 0, \; j \neq k.$$

In particular $[P_j, P_k]u \equiv 0$ in case $\lambda_j(x)$ are real and $\mu_j(x)$ are either real or purely imaginary and satisfy the conditions

$$\frac{\partial \lambda_k}{\partial x_j} - \frac{\partial \lambda_j}{\partial x_k} = 0, \quad \frac{\partial \mu_k}{\partial x_j} - \frac{\partial \mu_j}{\partial x_k} = 0, \; j \neq k, \tag{5.54}$$

which we shall assume.

We consider the system (5.53) with these conditions in a cylindrical domain $\Omega_t = \{(x,t) : x \in D, t \geq 0\}$, where D is a bounded domain in $\mathbb{R}^n$ with smooth boundary $\Gamma = \partial D$, assuming that $a_j(x,t)$, $b_j(x,t)$, $f_j(x,t)$, are sufficiently smooth functions vanishing as $t \to \infty$. Denote by

$$G_0 = \{(x_n, t) : a \le x_n \le b,\ t \ge 0\} = \Omega \cap (x' = 0)$$

a rectangular region, obtained from cylinder Ω_t by intersecting it with the two-dimensional plane $x' = 0$, $x' = (x_1, \dots, x_{n-1})$.

(a) First we consider the case $\mu_j(x) \equiv 0$, $j = 1, \dots, n$ and all $\lambda_j(x)$ are real and

$$\sum_{j=1}^{n} \lambda_j^2(x) \neq 0.$$

In this case the system (5.53) has one double multiplicity family characteristic $\omega(x, t) = t - \alpha(x) = \text{const.}$, where

$$\alpha(x) = \int_{\ell} (\lambda, \mathrm{d}x) = \sum_{j=1}^{n} \int_{\ell} \lambda_j(x)\, \mathrm{d}x_j$$

where ℓ is an arbitrary curve in $\mathbb{R}^n$, connecting the point $x(x_1, \dots, x_n)$ with the fixed point $x^0 = (x_1^0, \dots, x_n^0)$.

Now we formulate the following initial-boundary-value problem ($\mu_j \equiv 0$):

Problem *M*. Find in Ω_t a solution of (5.53), satisfying the conditions

$$u|_{x', t=0} = h_1(x_n), \quad a \le x_n \le b, \tag{5.55}$$

$$u|_{x'=0,\, x_n=a} = h_2(t), \quad t \ge 0, \tag{5.56}$$

where h_i are given complex-valued continuous functions.

Theorem 5.5. *Let* $\mu_j(x) \equiv 0$, $j = 1, \dots, n$, $\lambda_j(x)$ be real and

$$\frac{\partial \lambda_k}{\partial x_j} - \frac{\partial \lambda_j}{\partial x_k} = 0, \quad j \neq k,$$

$$\frac{\partial a_k}{\partial x_j} - \frac{\partial a_j}{\partial x_k} + \lambda_j \frac{\partial a_k}{\partial t} - \lambda_k \frac{\partial a_j}{\partial t} = 0,$$

$$\tag{5.57}$$

$$\frac{\partial b_k}{\partial x_j} - \frac{\partial b_j}{\partial x_k} + \lambda_j \frac{\partial b_k}{\partial x_j} - \lambda_k \frac{\partial b_j}{\partial t} + 2i\,(b_j \operatorname{Im} a_k - b_k \operatorname{I}m\, a_j) = 0, \quad \operatorname{Im}(b_j \bar{b}_k) = 0,$$

$$\frac{\partial f_k}{\partial x_j} - \frac{\partial f_j}{\partial x_k} + \lambda_j \frac{\partial b_k}{\partial x_j} - \lambda_k \frac{\partial b_j}{\partial t} + a_k f_j - a_j f_k + b_k \bar{f}_j - b_j \bar{f}_k = 0,$$

and let the functions h_j, f_j satisfy some compatibility conditions at points $x' = 0$, $x_n = a$, $t = 0$. Then Problem M has a unique solution, continuous in $\bar{\Omega}_t$.

Proof. Suppose for determinacy $\lambda_n(x) \neq 0$. Then making nondegenerate transformation

$$\xi_j = x_j, \ \ j = 1,\ldots,n-1, \ \ \xi_n = \alpha(x), \ \ \tau = t \tag{5.58}$$

we transform the system (5.53) into the system

$$\begin{aligned} &u_{\xi_j} - \tilde{a}_j(\xi,\tau)u - \tilde{b}_j(\xi,\tau)\bar{u} = f_j(\xi,\tau), \ \ j = 1,\ldots,n-1, \\ &u_{\xi_n} + u_\tau - \tilde{a}_n(\xi,\tau)u - \tilde{b}_n(\xi,\tau)\bar{u} = \tilde{f}_n(\xi,\tau), \end{aligned} \tag{5.59}$$

where

$$\tilde{a}_j = a_j - \lambda_j \frac{a_n}{\lambda_n}, \ \tilde{b}_j = b_j - \lambda_j \frac{b_n}{\lambda_n}, \ \tilde{f}_j = f_j - \lambda_j \frac{f_n}{\lambda_n},$$

$$\tilde{a}_n = \frac{a_n}{\lambda_n}, \ \tilde{b}_n = \frac{b_n}{\lambda_n}, \ \tilde{f}_n = \frac{f_n}{\lambda_n},$$

From (5.57) it follows that

$$\frac{\partial \tilde{a}_k}{\partial \xi_j} - \frac{\partial \tilde{a}_j}{\partial \xi_n} = 0, \ \ j \neq k.$$

Hence the curvilinear integral

$$\omega(\xi) = \int_\ell (\tilde{a}, \mathrm{d}\xi), \ \ \tilde{a} = (\tilde{a}_1,\ldots,\tilde{a}_n)$$

does not depend on the curve ℓ and so introducing a new function by the equality

$$v(\xi,\tau) = u(\xi,\tau)\, \mathrm{e}^{-\omega(\xi)},$$

we obtain instead (5.59) the following system:

$$\begin{aligned} &v_{\xi_j} - b_j^0(\xi,\tau)\,\bar{v} = f_j^0(\xi,\tau), \\ &v_{\xi_n} + v_\tau - b_n(\xi,\tau)\bar{v} = f_n(\xi,\tau), \end{aligned} \tag{5.60}$$

where

$$b_j^0(\xi, \tau) = b_j(\xi, \tau)\, e^{-2i\,\mathrm{Im}\,\omega(\xi)}, \quad f_j^0(\xi, \tau) = \tilde{f}_j(\xi, \tau)\, e^{-\omega(\xi)}.$$

It is easy to verify that, as follows from (5.57), for the system (5.60) the compatibility conditions are fulfilled. Hence the first $n - 1$ equations of the system (5.60) are equivalent to the following Volterra-type integral equation:

$$v(\xi, \tau) - \sum_{j=1}^{n-1} \int_\ell b_j^0(\xi', \xi_n, \tau)\, \overline{v(\xi'; \xi_n, \tau)}\, d\xi_j$$

$$= v^0(\xi_n, \tau) + \sum_{j=1}^{n-1} \int_\ell f_j^0(\xi'; \xi_n, \tau)\, d\xi_j, \tag{5.61}$$

where $v^0(\xi_n, \tau)$ is an arbitrary complex-valued continuously differentiable function of only two variables ξ_n, τ: $v^0(\xi_n, \tau) = v|_{\xi'=0}$.

From (5.61) by means of successive approximation we find

$$v(\xi, \tau) = \Gamma_1(\xi, \tau)\, v^0(\xi_n \tau) + \Gamma_2(\xi, \tau)\, \overline{v^0(\xi_n, \tau}$$

$$+ \sum_{j=1}^{n-1} \int_\ell \left[K_1(\xi', \eta'; \xi_n, \tau) f_j^0(\eta'; \xi_n, \tau) - K_2(\xi', \eta'; \xi_n, \tau)\overline{f_j^0(\eta'; \xi_n, \tau)}\right] d\eta_j, \tag{5.62}$$

where $\Gamma_1(\xi, \tau)$, $\Gamma_2(\xi, \tau)$ form a solution of the following system of integral equations:

$$\Gamma_1(\xi, \tau) - \sum_{j=1}^{n-1} \int_\ell b_j^0(\xi', \xi_n, \tau)\, \overline{\Gamma_2(\xi'; \xi_n, \tau)}\, d\xi_j = 1,$$

$$\Gamma_2(\xi, \tau) - \sum_{j=1}^{n-1} \int_\ell b_j^0(\xi', \xi_n, \tau)\, \overline{\Gamma_1(\xi'; \xi_n, \tau)}\, d\xi_j = 0,$$

$$K_1(\xi', \eta'; \xi_n, \tau) = \Gamma_1(\xi'; \xi_n, \tau)\, \overline{\Gamma_1(\eta'; \xi_n, \tau)} - \Gamma_2(\xi', \xi_n, \tau)\, \overline{\Gamma_2(\eta'; \xi_n, \tau)},$$

$$K_2(\xi', \eta'; \xi_n, \tau) = \Gamma_1(\xi'; \xi_n, \tau)\, \Gamma_2(\eta'; \xi_n, \tau) - \Gamma_2(\xi'; \xi_n, \tau)\, \Gamma_1(\eta'; \xi_n, \tau). \tag{5.63}$$

From (5.63) it follows that $|\Gamma_1(\xi, \tau)|^2 - |\Gamma_2(\xi, \tau)|^2 \equiv 1$. Indeed, since from (5.63) it follows that

$$\frac{\partial \Gamma_1}{\partial \xi_j} = b_j^0(\xi, \tau)\, \overline{\Gamma_2(\xi, \tau)}, \quad \frac{\partial \Gamma_2}{\partial \xi_j} = b_j^0(\xi, \tau)\, \overline{\Gamma_1(\xi, \tau)} \tag{5.64}$$

and

$$\Gamma_1|_{\xi'=0} = 1, \quad \Gamma_2|_{\xi'=0} = 0, \tag{5.65}$$

then the function $\Delta(\xi, \tau) = |\Gamma_1(\xi, \tau)|^2 - |\Gamma_2(\xi, \tau)|^2$ does not depend on $\xi' = (\xi_1, \ldots, \xi_{n-1})$, because from (5.64)

$$\frac{\partial \Delta}{\partial \xi_j} = \frac{\partial \Gamma_1}{\partial \xi_j} \overline{\Gamma_1(\xi, \tau)} - \frac{\partial \Gamma_2}{\partial \xi_j} \overline{\Gamma_2(\xi, \tau)} \frac{\partial \overline{\Gamma}_1}{\partial \xi_j} - \Gamma_2(\xi, \tau) \frac{\partial \overline{\Gamma}_2}{\partial \xi_j} = 0.$$

Now from (5.65) it follows that $\Delta|_{\xi'=0} = 1$, i.e. $\Delta(\xi, \tau) \equiv 1$. Substituting (5.62) into the last equation of the system (5.60) we see that the function $v^0(\xi_n, \tau)$ satisfies the equation

$$V^0_{\xi_n} + V^0_\tau - a_n(0, \xi_n; \tau)\, V^0 - b_n(0; \xi_n, \tau)\, \bar{V}^0 = f_n(0, \xi_n, \tau) \tag{5.66}$$

in a rectangular domain $G_0' = \{(\xi_n, \tau) : \alpha^0 \le \xi_n \le \beta^0, \tau \ge 0\}$, where

$$\alpha^0 = \int_{x_n^0}^{a} \lambda_n(0, \ldots, 0; x_n)\, dx_n, \quad \beta^0 = \int_{x_n^0}^{b} \lambda_n(0, \ldots, 0; x_n)\, dx_n.$$

Taking into account the conditions (5.55), (5.56), we have the initial condition

$$v^0|_{\tau=0} = h_1(\xi_n) = h_1[x_n(0, \ldots, 0, \xi_n)], \quad \alpha^0 \le \xi_n \le \beta^0 \tag{5.67}$$

and the boundary condition

$$v^0|_{\xi_n = \alpha^0} = h_2(\tau), \quad \tau \ge 0. \tag{5.68}$$

The general solution of the equation (5.66) takes the form

$$v^0(\xi_n, \tau) = \Gamma^1(\xi_n, \tau)\, \varphi(\xi_n - \tau) + \Gamma^2(\xi_n, \tau)\, \overline{\varphi(\xi_n - \tau)}$$

$$+ \int_0^{\xi_n + \tau} \left[k^1(\xi_n, \tau; \sigma) f_n(\xi_n, \tau; \sigma) - k^2(\xi_n, \tau; \sigma) \overline{f_n(\xi_n, \tau; \sigma)} \right] d\sigma, \tag{5.69}$$

where φ is an arbitrary complex-valued function of one variable. Now by means of (5.67), (5.68) we can determine the smooth value of the function φ for $\alpha^0 \le \xi_n \le \beta^0$, $\tau \ge 0$ if, for instance, $h_1(\alpha^0) = h_2(0)$.

(b) Let $\lambda_j(x) \equiv 0$, $j = 1,\ldots,n$ and $\mu_j(x)$ be either real or purely imaginary. In this case the system (5.53) has two different family characteristic $\omega(x, t) = t - \beta(x) =$ const. and $\omega(x, t) = t + \beta(x) =$ const., where

$$\beta(x) = \int_\ell (|\mu|, \mathrm{d}x) = \sum_{j=1}^{n} \int_\ell |\mu_j(x)| \, \mathrm{d}x_j. \tag{5.70}$$

Let Ω_t be the same cylindrical domain as before.

Problem *N*. Find in Ω_t a solution of the system (5.53) ($\lambda_j(x) \equiv 0$), satisfying the initial condition

$$u|_{x', t=0} = g_1(x_n), \quad a \le x_n \le b \tag{5.71}$$

and boundary conditions

$$\mathrm{Re}\, u|_{x'=0, x_n=a} = g_2(t), \quad \mathrm{Im}\, u|_{x'=0, x_n=b} = g_3(t), \quad t \ge 0 \tag{5.72}$$

in case $\mu_j(x)$ are real, and boundary conditions

$$[\mathrm{Re}\, u + \mathrm{Im}\, u]_{x'=0, x_n=a} = g_2(t), \; [\mathrm{Re}\, u - \mathrm{Im}\, u]_{x'=0, x_n=b} = g_3(t), \; t \ge 0 \tag{5.73}$$

in case $\mu_j(x)$ are purely imaginary.

Theorem 5.6. *Let* $\lambda_j(x) \equiv 0$, $\mu_j(x)$ *be either real or purely imaginary,* $\sum_{j=1}^{n} |\mu_j|^2 \neq 0$ *and*

$$\frac{\partial \mu_k}{\partial x_j} - \frac{\partial \mu_j}{\partial x_k} = 0, \quad j \neq k,$$

$$\frac{\partial a_k}{\partial x_j} - \frac{\partial a_j}{\partial x_k} + \mu_j \frac{\partial \bar{b}_k}{\partial t} - \mu_k \frac{\partial \bar{b}_j}{\partial t} = 0,$$

$$\frac{\partial b_k}{\partial x_j} - \frac{\partial b_j}{\partial x_k} + \mu_j \frac{\partial \bar{a}_k}{\partial t} - \mu_j \frac{\partial \bar{a}_j}{\partial t} + 2\mathrm{i}\,(b_j \,\mathrm{Im}\, a_k - b_k \,\mathrm{Im}\, a_j = 0,$$

$$\mathrm{Im}\,(b_j \bar{b}_k) = 0, \; \mathrm{Im}\,(\mu_j \bar{b}_k - \mu_k \bar{b}_j) = 0, \; \mu_j \,\mathrm{Im}\, a_k - \mu_k \,\mathrm{Im}\, a_j = 0,$$

$$\frac{\partial f_k}{\partial x_j} - \frac{\partial f_j}{\partial x_k} + \mu_j \frac{\partial \bar{f}_k}{\partial t} - \mu_k \frac{\partial \bar{f}_j}{\partial t} + a_k f_j - a_j f_k + b_k \bar{f}_j - b_j \bar{f}_k = 0, \quad j \neq k,$$

ınd at the points $x' = 0,\ x_n = a, x_n = b, t = 0$ *functions* f_j, g_j *satisfy some :ompatibility condition, then Problem N has a unique solution, continuous in* $\bar{\Omega}_t$.

Proof. Suppose $\mu_n(x) \neq 0$ and make the transformation $(x, t) \to (\xi, \tau)$:

$$\xi_j = x_j,\ j = 1,...,n,\ \xi_n = \beta(x),\ \tau = t. \tag{5.74}$$

Then we obtain the following system in variables ξ, τ :

$$u_{\xi_j} - a_j(\xi, \tau)u - b_j(\xi, \tau)\bar{u} = f_j(\xi, \tau),\ j = 1,...,n-1,$$

$$u_{\xi_n} + \varepsilon\ \bar{u}_\tau - a_n(\xi, \tau)u - b_n(\xi, \tau)\bar{u} = f_n(\xi, \tau), \tag{5.75}$$

where ε is equal to either 1 or i.

Now as before from the first $n - 1$ equations, we derive for $u(\xi, \tau)$ the following expression:

$$u(\xi, \tau) = \Gamma_1^*(\xi, \tau)u^0(\xi_n, \tau) + \Gamma_2^*(\xi, \tau)\,\overline{u^0(\xi_n, \tau)}$$

$$+ \sum_{j=1}^{n-1} \int_\ell \left[K_1^*(\xi', \eta'; \xi_n, \tau)f_j(\eta'; \xi_n, \tau) - K_2^*(\xi', \eta'; \xi_n, \tau)\overline{f_j(\eta'; \xi_n, \tau)}\right] d\eta_j, \tag{5.76}$$

where $u^0(\xi_n, \tau)$ is an arbitrary complex-valued continuously differentiable function of only two variables, $u^0(\xi_n, \tau) = u\mid_{\xi' = 0}$, which according to the last equation of the system (5.75) satisfies either the equation

$$u^0_{\xi_n} + \bar{u}^0_\tau - a_n(0; \xi_n, \tau)u^0 - b_n(0; \xi_n, \tau)\bar{u}^0 = f_n(0; \xi_n, \tau) \tag{5.77}$$

or the equation

$$u^0_{\xi_n} + \mathrm{i}\ \bar{u}^0_\tau - a_n(0; \xi_n, \tau)u^0 - b_n(0; \xi_n, \tau)\bar{u}^0 = f_n(0; \xi_n, \tau) \tag{5.78}$$

in the rectangular domain $G_0' = \{(\xi_n, \tau) : \alpha^0 \leq \xi_n \leq \beta^0, \tau \geq 0\}$.

According to (5.71)-(5.73) the function $u^0(\xi_n, \tau)$ satisfies the initial condition

$$u^0\mid_{\tau = 0} = g_1(\xi_n) = g_1[x_n(0,...,0, \xi_n)] \tag{5.79}$$

and either the boundary conditions

$$\operatorname{Re} u^0 |_{\xi_n = \alpha^0} = g_2(\tau), \quad \operatorname{Im} u^0 |_{\xi_n = \beta^0} = g_3(\tau), \quad \tau \geq 0 \tag{5.80}$$

in the case of equation (5.77), or the boundary conditions

$$[\operatorname{Re} u^0 + \operatorname{Im} u^0]_{\xi_n = \alpha^0} = g_2(\tau), \quad [\operatorname{Re} u^0 - \operatorname{Im} u^0]_{\xi_n = \beta^0} = g_3(\tau), \quad \tau \geq 0 \tag{5.81}$$

in the case of equation (5.78).

Now separating real and imaginary parts in (5.77) and (5.78) and then adding and subtracting we obtain an equation of the form (5.66) or a similar equation with $v^0_{\xi_n} - v^0_\tau$, instead of $v^0_{\xi_n} + v^0_\tau$ and hence for their solution we obtain the expression (5.69) or a similar expression with $\varphi(\xi_n + \tau)$ instead $\varphi(\xi_n - \tau)$, and then by means of (5.79)-(5.81) we determine their continuous value if, for instance, $g_1(\alpha^0) = g_2(0)$, $g_1(\beta^0) = g_3(0)$ etc.

(c) Here we show how nonelliptic overdetermined systems like (5.53) arise. Consider, for instance, in a three-dimensional cylinder

$$\Omega_t = \{(x_1, x_2, t) : (x_1, x_2) \in G, \ t \geq 0\}$$

the problem of determining in Ω_t a solution of the first-order hyperbolic system of equations with respect to four real functions u_1, u_2, u_3, u_4:

$$\frac{\partial u_1}{\partial t} + \frac{\partial u_3}{\partial x_1} + \frac{\partial u_4}{\partial x_2} = f_1, \quad \frac{\partial u_2}{\partial t} - \frac{\partial u_3}{\partial x_2} + \frac{\partial u_4}{\partial x_1} = f_2,$$

$$\frac{\partial u_3}{\partial t} + \frac{\partial u_1}{\partial x_1} - \frac{\partial u_2}{\partial x_2} = f_3, \quad \frac{\partial u_4}{\partial t} + \frac{\partial u_1}{\partial x_2} + \frac{\partial u_2}{\partial x_1} = f_4, \tag{5.82}$$

satisfying the initial conditions

$$u_j|_{t=0} = h_j(x), \quad x \in G, \quad j = 1,2,3,4 \tag{5.83}$$

and boundary conditions

$$[\alpha_1 u_1 + \alpha_2 u_2]_{s_t} = 0, \quad [\beta_1 u_1 + \beta_2 u_2]_{s_t} = 0 \tag{5.84}$$

on the surface $s_t = \{(x_1, x_2, t) : (x_1, x_2) \in \Gamma = \partial G, t \geq 0\}$, where α_j, β_j are given real numbers normalized such that

$$\alpha_1^2 + \alpha_2^2 = \beta_1^2 + \beta_2^2 = 1.$$

From (5.82) it follows that the functions $u_j(x, t)$ are a solution of the following inhomogeneous wave equations:

$$\frac{\partial^2 u_1}{\partial t^2} - \frac{\partial^2 u_1}{\partial x_1^2} - \frac{\partial^2 u_1}{\partial x_2^2} = \frac{\partial f_1}{\partial t} - \frac{\partial f_3}{\partial x_1} - \frac{\partial f_4}{\partial x_2},$$

$$\frac{\partial^2 u_2}{\partial t^2} - \frac{\partial^2 u_2}{\partial x_1^2} - \frac{\partial^2 u_2}{\partial x_2^2} = \frac{\partial f_2}{\partial t} + \frac{\partial f_3}{\partial x_2} - \frac{\partial f_4}{\partial x_1},$$

$$\frac{\partial^2 u_3}{\partial t^2} - \frac{\partial^2 u_3}{\partial x_1^2} - \frac{\partial^2 u_3}{\partial x_2^2} = \frac{\partial f_3}{\partial t} - \frac{\partial f_1}{\partial x_1} + \frac{\partial f_2}{\partial x_2},$$

$$\frac{\partial^2 u_4}{\partial t^2} - \frac{\partial^2 u_4}{\partial x_1^2} - \frac{\partial^2 u_4}{\partial x_2^2} = \frac{\partial f_4}{\partial t} - \frac{\partial f_1}{\partial x_2} - \frac{\partial f_2}{\partial x_1} \tag{5.85}$$

and so the functions $v_1 = \alpha_1 u_1 + \alpha_2 u_2$, $v_2 = \beta_1 u_3 + \beta_2 u_4$ are a solution in Ω_t of the inhomogeneous wave equations

$$\frac{\partial^2 v_1}{\partial t^2} - \frac{\partial^2 v_1}{\partial x_1^2} - \frac{\partial^2 v_1}{\partial x_2^2} = F_1, \quad \frac{\partial^2 v_2}{\partial t^2} - \frac{\partial^2 v_2}{\partial x_1^2} - \frac{\partial^2 v_2}{\partial x_2^2} = F_2 \tag{5.86}$$

with

$$F_1 = \frac{\partial}{\partial t}(\alpha_1 f_1 + \alpha_2 f_2) - \frac{\partial}{\partial x_1}(\alpha_2 f_3 - \alpha_1 f_4) + \frac{\partial}{\partial x_2}(\alpha_2 f_3 - \alpha_1 f_4)$$

$$F_2 = \frac{\partial}{\partial t}(\beta_1 f_1 + \beta_2 f_2) - \frac{\partial}{\partial x_1}(\beta_1 f_1 - \beta_2 f_2) + \frac{\partial}{\partial x_3}(\beta_1 f_2 - \beta_2 f_1),$$

satisfying the initial conditions

$$v_1|_{t=0} = \alpha_1 h_1 + \alpha_2 h_2, \quad v_2|_{t=0} = \beta_1 h_3 + \beta_2 h_4 \tag{5.87}$$

and boundary conditions

$$v_1|_{S_t} = 0, \quad v_2|_{S_t} = 0. \tag{5.88}$$

Hence we can find $v_1(x, t)$, $v_2(x, t)$ in Ω_t uniquely through given conditions. The functions $u_j(x, t)$ can be found in Ω_t if we can find in Ω_t the functions $v_3 = -\alpha_2 u_1 + \alpha_1 u_2, v_4 = -\beta_2 u_3 + \beta_1 u_4$, because then we obtain

$$u_1 = \alpha_1 v_1 - \alpha_2 v_3,\ u_2 = -\alpha_1 v_1 + \alpha_2 v_3,$$

$$u_3 = \beta_1 v_2 - \beta_2 v_4,\ u_4 = -\beta_1 v_2 + \beta_2 v_4. \tag{5.89}$$

But the system (5.82), after determining v_1 and v_2, becomes the following over-determined system with respect to the function $u(x, t) = v_3 + \mathrm{i}\, v_4$:

$$\frac{\partial u}{\partial x_1} + \mathrm{i}\,(\alpha_1 \beta_1 + \alpha_2 \beta_2) \frac{\partial \bar{u}}{\partial t} = F_1^0,$$

$$\frac{\partial u}{\partial x_2} + \mathrm{i}\,(\alpha_1 \beta_2 + \alpha_2 \beta_1) \frac{\partial \bar{u}}{\partial t} = F_2^0, \tag{5.90}$$

which is the particular case of the system (5.53) with $n = 2$, $\lambda_j \equiv 0$, $\mu_1 = \mathrm{i}(\alpha_1\beta_1 + \alpha_2\beta_2)$, $\mu_2 = \mathrm{i}(\alpha_1\beta_2 - \alpha_2\beta_1)$, $a_j = b_j \equiv 0$.

Since a set of solutions of the system (5.90) has infinite dimension, then it follows that the problem (5.82), (5.83), (5.84) is not well posed. To obtain a well-posed problem for the system (5.82) it is necessary according to Theorem 5.6 to supply, on the lines $x_2 = a, x_2 = b$ of the rectangular domain

$$G' = \{(x_2, t) : a \le x_2 \le b, t \ge 0\} = \Omega \cap \{x_1 = 0\},$$

boundary conditions:

$$[v_3 + v_4\}_{x_1 = 0, x_2 = a} = h^1(t),\ [v_3 - v_4]_{x_1 = 0, x_2 = b}\, h^2(t),\ t \ge 0 \tag{5.91}$$

Now by means of Theorem 5.6 we obtain

Theorem 5.7. *If at points $x_1 = 0, x_2 = a, x_2 = b, t = 0$ the compatibility conditions for the functions $f_j, h_j, h^{(j)}$ are fulfilled, then the problem (5.82), (5.83), (5.84), (5.91) is well posed in the classical sense.*

(d) Now let us turn to the overdetermined system (5.8), which we consider in the cylinder

$$\Omega = \{(x_1, x_2, x_3) : (x_2, x_3) \in G,\ x_1 \ge 0\},$$

where G is a bounded domain in $\mathbb{R}^2$ with smooth boundary $\Gamma = \partial G$. Note that together with the homogeneous initial conditions $u_j \mid_{x_1 = 0} = 0$, this system is equivalent to the following (determined) hyperbolic system with respect to four real functions u_0, u_1, u_2, u_3:

$$\frac{\partial u_1}{\partial x_1} + \frac{\partial u_2}{\partial x_2} + \frac{\partial u_3}{\partial x_3} = 0, \quad \frac{\partial u_0}{\partial x_1} + \frac{\partial u_3}{\partial x_2} - \frac{\partial u_2}{\partial x_3} = 0,$$

$$\frac{\partial u_0}{\partial x_2} + \frac{\partial u_1}{\partial x_3} + \frac{\partial u_3}{\partial x_1} = 0, \quad \frac{\partial u_0}{\partial x_3} - \frac{\partial u_2}{\partial x_1} - \frac{\partial u_1}{\partial x_2} = 0 \tag{5.92}$$

supplied by the homogeneous initial conditions $u_j \mid_{x_1 = 0}$ and the homogeneous boundary conditions $u_0 \mid_{S_x} = 0$ on the surface

$$S_x = \{(x_1, x_2, x_3) : (x_1, x_3) \in \Gamma = \partial G,\ x_1 \geq 0\}.$$

Indeed, from (5.92) it follows that the function $u_0(x)$ is a solution in Ω of the wave equation

$$\frac{\partial^2 u_0}{\partial x_1^2} - \frac{\partial^2 u_0}{\partial x_2^2} - \frac{\partial^2 u_0}{\partial x_3^2} = 0, \tag{5.93}$$

satisfying the initial conditions

$$u_0 \mid_{x_1 = 0} = 0, \quad \frac{\partial u_0}{\partial x_1}\Big|_{x_1 = 0} = \left(\frac{\partial u_3}{\partial x_2} - \frac{\partial u_3}{\partial x_2}\right)_{x_1 = 0} = 0 \tag{5.94}$$

and boundary condition

$$u_0 \mid_{S_x} = 0, \tag{5.95}$$

and hence $u_0(x, t) \equiv 0$, i.e. the system (5.92) becomes (5.8) with conditions $u_j \mid_{x_1 = 0} = 0$. Since the system (5.92) coincides with the system (5.82) apart from notation, then by means of Theorem 5.7 it is easy to see that for the system (5.8) in the cylinder Ω, well-posed conditions are the initial conditions

$$u_j \mid_{x_1 = 0} = 0, \quad (x_2, x_3) \in G, \quad j = 1,2,3, \tag{5.96}$$

the boundary condition

$$[\beta_1 u_2 + \beta_2 u_3]_{S_x} = 0, \; x_1 \geq 0, \; \beta_1^2 + \beta_2^2 = 1 \tag{5.97}$$

on the surface and the boundary conditions

$$[u_1 + \beta_2 u_2 - \beta_1 u_3]_{x_2 = 0, x_3 = a} = h^{(1)}(x_1),$$

$$[u_1 - \beta_2 u_2 + \beta_1 u_3]_{x_2 = 0, x_3 = b} = h^{(2)}(x_1), \; x_1 \geq 0. \tag{5.98}$$

on the sides of the rectangular domain

$$G' = \{(x_3, x_1) : a \leq x_3 \leq b, x_1 \geq 0\} = \Omega \cap \{x_2 = 0\}.$$

Thus we have proved the following.

Theorem 5.8. *If the functions* $h^{(j)}(x_1)$ *satisfy equalities* $h^{(1)}(0) = h^{(2)}(0) = 0$, *then the problem (5.8), (5.96), (5.97), (5.98) has a unique solution continuous in* $\bar{\Omega}$.

Remark. As we have seen, a solution of a nonelliptic first-order overdetermined system (5.53) has a special wavelike structure. In particular all solutions of the system

$$\frac{\partial u}{\partial x_j} + \frac{x_j}{c|x|}\frac{\partial u}{\partial t} = 0, \; j = 1,\ldots,n, \tag{5.99}$$

where c is a real positive constant, are represented by the formula

$$u(x, t) = \varphi(|x| - c\,t), \; |x| = (x_1^2 + \cdots + x_n^2) \tag{5.100}$$

with arbitrary C^1-function φ, and may be presented as waves travelling outwards from the origin with speed c. All solutions of the system

$$\frac{\partial u}{\partial x_j} + \frac{x_j}{c|x|}\frac{\partial \bar{u}}{\partial t} = 0, \; j = 1,\ldots,n \tag{5.101}$$

are represented by the formula

$$u(x, t) = \varphi(|x| - c\,t) + i\,\psi(|x| + c\,t), \tag{5.102}$$

and all solutions of the system

$$\frac{\partial u}{\partial x_j} + \frac{i x_j}{c|x|} \frac{\partial \bar{u}}{\partial t} = 0, \quad j = 1,\dots,n \tag{5.103}$$

are represented by the formula

$$u(x, t) = e^{\pi i/4} \varphi(|x| - c t) + e^{-\pi i/4} \psi(|x| + c t) \tag{5.104}$$

with arbitrary C^1-functions $\varphi, \psi,$ and may be presented as a superposition of two waves, one approaching the origin with speed c and the other travelling outwards from the origin at the same speed.

6 The statement of well-posed problems for degenerate systems of partial differential equations

6.1 Introduction

Let $\mathbb{R}^n$ as usual denote the n-dimensional Cartesian space of points $x = (x_1,...,x_n)$. In some domain $\Omega \subset \mathbb{R}^n$ we consider a system of linear partial differential equations with respect to the vector function $u(x) = (u_1,...,u_m)$

$$\sum_{|\alpha| \leq N} A_\alpha(x) D^\alpha u = F(x), \tag{6.1}$$

where $A_\alpha(x)$ are $(r \times m)$ matrices and $F(x) = (F_1,...,F_m)$ is a vector function given in Ω. We presuppose for (6.1) to be degenerate in Ω, i.e. even in case $r = m$,

$$\det \sigma(x, \xi) \equiv 0, \quad \sigma(x, \xi) = \sum_{|\alpha| \leq N} A_\alpha(x) \xi^\alpha, \tag{6.2}$$

for any $x \in \Omega$ and any $\xi = (\xi_1,...,\xi_n) \in \mathbb{R}^n$.

The simplest examples of the first-order degenerate systems in the plane are the following $(r = m = 2, N = 1)$:

$$A_1 u_{x_1} + A_2 u_{x_2} + A_0 u = F, \tag{6.3}$$

where A_1 is a real square matrix

$$A_1 = \begin{pmatrix} 1 + \cos\omega & \sin\omega \\ \sin\omega & 1 - \cos\omega \end{pmatrix}$$

and A_2 is a real square matrix having the form either

$$A_2 = \begin{pmatrix} \sin\omega & -(1 + \cos\omega) \\ 1 - \cos\omega & -\sin\omega \end{pmatrix}$$

or

$$A_2 = \begin{pmatrix} -\sin w & -(1-\cos w) \\ 1+\cos w & \sin w \end{pmatrix}$$

where $w = w(x_1, x_2)$ is an arbitrary bounded real function.

The simplest example of a second-order degenerate system in the plane is the following ($r = m = 2$, $N = 2$):

$$\frac{\partial^2 u_1}{\partial x_1^2} + \frac{\partial^2 u_2}{\partial x_1 \partial x_2} + \frac{\partial u_2}{\partial x_1} = 0, \quad \frac{\partial^2 u_1}{\partial x_1 \partial x_2} + \frac{\partial^2 u_2}{\partial x_2^2} - \frac{\partial u_1}{\partial x_1} = 0 \qquad (6.4)$$

This system arises in variational problem (see [7], Chapter 2, Section 2.6).

Here is another example of a system with complex-valued coefficients: A_1 is a complex second-order square matrix

$$A_1 = \begin{pmatrix} 1 & \mu \\ \bar{\mu} & 1 \end{pmatrix}$$

and A_2 is a complex second-order square matrix having the form either

$$A_2 = i \begin{pmatrix} 1 & -\mu \\ \bar{\mu} & -1 \end{pmatrix}$$

or

$$A_2 = i \begin{pmatrix} 1 & \mu \\ -\bar{\mu} & -1 \end{pmatrix}$$

where $\mu = \mu(x_1, x_2)$ is an arbitrary complex-valued function with $|\mu| = 1$ and $\bar{\mu}$ is complex conjugate of μ.

Turning to three-dimensional space, we choose as an example the degenerate system[1]

1 Of course the system $\operatorname{div} u + (A, u) = f_0$, $\operatorname{grad} u_0 + a(x)u = f(x)$ with $\operatorname{curl} A(x) = 0$ is the simplest one.

$$\operatorname{curl} u + a(x)u = f(x) \tag{6.5}$$

where $a(x)$ is a given real (3×3) matrix. The principal symbol of (6.5),

$$\sigma(\xi) = \begin{pmatrix} 0 & -\xi_3 & \xi_2 \\ \xi_3 & 0 & -\xi_1 \\ -\xi_2 & \xi_1 & 0 \end{pmatrix} \tag{6.6}$$

is degenerate: $\det \sigma(\xi) \equiv 0,\ \xi \in \mathbb{R}^3$ and evidently has a rank equal to 2.

As an example of a second-order system we consider the following:

$$\operatorname{grad} \operatorname{div} u + \sum_{i=1}^{3} a_i(x) u_{x_i} + a_0(x) u = f(x) \tag{6.7}$$

with degenerate principal symbol: $\sigma(\xi) = (\xi_i \xi_j),\ 1 \le i, j \le 3$. Obviously the system (6.4) is a particular case of (6.7). Since the homogeneous system

$$\operatorname{curl} u + [\, b \times u \,] = 0 \tag{6.8}$$

with an irrotational vector field $\operatorname{curl} b(x) = 0,\ x \in \Omega$, has a solution

$$u(x) = e^{w(x)} \operatorname{grad} \Phi(x),\ x \in \Omega$$

with an arbitrary C^2-function $\Phi(x)$ of the variable $x = (x_1, x_2, x_3) \in \Omega$ as well as the homogeneous system

$$\operatorname{grad} \operatorname{div} u + \operatorname{grad} (b, u) = 0 \tag{6.9}$$

with $\operatorname{curl} b(x) = 0,\ x \in \Omega$ has a solution

$$u(x) = e^{\omega(x)} \operatorname{curl} \psi(x) + 1/3\, Cx,\ x \in \Omega$$

with an arbitrary C^2 vector-function of the variable $x \in \Omega$, where $\omega(x)$ is the solution of the system $\operatorname{grad} w(x) = -b(x), x \in \Omega$, then it follows that a solution of the system (6.8), (6.9) cannot be determined in a bounded domain Ω apart from the finite-dimensional subspace defined by any boundary condition on $\partial\Omega$.

The same is true for the homogeneous system (3) with special matrix A_0, in particular, with $A_0 \equiv 0$. Further we shall exclude this case, because we are going to derive the theory of boundary-value problems or initial-boundary-value problems for degenerate systems perturbed by lower terms rich enough in content. The main idea which we exploit here is simple: assume that the symbol $\sigma(x, \zeta)$ of (6.1) with $\cdot = m$ has constant rank in Ω equal m_0, then there are $m - m_0$ linearly independent eigenvectors $e^{(j)}(x, \zeta) = (e_1^{(j)},\dots,e_m^{(j)})$, satisfying $\sigma^{\mathrm{T}}(x, \zeta)e = 0$, where σ^{T} denotes the transpose of σ. By means of $e^{(j)}(x, D)$ the system (1) reduces to the following equivalent compatible overdetermined system:

$$e^{(j)}(x, D) \sum_{|\alpha|\le N-1} A_\alpha(x)D^\alpha u = e^{(j)}(x, D)F(x), \quad j = 1,\dots,m - m_0,$$

$$\sum_{|\alpha|\le N} A_\alpha(x)D^\alpha u = F(x) \tag{6.10}$$

and the problem is now to investigate this system.

6.2 Boundary-value problems for first-order systems

(a) First we consider the system (6.5) in a bounded domain $\Omega \subset \mathbb{R}^3$ with a sufficiently smooth boundary $\Gamma = \partial\Omega$ which is a closed surface in $\mathbb{R}^3$. Since the symbol (6.6) of this system has constant rank equal 2, then there is a unique eigenvector $e(\zeta) = (\zeta_1, \zeta_2, \zeta_3)$, satisfying the algebraic system $\sigma^{\mathrm{T}}(\zeta)e = 0$ and hence

$$e(D) := \left(\frac{\partial}{\partial x_1}, \frac{\partial}{\partial x_2}, \frac{\partial}{\partial x_3}\right) \equiv \nabla$$

is the Hamilton operator, so the system (6.10) in this case is

$$\sum_{i,j=1}^{3} a_{ij}(x)\frac{\partial u_j}{\partial x_i} + \frac{\partial a_{ij}}{\partial x_i} u_j = \operatorname{div} f,$$

$$\operatorname{curl} u + a(x)u = f, \tag{6.11}$$

because of

$$\operatorname{div} a(x)u = \sum_{i,j=1}^{3} a_{ij}(x)\frac{\partial u_j}{\partial x_i} + \frac{\partial a_{ij}}{\partial x_i} u_j .$$

The principal symbol of (6.11) is

$$\tilde{\sigma}(x,\xi) = \begin{pmatrix} \sigma_1(x,\xi) & \sigma_2(x,\xi) & \sigma_3(x,\xi) \\ 0 & -\xi_3 & \xi_2 \\ \xi_3 & 0 & -\xi_1 \\ -\xi_2 & \xi_1 & 0 \end{pmatrix}$$

$$\sigma_j(x,\xi) = \sum_{i=1}^{3} a_{ij}(x)\xi_i. \tag{6.12}$$

If we assume the real matrix $a(x)$ to be symmetric: $a_{ij} = a_{ji}$, then in case all its eigenvalues are strongly positive in Ω or strongly negative in Ω, the system (6.11) will be elliptic in Ω, because then the rank of the matrix (6.12) will be equal to 3: among three minors

$$\tilde{\sigma}^{(k)}(x,\xi) = (-1)^{k+1}\xi_k \sum_{i,j=1}^{3} a_{ij}(x)\xi_i\xi_j$$

of the matrix (6.12) at least one differs from zero, because of the determinacy of the quadratic form

$$\sum_{i,j=1}^{3} a_{ij}(x)\xi_i\xi_j .$$

Let us consider in Ω another first-order degenerate system in Ω for unknown function $u_0(x)$ and vector function $u(x) = (u_1, u_2, u_3)$:

$$\sum_{i,j=1}^{3} a_{ij}(x)\frac{\partial u_j}{\partial x_i} + (A(x), u) = f_0,$$

$$\text{grad}\, u_0 + \lambda(x)u = f(x), \tag{6.13}$$

where $\lambda(x)$, $f_0(x)$ are functions given in Ω such that $\lambda(x) \neq 0$ in the closure $\Omega = \Omega + \Gamma$, $A(x), f(x)$ are vector functions given in Ω such that $\text{curl}\, A(x) = 0$ in Ω and it is assumed that the quadratic form

$$\sum_{i,j=1}^{3} a_{ij}(x)\xi_i\xi_j$$

is determined in Ω. Since the principal symbol of (6.13),

$$\sigma(x,\xi) = \begin{pmatrix} 0 & \sigma_1(x,\xi) & \sigma_2(x,\xi) & \sigma_3(x,\xi) \\ \xi_1 & 0 & 0 & 0 \\ \xi_2 & 0 & 0 & 0 \\ \xi_3 & 0 & 0 & 0 \end{pmatrix},$$

has rank 1, then there are three linearly independent eigenvectors $e^{(1)}(\xi) = (0, 0, -\xi_3, \xi_2)$, $e^{(2)}(\xi) = (0, \xi_3, 0, -\xi_1)$, $e^{(3)}(\xi) = (0, -\xi_2, \xi_1, 0)$, satisfying the equation $\sigma^T(x,\xi)e = 0$, i.e.

$$(0, \xi_1, \xi_2, \xi_3)e = 0,$$

so the system (6.10) in this case is

$$\operatorname{curl}(\lambda(x)u) = \operatorname{curl} f(x),$$

$$\sum_{i,j=1}^{3} a_{ij}(x)\frac{\partial u_j}{\partial x_i} + (A(x), u) = f_0$$

$$\operatorname{grad} u_0 + \lambda(x)u = f(x)$$

or

$$\sum_{i,j=1}^{3} a_{ij}(x)\frac{\partial u_j}{\partial x_i} + (A(x), u) = f_0,$$

$$\operatorname{curl} u + [\operatorname{grad} \ln \lambda(x) \times u] = \lambda^{-1}(x) \operatorname{curl} f(x),$$

$$\operatorname{grad} u_0 + \lambda(x)u = f(x), \tag{6.14}$$

because of $\operatorname{curl} \lambda u = [\operatorname{grad} \lambda \times u] + \lambda \operatorname{curl} u$. The principal symbol of the overdetermined system (6.14)

$$\tilde{\sigma}(x,\xi)=\begin{pmatrix} 0 & \sigma_1(x,\xi) & \sigma_2(x,\xi) & \sigma_3(x,\xi) \\ 0 & 0 & -\xi_3 & \xi_2 \\ 0 & \xi_3 & 0 & -\xi_1 \\ 0 & \xi_2 & -\xi_1 & 0 \\ \xi_1 & 0 & 0 & 0 \\ \xi_2 & 0 & 0 & 0 \\ \xi_3 & 0 & 0 & 0 \end{pmatrix}$$

has rank 4, so the system (6.14) is elliptic in Ω.

Now let us consider the following problem.

Oblique component problem. Find a solution $u = (u_1, u_2, u_3) \in C^2$ of the system (6.5) in Ω or a solution $(u_0, u) = (u_0, u_1, u_2, u_3) \in C^2$ of the system (6.13) in Ω, continuous in Ω such that

$$(u, \ell)_{\Gamma} = 0, \tag{6.15}$$

where $\ell(x) = (\ell_1, \ell_2, \ell_3)$ is a continuous vector field given on Γ, $(u, \ell) = u_1\ell_1 + u_2\ell_2 + u_3\ell_3$ is the oblique component of the vector u.

Theorem 6.1. *Let the quadratic form*

$$\sum_{i,j=1}^{3} a_{ij}(x)\,\xi_i\,\xi_j$$

be determined for any $x \in \Omega$ and any $\xi \in \mathbb{R}^3\backslash 0$. Then in case $\cos(\ell, n) \neq 0$ on Γ, where $n = (n_1, n_2, n_3)$ is a normal vector on Γ, the oblique component problem is well posed.

To prove this theorem for the system (6.5) we just note that the overdetermined system (6.11) is equivalent (see [25], [40], [54]) to the following wider elliptic system with respect to the function $v_0(x)$ and vector function $u(x)$:

$$\sum_{i,j=1}^{3} a_{ij}(x)\,\frac{\partial u_j}{\partial x_i} + \frac{\partial a_{ij}}{\partial x_i}\,u_j = \operatorname{div} f,$$

$$\operatorname{grad} v_0 + \operatorname{curl} u + a(x)u = f \tag{6.16}$$

with complementary boundary condition

$$v_0 \,|\, _\Gamma = 0, \tag{6.17}$$

because acting to the second line in (6.16) by operator div and taking into account the first line we see that the function $v_0(x)$ will be harmonic in Ω and hence $v_0 \equiv 0$ by (6.17). In the case of the system (6.13) instead of (6.16) we get the following system:

$$\sum_{i,j=1}^{3} a_{ij}(x) \frac{\partial u_j}{\partial x_i} + (A, u) = f_0,$$

$$\operatorname{grad} v_0 + \operatorname{curl} u + [\operatorname{grad} \ln \lambda(x) \times u] = \lambda^{-1}(x) \operatorname{curl} f$$

$$\operatorname{grad} v_0 + \lambda(x) u = f \tag{6.18}$$

and the condition (6.17). Considering two first lines of (6.18) we get the following elliptic system with the same principal symbol as in system (6.16):

$$\sum_{i,j=1}^{3} a_{ij}(x) \frac{\partial u_j}{\partial x_i} + (A, u) = f_0,$$

$$\operatorname{grad} v_0 + \operatorname{curl} u + [\operatorname{grad} \ln \lambda(x) \times u] = \lambda^{-1} \operatorname{curl} f. \tag{6.19}$$

It is not difficult to see that the problem (6.15), (6.16), (6.17) as well the problem (6.15), (6.17), (6.19) is an elliptic one in the sense that they satisfy the so-called Shapiro-Lopatinski condition (see [47], [54], [58]), provided $\cos(\ell, n) \neq 0$ on Γ.

The theorem is proved in the case of the system (6.5). To complete the proof in the case of the system (6.13) it remains to find the unknown function $u_0(x)$ from the last line in (6.18). Since $\operatorname{curl}(f - \lambda u) = 0$ by the second line in (6.14), then it follows that

$$u_0(x) = C + \int_{\ell_x} (f - \lambda u)\, dx, \tag{6.20}$$

where C is an arbitrary constant and the curvilinear integral does not depend on the path ℓ_x of integration connecting the origin $x = 0$ with $x \in \Omega$, i.e. u_0 is determined uniquely apart from a constant. This completes the proof.

(b) Theorem 1 excludes important cases when $\cos(\ell, n)$ may be zero on some subset of surface Γ. This happens, for instance, when ℓ is a nonzero constant vector: then $\cos(\ell, n) = 0$ along the curve $\gamma = \partial G$, where G is a plane domain, obtained from Ω by intersection with the plane $(\ell, x) = 0 : G = \Omega \cap ((\ell, x) = 0)$. Unfortunately in this case the problem (6.5), (6.15) as well as the problem (6.13), (6.15) become ill-posed: no solution of the corresponding homogeneous problems is determined except for a finite-dimensional subspace. Here we try to find suitable complementary conditions to be imposed on γ to obtain a well-posed problem for the system (6.5) and (6.13). Assuming that Ω contains the origin $x = 0$ we also assume that $G = \Omega \cap ((\ell, x) = 0)$ is a plane domain with the boundary $\gamma = \partial G$ being a smooth closed curve without intersection points.

Problem E. Find in Ω a C^2-solution of (6.5) (or (6.13)), continuous in Ω, satisfying the condition (6.15) on Γ and the condition

$$(u, k)_\gamma = 0 \tag{6.21}$$

on γ, where $k = (k_1, k_2, k_3)$ is vector field Hölder continuous on γ.

To investigate this problem in the case of (6.5) we restrict ourselves to a constant matrix $a = (a_{ij})$. Then it is not difficult to see from (6.11) that the functions $u_k(x)$, $k = 1,2,3$ are solutions in Ω of the following second-order elliptic equation

$$\sum_{i,j=1}^{3} a_{ij} \frac{\partial^2 u_k}{\partial x_i \partial x_j} + \lambda_* u_k = f^{(k)}(x), \quad k = 1,2,3, \tag{6.22}$$

with

$$\lambda_* = a_{11}a_{22}a_{33} + 2a_{12}a_{13}a_{23} - (a_{11}a_{23}^2 + a_{22}a_{13}^2 + a_{33}a_{12}^2),$$

$$f^{(1)} = \operatorname{div} \frac{\partial f}{\partial x_1} + \sum_{i=1}^{3} \left(a_{i3} \frac{\partial f_2}{\partial x_i} - a_{i2} \frac{\partial f_3}{\partial x_i} \right) + a_{23}^2 - a_{22}a_{33})f_1$$

$$+ (a_{12}a_{33} - a_{13}a_{23})f_2 + (a_{12}a_{23} - a_{13}a_{22})f_3,$$

$$f^{(2)} = \operatorname{div} \frac{\partial f}{\partial x_2} + \sum_{i=1}^{3} \left(a_{i1} \frac{\partial f_3}{\partial x_i} - a_{i3} \frac{\partial f_1}{\partial x_i} \right) + (a_{12}a_{33} - a_{13}a_{23})f_1$$

$$+ (a_{13}^2 - a_{11}a_{33})f_2 + (a_{11}a_{23} - a_{12}a_{13})f_3,$$

$$f^{(3)} = \operatorname{div} \frac{\partial f}{\partial x_3} + \sum_{i=1}^{3} \left(a_{i2} \frac{\partial f_1}{\partial x_i} - a_{i1} \frac{\partial f_2}{\partial x_i} \right) + (a_{13}a_{22}\; a_{12}a_{33})f_1$$

$$+ (a_{11}a_{23} - a_{12}a_{13})f_2 + (a_{12}^2 - a_{11}a_{22})f_3.$$

First assume that[2] $\ell_1 = 1,\ \ell_2^2 + \ell_3^2 = 0$. Then (6.15) becomes $u_1|_\Gamma = 0$ and according to (6.22) $u_1(x)$ is expressed as

$$u_1(x) = -\int_\Omega G(x,\xi)u_1(\xi)\,d\xi + \int_\Omega G(x,\xi)f^{(k)}(\xi)\,d\xi, \tag{6.23}$$

where $G(x,\xi)$ is the Green's function of the equation

$$\sum_{i,j=1}^{3} a_{ij}\frac{\partial^2 u}{\partial x_i \partial x_j} = 0.$$

Since (6.23) is an integral equation of the Fredholm type, then because of the weak singularity of $G(x,\xi)$, it is uniquely solvable if λ is not the eigenvalue, and if λ is the eigenvalue the equation (6.23) has solution if and only if the function $f^{(\ell)}$ satisfies a finite number of conditions:

$$\int_\Omega f^{(\ell)}(x)v_j(x)\,dx = 0,\quad j = 1,\ldots,k^0 \tag{6.24}$$

and a solution $u_1(x)$ of the equation (6.23) is determined uniquely modulo of a finite-dimensional subspace by

$$u_1(x) = \int_\Omega \Gamma^{(1)}(x,\xi,\lambda)f^{(1)}(\xi)\,d\xi + \sum_{j=1}^{k^0} c_j u_1^{(j)}(x), \tag{6.25}$$

where $u_1^{(1)},\ldots,u_1^{(k_0)}$ are eigenvalues of the equation (6.23) with $f^{(1)} \equiv 0$ and $c_1,\ldots,c_{k_0}$ are arbitrary real constants.

Taking into account that $u_1(x)$ is already determined by (6.25) we rewrite the system (6.11) as a system with respect to unknowns u_2, u_3:

$$a_{22}\frac{\partial u_2}{\partial x_2} + a_{23}\frac{\partial u_3}{\partial x_2} + a_{32}\frac{\partial u_2}{\partial x_3} + a_{33}\frac{\partial u_3}{\partial x_3}$$

$$= -a_{12}\frac{\partial u_2}{\partial x_1} - a_{13}\frac{\partial u_3}{\partial x_1} - \sum_{i=1}^{3} a_{i1}\frac{\partial u_1}{\partial x_i} + \operatorname{div} f,$$

$$\frac{\partial u_3}{\partial x_2} - \frac{\partial u_2}{\partial x_3} = -\sum_{j=1}^{3} a_{1j}u_j + f_1,$$

2 Without loss of generality we may obviously assume in (6.13) that $|\ell| = 1$.

$$\frac{\partial u_2}{\partial x_1}=\frac{\partial u_1}{\partial x_2}-\sum_{j=1}^{3} a_{3j}u_j+f_3,$$

$$\frac{\partial u_3}{\partial x_1}=\frac{\partial u_1}{\partial x_3}+\sum_{j=1}^{3} a_{2j}u_j+f_2. \tag{6.26}$$

The last two equations of this system are

$$\frac{\partial w_0}{\partial x_1}+a_1^0 w_0+a_2^0 \bar{w}_0=\tilde{f}^0, \tag{6.27}$$

where $w_0(x)=u_2(x)+\mathrm{i}\,u_3(x)$

$$a_1^0=a_{23}+\mathrm{i}(a_{33}-a_{22})/2,\ a_2^0=(a_{22}+a_{33})/2\mathrm{i},$$

$$\tilde{f}^0=2\frac{\partial u_1}{\partial z}+\mathrm{i}(a_{12}+\mathrm{i}\,a_{13})u_1+f_3+\mathrm{i}\,f_2.$$

Let $w^0(x_2,x_3)=w_0|_{x_1=0}$. Then (6.27) is equivalent to the Volterra integral equation

$$w_0(x)+\int_0^{x_1}(a_1^0 w_0+a_2^0\bar{w}_0)(x_1',x_2,x_3)\,\mathrm{d}x_1'=w^0(x_2,x_3)+\tilde{f}_0, \tag{6.28}$$

where

$$\tilde{f}_0(x)=\int_0^{x_1}\tilde{f}_0(x_1',x_2,x_3)\,\mathrm{d}x_1'.$$

By successive approximation from (6.28) we obtain

$$w_0(x)=\Gamma_1^0(x)w^0(x_2,x_3)+\Gamma_2^0(x)\,\overline{w_0(x_2,x_3)}+\tilde{f}(x), \tag{6.29}$$

where $\tilde{f}(x)$ depends only on $\tilde{f}_0(x)$ and $\Gamma_j^0(x)$ are known functions such that $|\Gamma_1^0(x)|^2-|\Gamma_2^0(x)|^2\neq 0$ and $\Gamma_1^0|_{x_1=0}=1,\ \Gamma_2^0|_{x_1=0}=0.$

Substituting (6.29) into the first two equations of the system (6.26) we obtain the following system of equations for $\overset{0}{u}_2(x_2, x_3) = u_2 |_{x_1 = 0}$, $\overset{0}{u}_3(x_2, x_3) = u_3 |_{x_1 = 0}$:

$$a_{22} \frac{\partial \overset{0}{u}_2}{\partial x_2} + a_{23} \frac{\partial \overset{0}{u}_3}{\partial x_2} + a_{32} \frac{\partial \overset{0}{u}_2}{\partial x_3} + a_{33} \frac{\partial \overset{0}{u}_3}{\partial x_3} + A_{11} \overset{0}{u}_2 + A_{12} \overset{0}{u}_3 = \overset{0}{g}_1(x_2, x_3),$$

$$\frac{\partial \overset{0}{u}_3}{\partial x_2} - \frac{\partial \overset{0}{u}_2}{\partial x_3} + A_{21} \overset{0}{u}_2 + A_{22} \overset{0}{u}_3 = \overset{0}{g}_2(x_2, x_3). \tag{6.30}$$

in the plane domain $G = \Omega \cap (x_1 = 0)$.

Assuming $a_{22}a_{33} - a_{23}{}^2 > 0$ we conclude that (6.30) is an elliptic system which may be written in terms of the complex-valued function

$$w(z) = (a_{23} + \mathrm{i}\sqrt{(a_{22}\, a_{33} - a_{23}{}^2})\overset{0}{u}_2 + a_{33}\, \overset{0}{u}_3$$

as the single equation

$$\frac{\partial w}{\partial x_2} + \lambda^0 \frac{\partial w}{\partial x_3} + A^0 w + B^0 w = F^0(z),\ \lambda^0 = (a_{23} + \mathrm{i}\sqrt{a_{22}\, a_{33} - a_{23}{}^2})/a_{22} \tag{6.31}$$

in G.

Since

$$\overset{0}{u}_2 = (a_{22}\, a_{33} - a_{23}{}^2)^{-1/2}\, \mathrm{Im}\, w,$$

$$\overset{0}{u}_3 = a_{33}{}^{-1}\, \mathrm{Re}\, w - a_{23}\, a_{33}{}^{-1}\, (a_{22}\, a_{33} - a_{23}{}^2)^{-1/2}\, \mathrm{Im}\, w,$$

then

$$k_2 \overset{0}{u}_2 + k_3\, \overset{0}{u}_3 = a_{33}{}^{-1} (a_{22}\, a_{33} - a_{23}{}^2)^{-1/2}\, \mathrm{Re}\, \{ [\, k_3 \sqrt{(a_{22}\, a_{33} - a_{23}{}^2)}$$

$$- \mathrm{i}\, (k_2\, a_{33} - k_3\, a_{23})\,]\, w \}$$

and hence the boundary condition (6.21) reduces to the Riemann–Hilbert condition

$$\mathrm{Re}\, (a_0^*\, w) = h_0^* \tag{6.32}$$

on the curve $\gamma = \partial G$ with

$$a_0^* = k_3 \sqrt{(a_{22} a_{33} - a_{23}^2)} - i(k_2 a_{33} - k_3 a_{23}) \tag{6.33}$$

and h_0^* containing constants c_j and the right-hand side $f^{(1)}$:

$$h_0^* = \int_\Omega \Gamma^{*(1)}(x, \xi, \lambda) f^{(1)}(\xi)\, d\xi + \sum_{j=1}^{k^0} c_j h_j^* \tag{6.34}$$

Let us assume that $a^* \neq 0$ on γ and denote $\kappa^* = \mathrm{Ind}_\gamma a^*$. Then, according to the theory of the Riemann-Hilbert problem (see [58]) in case $\kappa^* \geq 0$, the problem (6.31), (6.32) is unconditionally solvable and the corresponding homogeneous problem ($F^0 = h_0^* \equiv 0$) has exactly $2\kappa^* + 1$ nontrivial solutions and in case $\kappa^* < 0$ the problem (6.31), (6.32) is solvable if and only if

> h_0^* satisfies $-2\kappa^* - 1$ orthogonality conditions and the corresponding homogeneous problem has no nontrivial solution.

The necessary and sufficient solvability conditions for the problem (6.32) gives ℓ' inhomogeneous algebraic equations for k^0 real constants $c_1,\ldots,c_{k^0}$:

$$\sum_{j=1}^{k^0} a_{ij} c_j = b_i, \quad i = 1,\ldots,\ell'.$$

If $r' = \mathrm{rank}\,(a_{ij})$, then we get from here $\ell' - r'$ independent solvability conditions imposed on the right-hand side of the problem (6.32), so the number k' of the solvability conditions of the original problem is equal to $\ell' + k^0 - r'$ and the number k of the solution of the original homogeneous problem is equal to $\ell + k^0 - r'$, where ℓ denotes the number of nontrivial solutions of the homogeneous problem (6.32). Since $\ell - \ell' = 2\kappa^* + 1$, then it follows that $k - k' = \ell + k^0 - r' - (\ell' + k^0 - r') = 2\kappa^* + 1$, i.e. we have proved the following theorem.

Theorem 6.2. *If* $\ell_1 = 1$, $\ell_2^2 + \ell_3^2 = 0$ and

$$a_0^* = k_3 \sqrt{(a_{22} a_{33} - a_{23}^2)} - i(k_2 a_{33} - k_3 a_{23}) \neq 0$$

on γ, then Problem E for the system (6.5) is well posed and has an index equal to $2\kappa_0^* + 1$, where $\kappa_0^* = \mathrm{Ind}_\gamma a_0^*$.

Now assume that $\ell_1 = 0$, $\ell_2^2 + \ell_3^2 = 1$. Then (6.15) becomes $\psi_1 |_\Gamma = 0$, where $\psi_1 = \ell_2 u_2 + \ell_3 u_3$ according to (6.22) is a solution of the elliptic equation

$$\sum_{i,j=1}^{3} a_{ij} \frac{\partial^2 \psi}{\partial x_i \partial x_j} + \lambda_* \psi_1 = f_*^1$$

with

$$f_*^1(x) = \ell_2 f^{(2)}(x) + \ell_3 f^{(3)}(x)$$

and is therefore expressed as

$$\psi_1(x) = \lambda \int_\Omega G(x, \xi) \psi_1(\xi)\, d\xi + \int_\Omega G(x, \xi) f^{*1}(\xi)\, d\xi. \tag{6.35}$$

Hence as before in case $f_*^1(x)$ satisfies a finite number orthogonality conditions such as (6.24), the function $\psi_1(x)$ is determined uniquely, except on a finite-dimensional subspace, by a formula such as (6.25). Introducing the new function $\psi_2(x)$ by $\psi_2(x) = -\ell_3 u_2(x) + \ell_2 u_3(x)$, we obtain

$$u_2(x) = -\ell_3 \psi_2(x) + \ell_2(x) \psi_1(x), \quad u_3(x) = \ell_2 \psi_2(x) + \ell_3 \psi_1(x). \tag{6.36}$$

Taking into account that $\psi_1(x)$ is already determined and substituting (6.36) into (6.11) we obtain the following system with respect to unknowns u_1, ψ_2:

$$a_{12} \frac{\partial u_1}{\partial x_2} + (a_{23}\ell_2 - a_{22}\ell_3) \frac{\partial \psi_2}{\partial x_2} + a_{13} \frac{\partial u_1}{\partial x_3} + (a_{33}\ell_2 - a_{23}\ell_3) \frac{\partial \psi_2}{\partial x_3}$$

$$= -a_{11} \frac{\partial u_1}{\partial x_1} + (a_{12}\ell_3 - a_{13}\ell_2) \frac{\partial \psi_2}{\partial x_1} + f_0^{(1)}(x),$$

$$\ell_2 \frac{\partial \psi_2}{\partial x_2} + \ell_3 \frac{\partial \psi_2}{\partial x_3} = -a_{11} u_1 + (a_{12}\ell_3 - a_{13}\ell_2) \psi_2(x) + f_0^{(2)}(x), \tag{6.37}$$

$$\frac{\partial u_1}{\partial x_2} = -\ell_3 \frac{\partial \psi_2}{\partial x_1} + a_{13} u_1 + (a_{33}\ell_2 - a_{23}\ell_3) \psi_2(x) + g_0^{(1)}(x),$$

$$\frac{\partial u_1}{\partial x_3} = \ell_2 \frac{\partial \psi_2}{\partial x_1} - a_{12} u_1 + (a_{22}\ell_3 - a_{23}\ell_2) \psi_2(x) + g_0^{(2)}(x),$$

where $f_0^{(k)}(x)$, $g_0^{(k)}(x)$ are expressed in terms of ψ_1 and its first derivatives and also in terms of the right-hand side of (6.11).

Considering (6.37) as inhomogeneous algebraic system of equations with respect

$$\frac{\partial u_1}{\partial x_2},\ \frac{\partial \psi_2}{\partial x_2}\ \frac{\partial u_1}{\partial x_3},\ \frac{\partial \psi_2}{\partial x_3}$$

and taking into account that its determinant $\Delta = a_{22}\ell_3{}^2 - 2a_{23}\ell_2\ell_3 + a_{33}\ell_2{}^2$ differs from zero, because obtained from

$$\sum_{i,j=1}^{3} a_{ij}\,\xi_i\,\xi_j$$

for $\xi = (0, -\ell_3, \ell_2) \neq 0$, we obtain

$$\frac{\partial u_1}{\partial x_2} + \ell_3\,\frac{\partial \psi_2}{\partial x_1} + A_{11}{}^0 u_1 + A_{12}{}^0\,\psi_2 = f_1^0,$$

$$\frac{\partial \psi_2}{\partial x_2} - a_{11}\,\ell_3\Delta^{-1}\,\frac{\partial u_1}{\partial x_1} - 2a_0\,\ell_3\Delta^{-1}\,\frac{\partial \psi_2}{\partial x_1} + A_{21}{}^0\,u_1 + A_{22}{}^0\,\psi_2 = f_2^0,$$

$$\frac{\partial u_1}{\partial x_3} - \ell_2\,\frac{\partial \psi_2}{\partial x_1} + B_{11}{}^0\,u_1 + B_{12}{}^0\,\psi_2 = g_1^0,$$

$$\frac{\partial \psi_2}{\partial x_3} + a_{11}\,\ell_2\Delta^{-1}\,\frac{\partial u_1}{\partial x_1} + 2a_0\,\ell_2\Delta^{-1}\,\frac{\partial \psi_2}{\partial x_1} + B_{21}{}^0\,u_1 + B_{22}{}^0\,\psi_2 = g_2^0 \qquad (6.38)$$

where $a_0 = a_{13}\ell_2 - a_{12}\ell_3$.

If $a_0 = 0$, then the system (6.38) may be written as two complex equations with respect to the unknown function $w(z) = \mathrm{i}\sqrt{(a_{11}\Delta^{-1})}\,u_1 + \psi_2$:

$$\frac{\partial w}{\partial x_2} + \mathrm{i}\,\ell_3\sqrt{(a_{11}\Delta^{-1})}\,\frac{\partial w}{\partial x_1} + A_2^0 w + B_2^0\,\bar{w} = F_2^0,$$

$$\frac{\partial w}{\partial x_3} - \mathrm{i}\,\ell_2\sqrt{(a_{11}\Delta^{-1})}\,\frac{\partial w}{\partial x_1} + A_3^0 w + B_3^0\,\bar{w} = F_3^0 \qquad (6.39)$$

If $a_0 \neq 0$, then the system (6.38) may be written as two equations with respect to the unknown complex-valued function $w(z) = \Delta^{-1}(a_0 - \mathrm{i}\sqrt{(a_{11}\Delta - a_0^2)})\, u_1 + \psi_2$:

$$\frac{\partial w}{\partial x_2} + \lambda_0 \frac{\partial w}{\partial x_1} + A_0^{(2)} w + B_0^{(2)} \bar{w} = F_0^{(2)},$$

$$\frac{\partial w}{\partial x_3} + \bar{\mu}_0 \frac{\partial w}{\partial x_1} + A_0^{(3)} w + B_0^{(3)} \bar{w} = F_0^{(3)}, \tag{6.40}$$

where

$$\lambda_0 = \Delta^{-1}\, \ell_3(a_0 + \mathrm{i}\sqrt{(a_{11}\Delta - a_0{}^2)}), \quad \bar{\mu}_0 = -\,\Delta^{-1}\, \ell_2(a_0 + \mathrm{i}\sqrt{(a_{11}\Delta - a_0{}^2)}).$$

If $a_0 = 0$ the boundary condition according to (6.36) takes the form

$$\mathrm{Re}\,(a_1^* \, w) = h_1^* \tag{6.41}$$

and in case $a_0 \neq 0$:

$$\mathrm{Re}\,(a_2^* \, w) = h_2^*, \tag{6.42}$$

where

$$a_1^* = k_3\, \ell_2 - k_2\, \ell_3 - \mathrm{i}\,(k_1 - k_2\, \ell_2 - k_3\, \ell_3)\sqrt{(\Delta^{-1}\, a_{11} - 1)}, \tag{6.43}$$

$$a_2^* = k_3\, \ell_2 - k_2\, \ell_3 - \mathrm{i}(a_{11}\Delta - a_0^2)^{-1/2}((k_1 - k_2\ell_2 - k_3\ell_3)$$

$$+ (k_3\ell_2 - k_2\ell_3)a_0) \tag{6.44}$$

By the nonsingular linear mapping $(x_1, x_2, x_3) \to (\xi_1, \xi_2, \xi_3)$

$$\xi_1 = x_1, \ \xi_2 == \ell_2 x_2 + \ell_3 x_3, \ \xi_3 = -\,\ell_3 x_2 + \ell_2 x_3, \tag{6.45}$$

we have

$$\frac{\partial w}{\partial \xi_1} = \frac{\partial w}{\partial x_1}, \ \frac{\partial w}{\partial \xi_2} = \ell_2 \frac{\partial w}{\partial x_2} + \ell_3 \frac{\partial w}{\partial x_3}, \ \frac{\partial w}{\partial \xi_3} = -\,\ell_3 \frac{\partial w}{\partial x_2} + \ell_2 \frac{\partial w}{\partial x_3},$$

and so by (6.39) and (6.40) we obtain, in case $a_0 = 0$,

$$\frac{\partial w}{\partial \xi_2} = \tilde{A}_2^0 w + B_2^0 \bar{w} + F_2^0,$$

$$\frac{\partial w}{\partial \xi_3} - i\sqrt{(a_{11}\Delta^{-1})}\frac{\partial w}{\partial \xi_1} + \tilde{A}_3^0 w + B_3^0 w = \tilde{F}_3^0, \tag{6.46}$$

and in case $a_0 \neq 0$: $\dfrac{\partial w}{\partial \xi_2} = \tilde{A}_0^{(2)} w + B_0^{(2)} \bar{w} + F_0^{(2)},$

$$\frac{\partial w}{\partial \xi_3} - \Delta^{-1}(a_0 + i\sqrt{(a_{11}\Delta - a_0^2)})\frac{\partial w}{\partial \xi_1} + \tilde{A}_0^{(3)} w + B_0^{(3)} \bar{w} = \tilde{F}_0^{(3)} \tag{6.47}$$

If $w \mid_{\xi_2 = 0} = w^0(\xi_1, \xi_3)$, then the first equations in (6.46), (6.47) reduces to Volterra's integral equation:

$$w(\xi) - \int_0^{\xi_2} [\tilde{A}(\xi_1, \xi_2', \xi_3) w(\xi_1, \xi_2', \xi_3)$$

$$+ \tilde{B}(\xi_1, \xi_2', \xi_3)\overline{w(\xi_1, \xi_2', \xi_3)}]\, d\xi_2' = w^0(\xi_1, \xi_3) + \tilde{F}(\xi), \tag{6.48}$$

so by successive approximation we obtain

$$w(\xi) = \tilde{\Gamma}_1^0(\xi) w^0(\xi_1, \xi_3) + \tilde{\Gamma}_2^0(\xi)\overline{w^0(\xi_1, \xi_3)} + \tilde{F}^0(\xi), \tag{6.49}$$

where $\tilde{F}^0(\xi)$ is expressed in terms of the right-hand sides and $\underset{\sim}{\Gamma}_j^0(\xi)$ are known functions such that $|\tilde{\Gamma}_1^0(\xi)|^2 - |\tilde{\Gamma}_2^0(\xi)|^2 \neq 0$ and $\Gamma_1^0|_{\xi_2 = 0} = 1$, $\tilde{\Gamma}_2^0|_{\xi_2 = 0} = 0$.

Substitution of (6.49) into the second equations of (6.46) and (6.47) gives the following equations for the unknown function $w_0(\xi_1, \xi_3)$ in case $a_0 = 0$:

$$\frac{\partial w^0}{\partial \xi_3} - i\sqrt{(a_{11}\Delta^{-1})}\frac{\partial w^0}{\partial \xi_1} + A_0(\xi_1, 0, \xi_3) w^0 + B^0(\xi_1, 0, \xi_3)\bar{w}^0 = \tilde{F}_3^0(\xi_1, 0, \xi_3), \tag{6.50}$$

and in case $a_0 \neq 0$:

$$\frac{\partial w^0}{\partial \xi_3} - \mathrm{i}\Delta^{-1}(a_0 + \mathrm{i}\sqrt{(a_{11}\Delta - a_0^2)})\frac{\partial w^0}{\partial \xi_1} + A_0(\xi_1, 0, \xi_3)w^0 + B^0(\xi_1, 0, \xi_3)\bar{w}^0$$

$$= \tilde{F}_0^{(3)}(\xi_1, 0, \xi_3) \qquad (6.51)$$

oth equations (6.50), (6.51) are elliptic. Moreover, since the image of the curve lies n the plane $\xi_2 = 0$, then the conditions (6.41), (6.42) remains the same for the nknown function $w^0(\xi_1, \xi_3)$. Therefore taking into account that the right-hand des of (6.41), (6.42), (6.50) and (6.51) besides that of (6.5) also contains finite umbers of constants, then by the same arguments used above, we conclude that roblem E has the same index as the Riemann-Hilbert problem (6.41) in case $_0 = 0$ and (6.42) in case $a_0 \neq 0$, i.e. we have proved the following:

'heorem 6.3. *If* $\ell_1 = 0$, $\ell_2^2 + \ell_3^2 = 1$ *and* $a_1^* \neq 0$ *on* γ *in case* $a_0 = 0$ *and* $a_2^* \neq 0$ *n* γ *in case* $a_0 \neq 0$, *then Problem E for the system (6.5) has an index equal to* $\kappa_1^* + 1$ *in case* $a_0 = 0$ *and equal to* $2\kappa_2^* + 1$ *in case* $a_0 \neq 0$, *where* $\kappa_k^* = \operatorname{Ind} \bar{a}_k^*$.

To investigate Problem E for the system (6.13) we also assume that its oefficients a_{ij}, A_j and λ are constant, so the system (6.14) in this case takes the orm

$$\sum_{i,j=1}^{3} a_{ij}\frac{\partial u_j}{\partial x_i} + A_1u_1 + A_2u_2 + A_3u_3 = f_0,$$

$$\frac{\partial u_3}{\partial x_2} - \frac{\partial u_2}{\partial x_3} = \lambda^{-1}\left(\frac{\partial f_3}{\partial x_2} - \frac{\partial f_2}{\partial x_3}\right),$$

$$\frac{\partial u_1}{\partial x_3} - \frac{\partial u_3}{\partial x_1} = \lambda^{-1}\left(\frac{\partial f_1}{\partial x_3} - \frac{\partial f_3}{\partial x_1}\right),$$

$$\frac{\partial u_2}{\partial x_1} - \frac{\partial u_1}{\partial x_2} = \lambda^{-1}\left(\frac{\partial f_2}{\partial x_1} - \frac{\partial f_1}{\partial x_2}\right). \qquad (6.52)$$

nd

$$\operatorname{grad} u_0 = f - \lambda u. \qquad (6.53)$$

From (6.52) it follows that the functions u_1, u_2, u_3 are solutions of the second-rder elliptic equation

$$\sum_{i,j=1}^{3} a_{ij} \frac{\partial^2 u_k}{\partial x_i \partial x_j} + \Sigma A_i \frac{\partial u_k}{\partial x_i} = \hat{f}^{(k)}(x) \tag{6.54}$$

in Ω with

$$\hat{f}^{(1)}(x) = \sum_{i=1}^{3} \left\{ \frac{\partial}{\partial x_i}\left(a_{i2}\frac{\partial}{\partial x_2} + a_{i3}\frac{\partial}{\partial x_3}\right) f_1 - \frac{\partial}{\partial x_1}\left(a_{i2}\frac{\partial f_2}{\partial x_1} + a_{i3}\frac{\partial f_3}{\partial x_1}\right)\right\}$$

$$+ \frac{\partial f_0}{\partial x_1} + \left(A_2\frac{\partial}{\partial x_2} + A_3\frac{\partial}{\partial x_3}\right) f_1 - \left(A_2\frac{\partial f_2}{\partial x_1} + A_3\frac{\partial f_3}{\partial x_1}\right),$$

$$\hat{f}^{(2)}(x) = \sum_{i=1}^{3} \left\{ \frac{\partial}{\partial x_i}\left(a_{i1}\frac{\partial}{\partial x_1} + a_{i3}\frac{\partial}{\partial x_3}\right) f_2 - \frac{\partial}{\partial x_i}\left(a_{i1}\frac{\partial f_1}{\partial x_2} + a_{i3}\frac{\partial f_3}{\partial x_2}\right)\right\}$$

$$+ \frac{\partial f_0}{\partial x_2} + \left(A_1\frac{\partial}{\partial x_1} + A_3\frac{\partial}{\partial x_3}\right) f_2 - \left(A_1\frac{\partial f_1}{\partial x_2} + A_3\frac{\partial f_3}{\partial x_2}\right),$$

$$\hat{f}^{(3)}(x) = \sum_{i=1}^{3} \left\{ \frac{\partial}{\partial x_i}\left(a_{i1}\frac{\partial}{\partial x_1} + a_{i2}\frac{\partial}{\partial x_2}\right) f_3 - \frac{\partial}{\partial x_i}\left(a_{i1}\frac{\partial f_1}{\partial x_3} + a_{i2}\frac{\partial f_2}{\partial x_3}\right)\right\}$$

$$+ \frac{\partial f_0}{\partial x_3} + \left(A_1\frac{\partial}{\partial x_1} + A_2\frac{\partial}{\partial x_2}\right) f_3 - \left(A_1\frac{\partial f_1}{\partial x_3} + A_2\frac{\partial f_2}{\partial x_3}\right);$$

Since by previous arguments in cases $\ell_1 = 1$, $\ell_2^2 + \ell_3^2 = 0$ and $\ell_1 = 0$, $\ell_2^2 + \ell_3^2 = 1$ we get Dirichlet's problem $u_1 \mid_\Gamma = 0$ and $\psi_1 \mid_\Gamma = 0$ correspondingly for u_1 the equation (6.54) with $k = 1$ and the equation

$$\sum_{i,j=1}^{3} a_{ij} \frac{\partial \psi_1}{\partial x_i \partial x_j} + \sum_{i=1}^{3} A_i \frac{\partial \psi_1}{\partial x_i} = \hat{f}(x) = \ell_2 \hat{f}^{(2)} + \ell_3 \hat{f}^{(3)} \tag{6.55}$$

for $\psi_1 = \ell_2 u_2 + \ell_3 u_3$, which is unconditionally and uniquely solvable, then we get not only the proof of Theorem 6.3 in the case of the system (6.13) with constant coefficients, but taking into account also equations (6.53), the following result, which is as precise as for the corresponding Riemann-Hilbert problem:

Theorem 4. *Let* $a_0^* \neq 0$ on γ *in case* $\ell_1 = 1$, $\ell_2^2 + \ell_3^2 = 0$, $a_1^* \neq 0$ *and* $a_2^* \neq 0$ *on* γ *correspondingly in case* $a_0 = 0$, $\ell_1 = 0$, $\ell_2^2 + \ell_3^2 = 1$ *and in case* $a_0 \neq 0$,

$\ell_1 = 0$, $\ell_2^2 + \ell_3^2 = 1$, *where* a_j^* *are determined by (6.33), (6.43) and (6.44). Then, if* $\kappa_j^* = \mathrm{Ind}_\gamma\, \bar{a}_j^*$, *Problem E for the system (6.13) is solvable unconditionally and the corresponding homogeneous problem has exactly* $2\kappa_j^* + 2$ *nontrivial solutions, but if* $\kappa_j^* < 0$, *then this problem is solvable if and only if its right-hand sides satisfy* $-2\kappa_j^* - 1$ *conditions and the corresponding homogeneous problem has no nontrivial solution.*

(c) As we have seen above an overdetermined system arose as an intermediate tool on the statement of the well-posed boundary value problem for degenerate systems. But the overdetermined systems above have independent interest. Therefore in this section we try to find well-posed problems for an overdetermined system itself. First we consider in the bounded domain $\Omega \subset \mathbb{R}^3$ the system

$$\operatorname{div} u + (A, u) = f_0, \quad \operatorname{curl} u + au = f \tag{6.56}$$

with real vector functions $A = (A_1, A_2, A_3)$, $f = (f_1, f_2, f_3)$, scalar function f_0 and (3×3) real matrix $a = (a_{ij})$ given in Ω. Note that if $au = [b \times u]$ the vector product of b and u, then in case (6.56) is compatible, i.e. $\operatorname{div} f - (b, f) = 0$, $\operatorname{curl} b = 0$ and also in case $A \equiv 0$, $a = \lambda I$, where I is the identity matrix and λ is a real nonzero constant and (6.56) is compatible, i.e. $\operatorname{div} f - \lambda f_0 = 0$ from the previous section it follows that the oblique component problem with $\cos(\ell, n) \neq 0$ on $\Gamma = \partial\Omega$ as well as the problem (6.15), (6.21) with constant $\ell = (\ell_1, \ell_2, \ell 3) \neq 0$ are well-posed for (6.56). Here we consider cases of the system (6.56) other than those mentioned above which present a quite different problem. Let us assume that the bounded domain Ω contains the origin and is such that the boundary γ of the plane domain $G = \Omega \cap (x_1 = 0)$ is a smooth closed curve without crossing points. We consider the system (6.56) in Ω with

$$A \equiv 0, \; a = \lambda \operatorname{diag}(1, -1, -1), \tag{6.57}$$

where $\lambda \neq 0$ is real constant.

Problem *e*. Find in Ω a solution $u = (u_1, u_2, u_3) \in C^2(\Omega)$ of the system (6.56) with (6.57), continuous in $\bar{\Omega}$, satisfying the conditions

$$u_1 \mid_\gamma = 0, \; [\, k_2 u_2 + k_3 u_3 \,] = 0, \tag{6.58}$$

where k_2, k_3 are Hölder-continuous functions given on γ.

Theorem 6.5. *If* $k_2^2 + k_3^2 \neq 0$ *on* γ, *then Problem e is well-posed and has an index equal to* $2\kappa + 1$, *where* $\kappa = \mathrm{Ind}_\gamma (k_2 - \mathrm{i}\, k_3)$. *Moreover, if* $\kappa \geq 0$, *then this problem* *is solvable unconditionally and the corresponding homogeneous problem* $(f_0 = f \equiv 0)$ *has exactly* $2\kappa + 1$ *nontrivial linearly independent solutions, but if* $\kappa < 0$, *then this problem is solvable if and only if its right-hand side* (f_0, f) *satisfies* $-2\kappa - 1$ *orthogonality conditions and the corresponding homogeneous problem has no nontrivial solution.*

Proof. In case (6.57) the system (6.56) is compatible if

$$\frac{\partial u_1}{\partial x_1} - \frac{\partial u_2}{\partial x_2} - \frac{\partial u_3}{\partial x_3} = \lambda^{-1} \operatorname{div} f$$

in Ω, i.e. the system (6.56) is equivalent to the following compatible overdetermined system of five equations:

$$\frac{\partial u_1}{\partial x_1} = \frac{1}{2}(f_0 + \lambda^{-1} \operatorname{div} f), \quad \frac{\partial u_2}{\partial x_2} + \frac{\partial u_3}{\partial x_3} = \frac{1}{2}(f_0 - \lambda^{-1} \operatorname{div} f),$$

$$\frac{\partial u_3}{\partial x_2} - \frac{\partial u_2}{\partial x_3} + \lambda u_1 = f_1, \quad \frac{\partial u_1}{\partial x_3} - \frac{\partial u_3}{\partial x_1} - \lambda u_2 = f_2, \quad \frac{\partial u_2}{\partial x_1} - \frac{\partial u_1}{\partial x_2} - \lambda u_3 = f_3, \tag{6.59}$$

from which it follows that the function $u_1(x)$ takes the form

$$u_1(x) = u_1^0(x_2, x_3) + \frac{1}{2}\int_0^{x_1} (f_0 + \lambda^{-1} \operatorname{div} f)\,(x_1'; x_2, x_3)\, dx_1', \tag{6.60}$$

where according to the first equality of (6.58) $u_1^0|_\gamma = 0$ and from (6.59) we derive that

$$\frac{\partial^2 u_1^0}{\partial x_1^2} + \frac{\partial^2 u_1^0}{\partial x_2^2} - \lambda^2 u_1^0 = f^0(x_2, x_3),$$

$$f^0(x_2, x_3) = \frac{1}{2}\left(\frac{\partial f_0}{\partial x_1} - \lambda^{-1} \operatorname{div} \frac{\partial f}{\partial x_1}\right)_{x_1 = 0} + \left(\frac{\partial f_2}{\partial x_3} - \frac{\partial f_3}{\partial x_1} - \lambda f_1\right)_{x_1 = 0}, \tag{6.61}$$

i.e. $u_1^0(x_2, x_3)$ is uniquely determined in G as a solution of the Dirichlet problem for the Helmholtz equation. From (6.56) it follows then that the function $w(x) = u_2(x) - iu_3(x)$ satisfies the system

$$\frac{\partial w}{\partial x_1} - \mathrm{i}\,\lambda \bar{w} = g_1(x), \quad \frac{\partial w}{\partial \bar{z}} = g_2(x), \tag{6.62}$$

vhere

$$g_1(x) = 2\frac{\partial u_1}{\partial z} + f_3 + \mathrm{i} f_2, \quad g_2(x) = \mathrm{i}\,\lambda/2 u_1(x) + \frac{1}{4}(f_0 - \lambda^{-1}\operatorname{div} f) + \mathrm{i}/2 f_1,$$

$$z = x_2 + \mathrm{i}x_3, \quad \bar{w} = u_2 + \mathrm{i}u_3 \quad \frac{\partial}{\partial \bar{z}} = \frac{1}{2}\left(\frac{\partial}{\partial x_2} + \mathrm{i}\frac{\partial}{\partial x_3}\right), \quad \frac{\partial}{\partial z} = \frac{1}{2}\left(\frac{\partial}{\partial x_2} - \mathrm{i}\frac{\partial}{\partial x_3}\right).$$

From the first equation (6.62) it follows that

$$w(x) = \sin \lambda x_1\, w^0(z) + \mathrm{i} \cosh \lambda x_1\, \overline{w^0(z)} + \int_0^{x_1} [\sinh \lambda (x_1 - x_1')g_1(x_1'; x_2; x_3)$$

$$+ \mathrm{i} \cosh (x_1 - x_1')g_1(x_1'; x_2; x_3)]\, dx_1, \tag{6.63}$$

vhere sinh, cosh are hyperbolic sine and cosine, $w^0(z) = w|_{x_1 = 0}$. Substitution of (6.63) into the second equation of (6.62) gives the inhomogeneous Cauchy–Riemann equation for $w^0(z)$:

$$\frac{\partial w^0}{\partial \bar{z}} = g_2(0, z)$$

n G and consequently

$$w^0(z) = \Phi(z) + T_G g_2, \quad T_G g_2 = -1/\pi \iint_G \frac{g_2(0, \zeta)\, dG_\zeta}{\zeta - z} \tag{6.64}$$

vhere $\Phi(z)$ is an arbitrary function, holomorphic in G. Now by (6.63), (6.64) the second equality of (6.58) reduces to the Riemann–Hilbert condition for holomorphic function Φ:

$$\operatorname{Re}(k_2 + \mathrm{i}\,k_3)\Phi = \operatorname{Re} T_G g_2|_\gamma = h \tag{6.65}$$

on γ, from the theory of which the assertions of the theorem follow.

(d) As we see the well-posedness or not of a problem for an overdetermined system depends on how the system is perturbed by lower terms. The same applies to the delicate question: why is there no well-posed boundary-value problem like the Riemann-Hilbert problem for holomorphic functions of several complex variables and what is the well-posed boundary-value problem for the Cauchy=Riemann equation or for a perturbed one in some bounded domain Ω in C^n with $n > 1$? Consider, for instance, in $\Omega \subset \Omega^n$ the Cauchy-Riemann system perturbed by constant coefficients a_i, b_j :

$$\frac{\partial w}{\partial \bar{z}_j} = a_j w + b_j \bar{w}, \quad j = 1,...,n. \tag{6.66}$$

If all $b_j = 0$, then the set of a solution of this system is similar to the set $H^n(\Omega)$ of functions of n variables holomorphic in Ω. Indeed, it is evident if all $a_j = 0$, but if at least one of a_j, say a_n, differs from zero, then by nonsingular linear mapping $z \to \zeta$,

$$\zeta_j = 1/a_n \sum_{i=1}^{j-1} \bar{a}_i z_i + z_j, \quad j = 1,...,n \tag{6.67}$$

the system (6.66) reduces to the system

$$\frac{\partial w}{\partial \bar{\zeta}_j} = 0, \quad j = 1,...,n-1, \quad \frac{\partial w}{\partial \bar{\zeta}_n} = a_n w,$$

from which it follows that

$$w = \Phi(\xi)\, e^{a_n \bar{\zeta}_n}, \tag{6.68}$$

where $\Phi(\xi) \in H^n(\Omega^*)$, Ω^* is the image of Ω under the mapping (6.67). If among b_j at least one differs from zero, then the set of solutions of the system (6.66) becomes poorer than $H^n(\Omega)$. Indeed, let, for instance, $b_n \neq 0$. Then by nonsingular mapping

$$\xi_j = 1/\bar{b}_n \sum_{i=1}^{j-1} \bar{b}_i z_i + z_j, \quad j = 1,...,n, \tag{6.69}$$

the system (6.66) reduces to the system

$$\frac{\partial w}{\partial \bar{\zeta}_j} = (a_j^*) w, \; j = 1,...,n-1, \; \frac{\partial w}{\partial \bar{\zeta}_n} = a_n w + b_n \bar{w} \tag{6.70}$$

in the domain Ω_*-image of Ω under the mapping (6.69).

The system (6.70) is compatible if

$$\frac{\partial w}{\partial \zeta_j} = \overline{\left(a_j^*\right)} w, \quad j = 1,\dots,n-1 \tag{6.71}$$

in Ω_*. From this equation and from the first $n-1$ equations of (6.70) it follows that[3]

$$w(\zeta) = \psi(\xi_n)\, e^{2\,\mathrm{Re} \sum_{j=1}^{n-1} a_j^* \zeta_j}, \tag{6.72}$$

and from the last equation of the system (6.70) it follows that the function of only one variable ξ_n is a generalized analytic function, i.e. (see for instance [55]):

$$\frac{\partial \psi}{\partial \bar{\zeta}_n} = a_n \psi + b_n \bar{\psi} \tag{6.73}$$

in the plane domain $G = \Omega_* \cap (\zeta_j = 0,\ j = 1,\dots,n-1)$.

The formula (6.72) can immediately be used to investigate boundary-value problems for the system (6.66). Consider, for instance, the following problem.

Problem *F*. Find in the bounded domain $\Omega \subset C^n$ a solution of the system (6.66) with $\Sigma_{j=1}^n |b_j| \neq 0$, continuous in Ω, satisfying the condition

$$\mathrm{Re}\,(\alpha w) = h$$

$$z_j + \bar{b}_n^{-1} \sum_{i=1}^{j-1} b_i z_i = 0, \quad j = 1,\dots,n-1,$$

where α, h are Hölder-continuous functions given on γ.

Theorem 6.6. *If $\alpha \neq 0$ on γ then Problem F is well posed and has an index equal to $2\kappa + 1$, where $\kappa = \mathrm{Ind}_\gamma \bar{\alpha}$. Moreover if $\kappa \geq 0$, then this problem is solvable unconditionally and the corresponding homogeneous problem ($h \equiv 0$) has exactly $2\kappa + 1$ nontrivial linearly independent (over the field of real numbers) solutions, but if $\kappa > 0$, then Problem F is solvable if and only if the function h satisfies $-2\kappa - 1$ orthogonality conditions and the corresponding homogeneous problem has no*

3 In the case of variable coefficients the local similarity principle of a solution of a perturbed Cauchy-Riemann equation in C^n, $n > 1$, was investigated maybe first by Koohara in [44] (see also [38], [48]).

nontrivial solution.

6.3 Initial-boundary-value problems for first-order systems

(a) In this section we consider again the degenerate system (6.5), but first with the following special matrix a:

$$a = \lambda\,\mathrm{diag}(-1, 1, 1) \tag{6.74}$$

and $\lambda \neq 0$ a real constant.

Let Ω_x be a cylindrical domain $\Omega_x - \{(x_1, x_2, x_3) : (x_2, x_3) \in G,\ x_1 \geq 0\}$, where G is a bounded domain, containing the origin, on the plane of variables (x_2, x_3), $S_x = \{(x_1, x_2, x_3) : (x_2, x_3) \in \Gamma = \partial G,\ x_1 \geq 0\}$ the surface of Ω_x. If we intersect Ω_x by the plane

$$\ell_2 x_2 + \ell_3 x_3 = \text{const.} \tag{6.75}$$

where ℓ_2, ℓ_3 are real numbers $\ell_2^2 + \ell_3^2 \neq 0$, then we get the rectangular region $R = \{(x_1, x_2, x_3) : \ell_2 x_2 + \ell_3 x_3 = 0,\ a^0 \leq -\ell_3 x_2 + \ell_2 x_3 \leq b^0,\ x_1 \geq 0\}$, where $-\ell_3 x_2 + \ell_2 x_3 = a^0$, $-\ell_3 x_2 + \ell_2 x_3 = b^0$ are the element of the cylinder Ω_x, which lies on the plane $\ell_2 x_2 + \ell_3 x_3 = 0$.

Problem *H*. Find in Ω_x a solution $u \in C^2(\Omega_x)$ of the system (6.5) with matrix a given by (6.74), continuous and bounded in Ω_x, satisfying the initial condition

$$u|_{x_1 = 0} = 0, \quad x \in G, \tag{6.76}$$

the boundary condition

$$\ell_2 u_2 + \ell_3 u_3 = 0 \tag{6.77}$$

on S_x and the conditions

$$[u_1 + \ell_3 u_2 - \ell_2 u_3]_{a^0} = 0, \quad [u_1 - \ell_3 u_2 - \ell_2 u_3]_{b^0} = 0 \tag{6.78}$$

on the elements $-\ell_3 x_2 + \ell_2 x_3 = a^0$, $-\ell_3 x_2 + \ell_2 x_3 = b^0$, $\ell_2 x_2 + \ell_3 x_3 = 0$ of the cylinder Ω_x.

Theorem 6.7. *If the right-hand side $f(x)$ of the system (6.5) with matrix a given by (6.74) satisfies some consistent conditions on the intersection of lines $\ell_2 x_2 + \ell_3 x_3 = 0$, $-\ell_3 x_2 + \ell_2 x_3 = a^0$, b^0 with line $\ell_2 x_2 + \ell_3 x_3 = 0,$, $x_1 = 0$, then Problem H has a unique solution.*

Proof. Applying the operator div to both sides of (6.5) and taking into account (6.74), we obtain an equivalent compatible overdetermined system

$$\frac{\partial u_1}{\partial x_1} - \frac{\partial u_2}{\partial x_2} - \frac{\partial u_3}{\partial x_3} = -\operatorname{div} f,$$

$$\frac{\partial u_3}{\partial x_2} - \frac{\partial u_2}{\partial x_3} - \lambda u_1 = f_1,$$

$$\frac{\partial u_1}{\partial x_3} - \frac{\partial u_3}{\partial x_1} + \lambda u_2 = f_2,$$

$$\frac{\partial u_2}{\partial x_1} - \frac{\partial u_1}{\partial x_2} + u_3 = f_3. \tag{6.79}$$

It is easy to see from (6.69) that the functions $u_k(x)$, $k = 1,2,3$ satisfy the inhomogeneous Klein-Gordon equation

$$\left(\frac{\partial^2}{\partial x_1^2} - \frac{\partial^2}{\partial x_2^2} - \frac{\partial^2}{\partial x_3^2} + \lambda^2\right) u_k = f^{(k)}(x), \tag{6.80}$$

where

$$f^{(1)}(x) = \frac{\partial f_3}{\partial x_2} - \frac{\partial f_2}{\partial x_3} - \operatorname{div} \frac{\partial f}{\partial x_1} - \lambda f_1, \quad f^{(2)}(x) = \frac{\partial f_1}{\partial x_1} + \frac{\partial f_1}{\partial x_3} - \operatorname{div} \frac{\partial f}{\partial x_2} + \lambda f_2,$$

$$f^{(3)}(x) = -\frac{\partial f_1}{\partial x_2} - \frac{\partial f_2}{\partial x_1} - \operatorname{div} \frac{\partial f}{\partial x_3} + \lambda f_3.$$

Hence the function $\psi_1(x) = \ell_2 u_2 + \ell_2 + \ell_3 u_3$ is uniquely determined in Ω_x as a solution of an initial-boundary-value problem: ψ_1 is a solution of the equation

$$\frac{\partial^2 \psi_1}{\partial x_1^2} - \frac{\partial^2 \psi_1}{\partial x_2^2} - \frac{\partial^2 \psi_1}{\partial x_3^2} + \lambda^2 \psi_1 = \ell_2 f^{(2)}(x) + \ell_3 f^{(3)}(x) \tag{6.81}$$

in Ω_x, satisfying the initial conditions

$$\psi_1|_{x_1=0}, \quad \frac{\partial \psi_1}{\partial x_1}\Big|_{x_1=0} = \ell_2 f_3(0, x_2, x_3) - \ell_3 f_2(0, x_2, x_3) \tag{6.82}$$

on G and the boundary condition

$$\psi_1 = 0 \tag{6.83}$$

on the surface S_x.

Therefore Problem H will be solved if we find the functions $u_1(x)$ and $\psi_2(x) = -\ell_3 u_2(x) + \ell_2 u_3(x)$, because then we find also $u_2(x)$ and $u_3(x)$ by

$$u_2(x) = \ell_2 \psi_1(x) - \ell_3 \psi_2(x), \quad u_3(x) = \ell_3 \psi_1(x) + \ell_2 \psi_2(x). \tag{6.84}$$

Substitutions of (6.84) into (6.79) gives the following system for u_1, ψ_2:

$$\ell_3 \frac{\partial \psi_2}{\partial x_2} - \ell_2 \frac{\partial \psi_3}{\partial x_3} = -\frac{\partial u_1}{\partial x_1} + f_0^{(0)}$$

$$\ell_2 \frac{\partial \psi_2}{\partial x_2} + \ell_3 \frac{\partial \psi_2}{\partial x_3} = \lambda u_1 + f_1^0, \tag{6.85}$$

$$\frac{\partial u_1}{\partial x_2} + \ell_3 \frac{\partial \psi_2}{\partial x_1} = \lambda \ell_2 \psi_2 + f_2^0,$$

$$\frac{\partial u_1}{\partial x_3} - \ell_2 \frac{\partial \psi_2}{\partial x_1} = \lambda \ell_3 \psi_2 + f_3^0, \tag{6.86}$$

where

$$f_0^{(0)} = \ell_2 \frac{\partial \psi_1}{\partial x_2} + \ell_3 \frac{\partial \psi_1}{\partial x_3} - \operatorname{div} f, \quad f_1^0 = -\ell_3 \frac{\partial \psi_1}{\partial x_1} + \ell_2 \psi_1 + f_1,$$

$$f_2^0 = \ell_3 \frac{\partial \psi_1}{\partial x_1} - \ell_2 \psi_1 + f_2, \quad f_3^0 = \ell_2 \frac{\partial \psi_1}{\partial x_1} + \lambda \ell_3 \psi_1 - f_3;$$

From (6.85) since $\ell_2^2 + \ell_3^2 = 1$ we have

$$\frac{\partial \psi_2}{\partial x_2} + \ell_3 \frac{\partial u_1}{\partial x_1} - \lambda \ell_2 u_1 = f_0^{(2)},$$

$$\frac{\partial \psi_2}{\partial x_3} - \ell_2 \frac{\partial u_1}{\partial x_1} - \lambda \ell_3 u_1 = f_0^{(3)}, \tag{6.87}$$

where

$$f_0^{(2)} = \ell_3 f_0^{(0)} + \ell_2 f_1^0, \; f_0^{(3)} = \ell_3 f_1^0 - \ell_2 f_0^{(0)}.$$

We may combine the first equations in (6.85) and (6.87) into the following single complex equation with respect to the complex-valued unknown function $w(z) = u_1 + \mathrm{i}\,\psi_2$:

$$\frac{\partial w}{\partial x_2} + \mathrm{i}\ell_3 \frac{\partial \bar{w}}{\partial x_1} - \mathrm{i}\,\lambda\,\ell_2\,\bar{w} = g_1^0 \tag{6.88}$$

In addition, we may combine the second equations in (6.85) and (6.87) into the following:

$$\frac{\partial w}{\partial x_3} - \mathrm{i}\,\ell_3 \frac{\partial \bar{w}}{\partial x_1} - \mathrm{i}\,\lambda\,\ell_3\,\bar{w} = q_3^0, \tag{6.89}$$

where

$$q_2^0 = f_2^0 + \mathrm{i} f_0^{(2)}, \; q_3^0 = f_3^0 + \mathrm{i} f_0^{(3)}.$$

Now by the nonsingular linear mapping $(x_1, x_2, x_3) \to (\xi_1, \xi_2, \xi_3)$:

$$\xi_1 = x_1, \; \xi_2 = \ell_2 x_2 + \ell_3 x_3, \; \xi_2 = -\ell_3 x_2 + \ell_2 x_3,$$

the system (6.88), (6.89) reduces to the following:

$$\frac{\partial w}{\partial \xi_2} - \mathrm{i}\,\lambda \bar{w} = q^0, \; \frac{\partial w}{\partial \xi_3} - \mathrm{i} \frac{\partial \bar{w}}{\partial \xi_1} = q_0, \tag{6.90}$$

where

$$q^0 = \ell_2 q_2^0 + \ell_3 q_3^0, \quad q_0 = \ell_2 q_2^0 - \ell_3 q_2^0.$$

From the first equation of (6.90) we obtain the representation

$$w(\xi) = \sinh \lambda\xi_2\, w^0(\xi_1, \xi_3) + i \cosh \lambda\, \xi_2\, \overline{w^0(\xi_1, \xi_3)}$$
$$+ \int_0^{\xi_2} \Big[\sinh \lambda(\xi_2 - \xi_2') q^0(\xi_1, \xi_2', \xi_3)$$
$$+ i \cosh (\xi_2 - \xi_2')\, \overline{q^0(\xi_1, \xi_2', \xi_3)} \Big] d\xi_2' \tag{6.91}$$

where $w^0(\xi_1, \xi_3) = w \mid_{\xi_2 = 0}$. Substituting this representation into the second equation of (6.90) gives the following equation for $w^0(\xi_1, \xi_3)$:

$$\frac{\partial w^0}{\partial \xi_3} - i \frac{\overline{\partial w^0}}{\partial \xi_1} = q_0(\xi_1, 0, \xi_3) \tag{6.92}$$

in the rectangular region $R' = \{(\xi_1, \xi_3) : a^0 \leq \xi_3 \leq b^0, \xi_1 \geq 0\}$. This is a simple hyperbolic system and its general solution in R' can be found by the method of characteristics:

$$w^0(\xi_1, \xi_3) = e^{\pi i/4} \varphi(\xi_3 + \xi_1) + e^{-\pi i/4} \psi(\xi_3 - \xi_1)$$
$$+ \frac{1}{4} \int_0^{\xi_3 - \xi_1} (g_0 + i g_0) d\xi' + \frac{1}{4} \int_0^{\xi_3 - \xi_1} (g_0 + i g_0) d\eta', \tag{6.93}$$

where φ and ψ are real functions of one variable. From (6.76) and (6.78) it follows that the function $w^0(\xi_1, \xi_3)$ satisfies the initial condition

$$w^0(0, \xi_3) = 0, \quad a{=}^0 \leq \xi_3 \leq b^0 \tag{6.94}$$

and the following boundary conditions

$$\operatorname{Re} w^0(\xi_1, a^0) - \operatorname{Im} w^0(\xi_1, a^0) = 0, \quad \xi_1 \geq 0,$$
$$\operatorname{Re} w^0(\xi_1, b^0) - \operatorname{Im} w^0(\xi_1, b^0) = 0, \quad \xi_1 \geq 0. \tag{6.95}$$

Taking into account (6.93) we easily see that the function $\varphi(\xi_3 + \xi_1)$ is determined in strip $a^0 \le \xi_3 + \xi_1 \le b^0$ by means of the initial condition (6.94) and in the half-plane $\xi_3 + \xi_1 \ge b^0$ by means of the second boundary condition (6.95); also the function $\psi(\xi_3 - \xi_1)$ is determined in the strip $a^0 \le \xi_3 - \xi_1 \le b^0$ by means of the initial condition (6.94) and in the half-plane $\xi_3 - \xi_1 \le a^0$ by mens of the first boundary condition (6.95), i.e. both functions $\varphi(\xi_3 + \xi_1)$ and $\psi(\xi_3 - \xi_1)$ in (6.93) are determined in the rectangular region $R' = \{(\xi_1, \xi_3) : a^0 \le \xi_3 \le b^0, \xi_1 \ge 0\}$.

It is evident now what consistent conditions we need to impose for the right-hand side $g_0(\xi_1, 0, \xi_3)$ of the equation (6.92) at the points $(\xi_1, \xi_3) = (0, a^0)$, $(\xi_1, \xi_3) = (0, b^0)$ in order to obtain continuous values of $\varphi(\xi_3 + \xi_1)$ and $\psi(\xi_3 - \xi_1)$ in the rectangle R'.

(b) Let us consider on the cylindrical domain Ω_x, described above, the following overdetermined system:

$$\frac{\partial u_1}{\partial x_1} - \frac{\partial u_2}{\partial x_2} - \frac{\partial u_3}{\partial x_3} = f_0,$$

$$\operatorname{curl} u + \lambda \operatorname{diag}(-1, -1, 1) u = f \tag{6.96}$$

with $\lambda \ne 0$ a real constant.

For this system we state the following

Problem *h*. Find in Ω_x a solution $u \in C^2(\Omega_x)$ of the system (6.96), satisfying the initial condition

$$u|_{x_1 = 0,\, x_2 = 0} = 0, \quad a^0 \le x_3 \le b_0, \tag{6.97}$$

where $x_3 = a^0$, $x_3 = b^0$ are the elements of the cylinder Ω_x which lies on the plane $x_2 = 0$ and satisfying the following boundary conditions:

$$u_2|_{x_2 = 0,\, x_3 = a^0} = 0, \quad u_2|_{x_2 = 0,\, x_3 = b^0} = 0,$$

$$[u_1 - u_3]_{x_2 = 0,\, x_3 = a^0} = 0, \quad [u_1 + u_3]_{x_2 = 0,\, x_3 = b^0} = 0 \tag{6.98}$$

Theorem 6.8. *If f_0, f satisfy some consistent conditions at points $x = (0, 0, a_0)$ and $x = (0, 0, b_0)$, then Problem h has a unique solution, continuous in $\bar{\Omega}_x$.*

Proof. Since $\lambda \ne 0$, then applying the operator div to the second line in (6.96) we see that the system (6.96) is compatible if

$$\frac{\partial u_1}{\partial x_1} + \frac{\partial u_2}{\partial x_2} \frac{\partial u_3}{\partial x_3} = -\lambda^{-1} \operatorname{div} f$$

in Ω_x. As a consequence the system (6.96) is equivalent to the following compatible overdetermined system of five equations

$$\frac{\partial u_2}{\partial x_2} = -\frac{1}{2}(f_0 + \lambda^{-1}\operatorname{div} f), \quad \frac{\partial u_1}{\partial x_1} - \frac{\partial u_3}{\partial x_3} = \frac{1}{2}(f_0 + \lambda^{-1}\operatorname{div} f),$$

$$\frac{\partial u_3}{\partial x_2} - \frac{\partial u_2}{\partial x_3} - \lambda u_1 = f_1, \frac{\partial u_1}{\partial x_3} - \frac{\partial u_3}{\partial x_1} - \lambda u_2 = f_2, \frac{\partial u_2}{\partial x_1} - \frac{\partial u_1}{\partial x_2} + \lambda u_3 = f_3 \quad (6.99)$$

From the first equation it follows that

$$u_2(x) = u_2^0(x_1, x_3) - \frac{1}{2}\int_0^{x_2} (f_0 + \lambda^{-1}\operatorname{div} f)(x_1, x_2', x_3)\,dx_2'$$

in Ω_x, where $u_2^0(x_1, x_3) = u_2|_{x_2=0}$.

Now by differentiation we derive from (6.99) that the function $u_2^0(x_1, x_3)$ will be a solution of the equation

$$\frac{\partial^2 u_2^0}{\partial x_1^2} - \frac{\partial^2 u_2^0}{\partial x_3^2} - \lambda^2 u_2^0 = f^0(x_1, x_3) \quad (6.100)$$

in the rectangular region $R_0 = \{(x_1, x_2, x_3) : x_2 = 0,\ a^0 \le x_3 \le b^0, x_1 \ge 0\}$, where

$$f_0(x_1, x_3) = \frac{1}{2}\left(\frac{\partial f_0}{\partial x_2} - \lambda^{-1}\operatorname{div}\frac{\partial f}{\partial x_2}\right)_{x_2=0} + \left(\frac{\partial f_1}{\partial x_3} + \frac{\partial f_3}{\partial x_1} + \lambda f_2\right)_{x_2=0}.$$

Moreover from the last equation of (6.99) from (6.97) and the first two equations of (6.98) it follows that $u_2^0(x_1, x_3)$ satisfies the initial conditions

$$u_2^0(0, x_3) = 0, \quad \frac{\partial u_2^0}{\partial x_3}\Big|_{x_1=0} = f_3(0, 0, x_3), \quad a^0 \le x_3 \le b^0 \quad (6.101)$$

and the boundary conditions

$$u_2^0(x_1, 0, a^0) = 0, \quad u_2^0(x_1, 0, b^0) = 0, \quad x_1 \ge 0 \quad (6.102)$$

Therefore (at least for those λ^2, which are not eigenvalues) the function $u_2^0(x_1, x_3)$ and hence the function $u_2(x)$ are uniquely determined in Ω_x. Consequently the system (6.99) reduces to the following system with respect to unknowns $v(x) = u_1 + u_3$, $w(x) = u_1 - u_3$:

$$\frac{\partial v}{\partial x_1} - \frac{\partial v}{\partial x_3} = g^{(1)}(x), \quad \frac{\partial w}{\partial x_1} + \frac{\partial w}{\partial x_3} = g^{(2)}(x),$$

$$\frac{\partial v}{\partial x_2} - \lambda v = g_1(x), \quad \frac{\partial w}{\partial x_2} + \lambda w = g_2(x) \tag{6.103}$$

in Ω_x, where

$$g^{(1)}(x) = -\lambda u_2 + \frac{1}{2}(f_0 - \lambda^{-1}\operatorname{div} f) - f_2,$$

$$g^{(2)}(x) = \lambda u_2 + \frac{1}{2}(f_0 - \lambda^{-1}\operatorname{div} f)$$

$$g_1(x) = \frac{\partial u_2}{\partial x_1} + \frac{\partial u_2}{\partial x_3} + f_1 - f_3, \quad g_2(x) = \frac{\partial u_2}{\partial x_1} - \frac{\partial u_2}{\partial x_3} - (f_1 + f_3).$$

From the last two equations of (6.103) it follows that

$$v(x) = v^0(x_1, x_3)e^{\lambda x} + \int_0^{x_2} g_1(x_1, x_2', x_3)e^{\lambda(x_2 - x_2')}\,dx_2',$$

$$w(x) = w^0(x_1, x_3)e^{-\lambda x} + \int_0^{x_2} g_2(x_1, x_2', x_3)e^{\lambda(x_2 - x_2')}\,dx_2', \tag{6.104}$$

where $v^0(x_1, x_3) = v|_{x_2 = 0}$, $w^0(x_1, x_3) = w|_{x_2 = 0}$. Substitution of (6.104) into the first two equations of (6.103) gives the following equations for v^0, w^0:

$$\frac{\partial v^0}{\partial x_1} - \frac{\partial v^0}{\partial x_3} = g^1(x_1, 0, x_3), \quad \frac{\partial w^0}{\partial x_1} + \frac{\partial w^0}{\partial x_3} = g^{(2)}(x_1, 0, x_3) \tag{6.105}$$

in rectangular region $R_0\{(x_1, x_3) : a^0 \le x_3 \le b^0, x_1 \ge 0\}$.

From (6.97) and (6.104) we obtain the following initial conditions:

$$v^0|_{x_1=0} = 0,\ w^0|_{x_1=0} = 0,\ a^0 \le x_3 \le b^0, \tag{6.106}$$

and from (6.98) and (6.104) the following boundary conditions:

$$v^0|_{x_3=b_0} = 0,\ w^0|_{x_3=a_0} = 0,\ x_1 \ge 0 \tag{6.107}$$

integration of (6.105) gives

$$v^0(x_1,x_3) = \varphi^0(x_3+x_1) - \frac{1}{2}\int_0^{x_3-x} g^{(1)}\left(\frac{\xi_3+x_1-\eta}{2}, 0, \frac{x_3+x_1+\eta}{2}\right) d\eta \tag{6.108}$$

$$w^0(x_1,x_3) = \psi^0(x_3-x_1) - \frac{1}{2}\int_0^{x_3+x_1} g^{(2)}\left(\frac{\xi-x_3+x_1}{2}; 0; \frac{\xi+x_3-x_1}{2}\right) d\xi \tag{6.109}$$

in R_0, φ^0 and ψ^0 are arbitrary real functions of one variable. By the first condition (6.106) and by (6.108) we determine the function $\varphi^0(x_3+x_1)$ in the strip $a^0 \le x_3 + x_1 \le b^0$:

$$\varphi^0(x_3+x_1) = \frac{1}{2}\int_0^{x_3+x_1} g^{(1)}\left(\frac{x_3+x_1-\eta}{2}, 0, \frac{\xi_3+x_1-\eta}{2}\right) d\eta, \tag{6.110}$$

and by the first condition of (6.107) and by (6.108) we determine $\varphi^0(x_3+x_1)$ in the half-plane $x_3+x_1 \ge b^0$:

$$\varphi^0(x_3+x_1) = \frac{1}{2}\int_0^{x_3+x_1} g^{(1)}\left(\frac{x_3+x_1-\eta}{2}, 0, \frac{x_3+x_1+\eta}{2}\right) d\eta \tag{6.111}$$

By the second condition of (6.106) and by (6.109) we determine the function $\psi^0(x_3+x_1)$ in the strip $a^0 \le x_3 - x_1 \le b^0$:

$$\psi^0(x_3+x_1) = -\frac{1}{2}\int_0^{x_3+x_1} g^{(2)}\left(\frac{\xi-x_3+x_1}{2}; 0; \frac{\xi+x_3-x_1}{2}\right) d\xi \tag{6.112}$$

and by the second condition of (6.107) and by (6.109) we determine $\psi^0(x_3+x_1)$ in the half-plane $x_3 - x_1 \le a^0$:

$$\psi^0(x_3+x_1) = -\frac{1}{2}\int_0^{x_3+x_1} g^{(2)}\left(\frac{\xi-x_3+x_1}{2}; 0; \frac{\xi+x_3-x_1}{2}\right) d\xi, \tag{6.113}$$

e. both functions $\psi^0(x_3+x_1)$, $\psi^0(x_3-x_1)$, presented in (6.108) and (6.109) are etermined in the rectangular region R_0. Now it is easy to see from (6.110) and 5.111) that ψ^0 will be continuous if $g^{(1)}$ satisfies the condition

$$\int_0^{x_1} g^{(1)}(\eta, 0, b^0-\eta)\,d\eta = 0, \quad 0 \le x_1 \le b^0/2.$$

t also follows from (6.112) and (6.113) that ψ^0 will be continuous if $g^{(2)}$ satisfies 1e condition

$$\int_0^{x_1} g^{(2)}(\xi, 0, \xi + a_0)\,d\xi = 0, \quad 0 \le x_1 \le a_0/2.$$

ιt the end of this section we give a definition of hyperbolicity for a general first-rder overdetermined system with respect to the vector function $u(x) = (u_1,\ldots,u_m)$:

$$Au := \sum_{k=1}^{n} A_k(x)\frac{\partial u}{\partial x_k} + A_0(x)u = F(x) \tag{6.114}$$

vith $(r \times m)$ matrices $A_i(x)$ and vector function $F(x) = (F_1,\ldots,F_m)$ given in the omain $\Omega \subset \mathbb{R}^n$, $r > m$.

Let $\Delta_j(x, \xi)$ denote the minors of order m of the characteristic matrix of (6.114)

$$\sigma(x, \xi) = \sum_{k=1}^{n} A_k(x)\,\xi_k$$

nd let

$$\text{char}\, A := = \{(x, \xi) : x \in \Omega,\ \xi \in \mathbb{R}^n\backslash 0,\ \Delta_j(x, \xi) = 0,\ \forall j\}.$$

Ve consider the system (6.114) as hyperbolic in Ω if the set char A is not empty nd, at least for one value of j, $\text{grad}_\xi\, \Delta_j(x, \xi) \ne 0$, $x \in \Omega$, $\forall \xi \in \mathbb{R}^n\backslash 0$. It is easy to heck that the overdetermined systems (6.79) and (6.96) are hyperbolic in the whole pace $\mathbb{R}^3$.

;.4 The boundary-value problem for second-order systems

.et us go back to the second-order system (6.7). Since for its principal symbol $\sigma^T(\xi) = (\xi_i\,\xi_j)$, $1 \le i, j \le 3$, then the equation $e^T(\xi)e = 0$ for eigenvectors $= (e_1, e_2, e_3))$ becomes $\xi_1 e_1 + \xi_2 e_2 + \xi_3 e_3 = 0$, which has three linearly ndependent solutions $e^{(1)} = (0, -\xi_3, \xi_2)$, $e^{(2)} = (\xi_3, 0, -\xi_1)$, $e^{(3)} = (\xi_2, -\xi_1, 0)$ and hence there are three differential operators

$$e^{(1)}(D) = (0, -\frac{\partial}{\partial x_3}, \frac{\partial}{\partial x_2}),\ e^{(2)} = (\frac{\partial}{\partial x_3}, 0, -\frac{\partial}{\partial x_1}),\ e^{(3)}(D) = (\frac{\partial}{\partial x_2}, -\frac{\partial}{\partial x_1}, 0),$$

so the overdetermined system (6.10) in this case becomes

$$e^{(j)}(D) \sum_{i=1}^{3} a_i(x)u_{x_i} + a_0(x)u) = e^{(i)}(D)f(x), \quad j = 1,2,3.$$

$$\text{grad div } u + \sum_{i=1}^{3} a_i(x)u_{x_i} + a_0(x)u = f(x)$$

or

$$\text{curl } (\sum_{i=1}^{3} a_i(x)u_{x_i} + a_0(x)u) = \text{curl } f(x),$$

$$\text{grad div } u + \sum_{i=1}^{3} a_i(x)u_{x_i} + a_0(x)u = f(x) \qquad (6.115)$$

Here we state the following

Problem *I*. Find in bounded domain $\Omega \subset R^3$ a solution $u \in C^2(\Omega)$ of the degenerate second-order system

$$\text{grad div } u + \lambda \text{ curl } u = f(x) \qquad (6.116)$$

with $\lambda \neq 0$ a real constant continuously differentiable in $\bar{\Omega} = \Omega + \Gamma$, satisfying the boundary conditions

$$(\nabla \times u, n)_{\Gamma} = 0, \quad (u, n)_{\Gamma} = 0 \qquad (6.117)$$

where n is the outward normal to $\Gamma = \partial\Omega$.

Theorem 6.9. *If the vector function $f(x)$ and its first-order derivatives belong to the space $L_p(\bar{\Omega})$, $p > 3$, then Problem I has a unique solution.*

Proof. According to (6.115) the equation (6.116) is equivalent to the overdetermined system

$$\operatorname{curl} \operatorname{curl} u = \lambda^{-1} \operatorname{curl} f,$$

$$\operatorname{grad} \operatorname{div} u + \lambda \operatorname{curl} u = f, \tag{6.118}$$

which is equivalent to the system

$$\Delta u + \lambda \operatorname{curl} u = f - \lambda^{-1} \operatorname{curl} f,$$

$$\operatorname{grad} \operatorname{div} u + \lambda \operatorname{curl} u = f, \tag{6.119}$$

because of $\operatorname{curl} \operatorname{curl} u = \operatorname{grad} \operatorname{div} u - \Delta u = -\Delta u - \lambda \operatorname{curl} u + f$. Introducing the function $\operatorname{div} u = \varphi$ and the vector function $\operatorname{curl} u = \psi$ and taking into account (6.119) we get the following first-order system for φ, ψ:

$$\operatorname{div} \psi = 0,$$

$$\operatorname{grad} \varphi - \operatorname{curl} \psi + \lambda\psi = f - \lambda^{-1} \operatorname{curl} f,$$

$$\operatorname{grad} \varphi + \lambda\psi = f. \tag{6.120}$$

Eliminating φ (6.120) leads to the following system only for ψ:

$$\operatorname{div} \psi = 0,$$

$$\operatorname{curl} \psi = \lambda^{-1} \operatorname{curl} f \tag{6.121}$$

From the first equality of (6.117) we have the boundary condition

$$(\psi, n)_\Gamma = 0 \tag{6.122}$$

The problem (6.121), (6.122) has a unique solution $\psi \in C^1(\Omega) \cap C(\bar{\Omega})$ for any given vector function $f(x)$, described above. Indeed, since the second equation (6.121) is $\operatorname{curl}(\psi - \lambda f) = 0$, then

$$\psi(x) = w(x) + \lambda^{-1} f(x)$$

and by means of the first equation (6.121) $w(x)$ is the solution of the Poisson equation

$$\Delta w = -\lambda^{-1} \operatorname{div} f \tag{6.123}$$

in Ω and by (6.122) we get Neumann's condition

$$\frac{\partial w}{\partial n} = -\lambda^{-1}(f, n) \tag{6.124}$$

on $\Gamma = \partial\Omega$. This Neumann problem is always solvable, because of

$$\int_\Gamma \Delta w \, dx = -\lambda^{-1} \int_\Omega \operatorname{div} f \, dx = -\lambda^{-1} \int_\Gamma (f, n) \, d\Gamma = \int_\Gamma \frac{\partial w}{\partial n} \, d\Gamma.$$

Hence the vector function ψ is uniquely determined in Ω. From the last equation of (6.120) we determine $\varphi(x)$ uniquely in terms of ψ and f apart from a constant C:

$$\varphi(x) = C + \int_\ell (f - \lambda\varphi, dx) \tag{6.125}$$

and the curvilinear integral does not depend on the path ℓ of integration connecting the origin $x = 0$ with point $x \in \Omega$, because of the second relation of (6.121). To find the desired vector function $u(x)$ we have the system

$$\operatorname{div} u = \varphi(x), \quad \operatorname{curl} u = \psi(x) \tag{6.126}$$

with the second boundary condition

$$(u, n)_\Gamma = 0. \tag{6.127}$$

The problem (6.126), (6.127) is solvable if

$$\int_\Omega \varphi(x) \, dx = \int_\Omega \operatorname{div} u \, dx = \int_\Gamma (u, n) \, d\Gamma = 0.$$

But according to (6.125) this equality is always fulfilled in case the constant C we choose as

$$C = \frac{1}{\operatorname{meas}\Omega} \int_\Omega \int_\ell (\lambda\psi - f, dx) \, d\xi.$$

Thus the vector function $u(x)$ is determined uniquely in Ω. Taking into account the previous results it is not difficult to extend the investigation of Problem I, replacing there the outward normal n to the arbitrary vector field $\ell = (\ell_1, \ell_2, \ell_3)$ with $\cos(\ell, n) \neq 0$.

References

[1] Adams J.F., Lax P.D. and Philips R.S. On matrices, whose real linear combinations are not singular, *Proc. Amer. Math. Soc.*, **16**, 318–322, 1965.

[2] Ahlfors L. On quasi conformal mappings, *J. Anal. Math.* 5, 1–58, 1954.

[3] Bedford E. The Dirichlet problem for some overdetermined system on the unit ball in $\mathbb{C}^n$ *Pacific J. Math.* **51** (1), 1974.

[4] Bedford E. and Federbush P. Pluriharmonic boundary values, *Tohoky Math. J.* **26** 505–11, 1974.

[5] Begehr H. and Hile G.N. Nonlinear boundary value problems for semilinear elliptic system in plane, *Math. Zeit.*, **179**, 241–261, 1982.

[6] Bergman S. The Kernel function and conformal mapping, *Amer. Math. Soc.*, Providence, RI, 1970.

[7] Bergman S. and Schiffer M. *Kernel Functions and Elliptic Differential Equations in Mathematical Physics*, Academic press, New York, 1953.

[8] Bers L. Theory of Pseudo-analytic Functions. *Courant Institute*, New York, 1953.

[9] Bitsadze A.V. On uniqueness of Dirichlet problem for elliptic partial differential equations. *Uspechy Mathem. Nauk*, **3** (6), 211–212 1948 (Russian).

[10] Bitsadze A.V. *Boundary Value Problems for Second Order Elliptic Equations*, Nauka, Moscow, 1966 (Russian) (English translation North-Holland Publishing Company, Amsterdam (1968)).

[11] Bitsadze A.V. *Some Classes of Partial Differential Equations*, Nauka, Moscow, 1981 (Russian) (English translation Gordon and Breach Science Publishers (1988)).

[12] Bogoliubov N.N. and Shirkov D.V. *Introduction to the Theory of Quantized Fields*, GITTL Moscow 1957; English translation, Interscience, New York, 1959.

[13] Calderon A.P. and Zygmund A. On the existence of certain singular integrals, *Acta Math.* **88**, 85–139, 1952.

[14] Calderon A.P. and Zygmund A. On singular integrals, *Amer. J. Math.*, β78, 289–309, 1956.

[15] Dzhuraev A. A system of Beltrami equations which are degenerate on a curve, *Dokl. Akad. Nauk SSSR*, **185**, 984–6, 1969. (Russian)

[16] Dzhuraev A. A certain method of investigating singular integral equations on a bounded plane region, *Dokl. Akad. Nauk. SSSR*, **197**, 1251–4, 1971 (Russian).

[17] Dzhuraev A. A certain system of singular integral equations on a bounded four-dimensional domain, *Dokl. Akad. Nauk SSSR*, **211**, 516-19, 1973. (Russian)

[18] Dzhuraev A. The properties of certain first order degenerate elliptic systems on the plane, *Dokl. Akad. Nauk SSSR*, **223** (3), 533-6, 1975. (Russian)

[19] Dzhuraev A. Hilbert formulas for the Moisil-Theodorescu system, Dokl. Akad. Nauk Tazhik SSR, **20** (10), 3-5, 1977. (Russian)

[20] Dzhuraev A. On a class of systems of singular integral equations on a bounded domain, *Proc. Roy. Soc. Edinburgh Sect. A.*, **83** (3-4), 347-64, 1979.

[21] Dzhuraev A. Structure of the solution of certain first order systems, *Trudy Mat. Inst. Steklov.*, **163**, 81-84, 1984. (Russian)

[22] Dzhuraev A. Two boundary value problems in a bounded plane domain for a general linear elliptic systems of two second-order equations, *Dokl. Akad. Nauk SSSR*, **287** (6), 1295-8, 1986. (Russian)

[23] Dzhuraev A. Two dimensional singular integral equations that degenerate along a curve, *Dokl. Akad. Nauk SSSR*, 290 (3), 542-5, 1986. (Russian)

[24] Dzhuraev A. Generalization of Bergman's Kernel function. *Complex Variables. Theory and Application. An International J.* **6** (2-4) 347-61, 1986.

[25] Dzhuraev A. *The Method of Singular Integral Equations*, Nauka, Moscow, 1987, 416 p. (Russian).

[26] Dzhuraev A. Degenerate elliptic and first-order mixed-type systems in a plane, *Soviet Math. Doklady*, **37** (1) 191-4, 1988 *Dokl. Akad. Nauk SSSR*, **298** (5), 1047-51, 1988.

[27] Dzhuraev A. Some boundary value problems for second-order systems of mixed type in a plane, *Soviet Math. Doklady*, **37** (1), 152-6, 1988; *Dokl. Akad. Nauk SSSR*, **298** (4) 797-802, 1988.

[28] Dzhuraev A. Mixed problems for first-order hyperbolic systems, *Soviet Math. Doklady*, **38** (2), 418-22, 1989; *Dokl. Akad. Nauk SSR*, β302 (2), 418-22, 1988. (Russian)

[29] Dzhuraev A. *Systems of Equations of Composite Type*, Longman Scientific and Technical Harlow; Wiley, New York, 1989.

[30] Dzhuraev A. On the theory of first-order elliptic systems in the plane and application. *Complex Variables. Theory and Application. An Internal J.* β14 (1-4), 109-15, 1990.

[31] Dzhuraev A. On elliptic systems in plane with discontinuous coefficients and their application. *Dokl. Akad. Nauk SSSR*, **312** (4), 790-95, 1990. (Russian).

[32] Dzhuraev A. On Cauchy and Schwartz problems for elliptic systems. *Complex Variables. Theory and Application. An International J,* **16**, 58-69.

[33] Dzhuraev A. On Kernel Matrices and Holomorphic Vectors. *Complex Variables. Theory and Application. An International J*, **16**, 43-57.

[34] Dzhuraev A. On Dirichlet problem for second order elliptic systems in multiply-connected plane domain. *Complex Variables. Theory and Application. An International J.* (to appear)

[35] Dzhuraev A. On elliptic systems degenerated at the boundary. *Proc. Roy. Soc. Edinburgh.* (to appear)

[36] Fichera G. *Linear Elliptic Differential Systems and Eigenvalue Problems.* Lecture Notes in Mathematics, 8, Springer-Verlag, Berlin, 1965.

[37] Gakhov F.D. *Boundary Value Problems*, Pergamon, Oxford, 1962.

[38] Garding L. Dirichlet's problem for linear partial differential equations, *Math. Scand.* **1**, 55-72, 1953.

[39] Gilbert R.P. and Buhanan J.L., *First Order Elliptic Systems. A Function Theoretic Approach.* Academic Press, New York, 1983.

[40] Gudovich I.S. and Krein, S.G, *Math. Sbornik* **72** (114), N4, 1971.

[41] Hörmander L. *An Introduction to Complex Analysis in Several Variables*, D. van Nostrand Company Inc. Princeton, New Jersey, Toronto, London, 1966.

[42] Ianushauskas A. *The Space Boundary Value Problems for Elliptic Equations*, Lithuanian Academy of Sciences, Vilnius, 1990 (Russian).

[43] Keng, Hua Loo, Lin Wei and Wu Ci-Quian, *Second-order Systems of Partial Differential Equations in the Plane*, Pitman, 1985.

[44] Koohara A. Similarity principle of the generalized Cauchy-Riemann equations for several complex variables, *J. Math. Soc. Japan*, **23**(2), 213-249, 1970.

[45] Kratz St.G. *Function Theory of Several Complex Variables*, Wiley, New York, 1982.

[46] Lewy H. An example of smooth linear partial differential equation without solution, *Ann. Math.*, **66**, 155-8, 1957.

[47] Lopatinski J.B. *Ukrain Math. I.* **5**(2), 123-51, 1953 (Russian).

[48] Magomedov G.A., Palamodov V.P., *Math. Sbornik*, **106** (148), N4(8), 1978 (Russian).

[49] Mihlin S.G., *Multidimensional Singular Integrals and Integral Equations*, Fizmatgiz, Moscow, 1962 (Russian).

[50] Miranda C., *Partial Differential Equations of Elliptic Type*, Springer-Verlag, Berlin and New York, 1970.

[51] Moisil G.G. and Theodorescu N., Fonction holomorphic dans l'espace, *Bull. Soc. St. Cluj*, **6**, 177-94, 1931.

[52] Muskhelishvili N.I. *Singular Integral Equations*, Noordhoff Int. Gröningen, Leyden, Netherlands, 1953.

[53] Olver P.J. *Application of Lie Groups to Differential Equations*, Springer-Verlag, N.Y., Berlin, Heidelberg, 1986.

[54] Saks R.S. *Differential Uravnenyia*, **8**, 126-33, 1972 (Russian).

[55] Shapiro Z.I. *Izvestiya Akad Nauk SSSR, Ser. Mathem.* **17**, 539-62, 1953.

[56] Shevchenko,V.T., On some criterion of solvability of boundary value problems for holomorphic vectors, *Review Roumaine de Mathematics et applic*. **13** (9) 1461-6, 1968.

[57] Tricomi F., Formula de inversione dell'ordine di due integratione doppie 'con arterisco'. *Rend. Acad. Naz. Lincei*, No. 9, 535-9, 1926.

[58] Vekua I.N., *Generalized Analytic Functions*, Pergamon, Oxford, 1962.

[59] Vinogradov V.S., Boundary value problems for first order elliptic systems on plane, *Differen. uravneniya*, **8** (7), 1440-48, 1972 (Russian).

[60] Visik M.I., On strongly elliptic systems diff. equations, *Mat. Sbornik*, **29** (3), 615-76, 1951. (Russian)

[61] Volevich L.R., *Mat. Sbornik* **68** (110), N3, 1965 (Russian).

[62] Weinstock B., Continuous boundary values of analytic functions of several complex variables, *Proc. A.M.S.*, 463-6, 1969.

[63] Wen G.C. and Begehr, H., *Boundary Value Problems for Elliptic Equations and Systems*, Longman, Harlow, 1990.

[64] Wendland W., *Elliptic Systems in the Plane*, Pitman, London, 1979.

Milton Keynes UK
Ingram Content Group UK Ltd.
UKHW021623071024
449327UK00020BA/1166

9 780367 450168